AN INTRODUCTION TO DIFFERENTIAL GEOMETRY

PRINCETON MATHEMATICAL SERIES

Editors: Marston Morse, H. P. Robertson, A. W. Tucker

———

1. THE CLASSICAL GROUPS
THEIR INVARIANTS AND REPRESENTATIONS
By Hermann Weyl

2. TOPOLOGICAL GROUPS
By L. Pontrjagin

3. AN INTRODUCTION TO DIFFERENTIAL GEOMETRY
WITH USE OF THE TENSOR CALCULUS
By Luther Pfahler Eisenhart

DIMENSION THEORY
By Witold Hurewicz and Henry Wallman
(In Press)

THE ANALYTICAL FOUNDATIONS
OF CELESTIAL MECHANICS
By Aurel Wintner
(In Preparation)

AN INTRODUCTION TO
DIFFERENTIAL
GEOMETRY
WITH USE OF THE TENSOR
CALCULUS

By LUTHER PFAHLER EISENHART

DOD PROFESSOR OF MATHEMATICS
PRINCETON UNIVERSITY

PRINCETON

PRINCETON UNIVERSITY PRESS

LONDON : HUMPHREY MILFORD : OXFORD UNIVERSITY PRESS

1940

Printed in the United States of America

TO

K . S . E .

Preface

Since 1909, when my *Differential Geometry of Curves and Surfaces* was published, the tensor calculus, which had previously been invented by Ricci, was adopted by Einstein in his General Theory of Relativity, and has been developed further in the study of Riemannian Geometry and various generalizations of the latter. In the present book the tensor calculus of euclidean 3-space is developed and then generalized so as to apply to a Riemannian space of any number of dimensions. The tensor calculus as here developed is applied in Chapters III and IV to the study of differential geometry of surfaces in 3-space, the material treated being equivalent to what appears in general in the first eight chapters of my former book with such additions as follow from the introduction of the concept of parallelism of Levi-Civita and the content of the tensor calculus.

Of the many exercises in the book some involve merely direct application of the text, but most of them constitute an extension of it.

In the writing of the book I have received valuable assistance and criticism from Professor H. P. Robertson and from my students, Messrs. Isaac Battin, Albert J. Coleman, Douglas R. Crosby, John Giese, Donald C. May, and in particular, Wayne Johnson.

The excellent line drawings and half-tone illustrations were conceived and executed by Mr. John H. Lewis.

Princeton, September 27, 1940   LUTHER PFAHLER EISENHART.

vii

Contents

CHAPTER IV

SURFACES IN SPACE

CHAPTER I

Curves in Space

1. CURVES AND SURFACES. THE SUMMATION CONVENTION

Consider space referred to a set of rectangular axes. Instead of denoting, as usual, the coordinates with respect to these axes by x, y, z we use x^1, x^2, x^3, since by using the same letter x with different superscripts to distinguish the coordinates we are able often to write equations in a condensed form. Thus we refer to the point of coordinates x^1, x^2, x^3 as the point x^i, where i takes the values 1, 2, 3. We indicate a particular point by a subscript as for example x_1^i, and when a point is a general or representative point we use x^i without a subscript. When the axes are rectangular, we call the coordinates *cartesian*.

In this notation parametric equations of the line through the point x_1^i and with direction numbers u^1, u^2, u^3 are*

$$(1.1) \qquad\qquad x^i = x_1^i + u^i t \qquad\qquad (i = 1, 2, 3).$$

This means that (1.1) constitute three equations as i takes the values 1, 2, 3. Here t is a parameter proportional to the distance between the points x_1^i and x^i, and t is the distance when u^i are direction cosines, that is, when

$$(1.2) \qquad\qquad \sum_i (u^i)^2 = 1,$$

which we write also at times in the form

$$\sum_i u^i u^i = 1.$$

An equation of a plane is

$$(1.3) \qquad\qquad a_1 x^1 + a_2 x^2 + a_3 x^3 + a = 0,$$

where the a's are constants. In order to write this equation in condensed form we make use of the so-called *summation convention* that when the same index appears in a term as a subscript and a superscript this term stands for the sum of the terms obtained by giving the index

* See C. G., p. 85. A reference of this type is to the author's *Coordinate Geometry*, Ginn and Company, 1939.

each of its values, in the present case the values 1, 2, 3. By means of this convention equation (1.3) is written

$$(1.4) \qquad a_i x^i + a = 0.$$

This convention is used throughout the book. At first it may be troublesome for the reader, but in a short time he will find it to be preferable to using the summation sign in such cases. A repeated index indicating summation is called a *dummy index*. Any letter may be used as a dummy index, but when a term involves more than one such index it is necessary to use different dummy indices. Any index which is not a dummy index and thus appears only once in a term is called a *free index*.

Since two intersecting planes meet in a line, the two equations

$$(1.5) \qquad a_i x^i + a = 0, \qquad b_i x^i + b = 0,$$

that is,

$$a_1 x^1 + a_2 x^2 + a_3 x^3 + a = 0, \qquad b_1 x^1 + b_2 x^2 + b_3 x^3 + b = 0,$$

are equations of a line, provided that the ratios a_1/b_1, a_2/b_2, a_3/b_3 are not equal; if these ratios are equal the planes are coincident or parallel according as a/b is equal to the above ratios or not.*

An equation (1.4) is an equation of a plane in the sense that it picks out of space a two dimensional set of points, this set having the property that every point of a line joining any two points of the set is a point of the set; this is Euclid's geometric definition of a plane. In like manner any functional relation between the coordinates, denoted by

$$(1.6) \qquad f(x^1, x^2, x^3) = 0,$$

picks out a two dimensional set of points, by which we mean that only two of the coordinates of a point of the locus may be chosen arbitrarily. The locus of points whose coordinates satisfy an equation of the form (1.6) is called a *surface*. Thus

$$(1.7) \qquad \sum_i x^i x^i + 2 a_j x^j + b = 0,$$

where the a's and b are constants, is an equation of a sphere with center at the point $-a_i$ and radius r given by

$$r^2 = \sum_i a_i a_i - b.\dagger$$

* C. G., pp. 100, 101.
† C. G., p. 128, Ex. 14.

Whenever throughout this book we consider any function, it is under-stood that the function is considered in a domain within which it is continuous in all its variables, together with such of its derivatives as are involved in the discussion.

Since an equation which does not involve one of the coordinates does not impose any restriction on this coordinate, such an equation is an equation of a *cylinder*. Thus

$$(1.8) \qquad\qquad f(x^1, x^2) = 0$$

is an equation of a cylinder whose generators, or elements, are parallel to the x^3-axis, each generator being determined by a pair of values satisfying (1.8). If x_1^1, x_1^2 are two such values, the generator is defined by the two equations

$$x^1 = x_1^1, \qquad x^2 = x_1^2,$$

these being a special form of (1.5) in this case. It does not follow that when all three coordinates enter as in (1.6) that the surface is not a cylinder, but that if it is a cylinder the generators are not parallel to one of the coordinate axes. Later (§12) there will be given a means of determining whether an equation (1.6) is an equation of a cylinder.

Two independent equations

$$(1.9) \qquad\quad f_1(x^1, x^2, x^3) = 0, \qquad f_2(x^1, x^2, x^3) = 0$$

define a *curve*, a one dimensional locus, for, only one of the coordinates of a point on the locus may be chosen arbitrarily. A line is a curve, its two equations being linear, as for example in (1.5). A curve, being one dimensional, may be defined also by three equations involving a parameter, as

$$(1.10) \qquad\qquad x^i = f^i(t),$$

which are called *parametric equations of the curve*. These are a gen-eralization of equations (1.1). The functions f^i in (1.10) are under-stood to be single-valued and such that for no value of t are all the first derivatives of f^i equal to zero; the significance of this requirement appears in §3.

If $\varphi(u)$ is a single-valued function of u, and one replaces t in equations (1.10) by $t = \varphi(u)$, there is obtained another set of parametric equa-tions of the curve, namely

$$(1.11) \qquad\qquad x^i = f^i(\varphi(u)) = \varphi^i(u).$$

Since all the first derivatives $\dfrac{dx^i}{du}$ are not to be zero for a value of u, there is the added condition on φ that $\dfrac{d\varphi}{du} \neq 0$; this means that the equation $t = \varphi(u)$ has a unique inverse.*

Thus the number of sets of parametric equations of a particular curve is of the order of any function satisfying the above condition. When, in particular, one of the coordinates, say x^3, is taken as parameter, the equations are

(1.12) $$x^1 = f^1(x^3), \qquad x^2 = f^2(x^3), \qquad x^3 = x^3,$$

the forms of the f's depending, of course, upon the curve. From the form of (1.12) it follows that the curve is the intersection of the two cylinders whose respective equations are the first two of (1.12).

When all the points of a curve do not lie in a plane, the curve is said to be *skew* or *twisted*. The condition that a curve with equations (1.10) be a *plane* curve, that is, all of its points lie in a plane, is, as follows from (1.4), that the functions f^i be such that

(1.13) $$a_i f^i + a = 0,$$

that is

$$a_1 f^1 + a_2 f^2 + a_3 f^3 + a = 0,$$

where the a's are constants. Differentiating equation (1.13) three times with respect to t and denoting differentiation by primes, we obtain the three equations

(1.14) $$a_i f^{i'} = 0, \qquad a_i f^{i''} = 0, \qquad a_i f^{i'''} = 0.$$

In order that the a's be not all zero, we must have†

(1.15) $$\begin{vmatrix} f^{1'} & f^{2'} & f^{3'} \\ f^{1''} & f^{2''} & f^{3''} \\ f^{1'''} & f^{2'''} & f^{3'''} \end{vmatrix} = 0.$$

Conversely, we shall show that if three functions $f^i(t)$ satisfy this condition, constants a_i and a can be found satisfying (1.13); and consequently that the curve $x^i = f^i(t)$ is plane. If (1.15) is satisfied there exist quantities b_i, ordinarily functions of t, such that

(1.16) $$b_i f^{i''} = 0, \qquad b_i f^{i'''} = 0, \qquad b_i f^{i''''} = 0.‡$$

* Fine, 1927, 1, p. 55. References of this type are to the Bibliography at the end of the book.
† C. G., p. 114.
‡ C. G., p. 116.

Assuming that the b's in these equations are functions of t and differentiating the first two of these equations with respect to t, the resulting equations are reducible to

(1.16′) $b_i' f^{i'} = 0, \qquad b_i' f^{i''} = 0.$

The b's in the first two of (1.16) are proportional respectively to the cofactors of the elements of the last row in the determinant (1.15).* The same is true of the b''s in (1.16′). Consequently we have

$$\frac{b_1'}{b_1} = \frac{b_2'}{b_2} = \frac{b_3'}{b_3}.$$

If we denote the common value of these ratios by $\varphi'(t)$, we have on integration

$$b_i = a_i e^{\varphi},$$

where the a's are constants. Substituting in the first of (1.16) and discarding the factor e^{φ}, we obtain

$$a_i f^{i'} = 0.$$

On integrating this equation with respect to t, we obtain an equation of the form (1.13). Hence we have†

[1.1] *A curve with equations* (1.10) *is a plane curve, if and only if the functions f^i satisfy equation* (1.15).

When for the curve with the equations

(1.17) $x^1 = c_1 t, \qquad x^2 = c_2 t^2, \qquad x^3 = c_3 t^3$

the expressions for x^i are substituted in an equation of a plane (1.4), we obtain a cubic equation in t for each of whose roots the corresponding point of the curve lies in the given plane. The curve (1.17) is called a *twisted cubic*. When a curve meets a general plane in n points, it is called a *twisted curve of the nth order*.

---

* C. G., p. 104.

† In the numbering of an equation or equations, as in (1.3), the number preceding the period is that of the section and the second number specifies the particular equation or equations. The same applies to the number of a theorem but in this case brackets are used in place of parentheses. This notation is used throughout the book.

From (1.17) we have that the projections of the curve upon the coordinate planes have the respective equations

(1.18)
$$c_1^2 x^2 - c_2(x^1)^2 = 0, \qquad x^3 = 0,$$
$$c_1^3 x^3 - c_3(x^1)^3 = 0, \qquad x^2 = 0,$$
$$c_3^2 (x^2)^3 - c_2^3(x^3)^2 = 0, \qquad x^1 = 0.$$

These curves are shown schematically as follows for positive values of the c's:

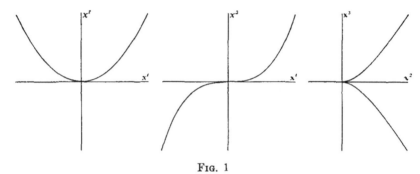

Fɪɢ. 1

From (1.18) it is seen that the curve is the intersection of the three cylinders whose equations are the first of each pair of equations (1.18), the generators of these cylinders being parallel to the x^3-, x^2- and x^1-axes respectively.

It will be found that many formulas and equations can be put in simpler form by means of quantities e_{ijk} and e^{ijk} defined as follows:

(1.19) $\left. \begin{matrix} e_{ijk} \\ e^{ijk} \end{matrix} \right\} = \begin{cases} 0 \text{ when two or three of the indices have the same} \\ \quad \text{values;} \\ 1 \text{ when the respective indices have the values } 1, 2, 3; \\ 2, 3, 1 \text{ or } 3, 1, 2; \\ -1 \text{ when the respective indices have the values } 1, 3, 2; \\ 3, 2, 1 \text{ or } 2, 1, 3. \end{cases}$

Consider, for example, the two equations

$$a_i x^i = 0, \qquad a_i y^i = 0,$$

from which it follows that*

(1.20) $\qquad a_1 : a_2 : a_3 = \begin{vmatrix} x^2 & x^3 \\ y^2 & y^3 \end{vmatrix} : \begin{vmatrix} x^3 & x^1 \\ y^3 & y^1 \end{vmatrix} : \begin{vmatrix} x^1 & x^2 \\ y^1 & y^2 \end{vmatrix}.$

* C. G., p. 104.

Thus, denoting by $1/r$ the factor of proportionality, we have

$$ra_1 = e_{1jk}x^jy^k = e_{123}x^2y^3 + e_{132}x^3y^2 = x^2y^3 - x^3y^2,$$

and consequently (1.20) may be written

$$ra_i = e_{ijk}x^jy^k.$$

Applying this process to the last two of equations (1.14) we have

$$ra_i = e_{ijk}f^{j\prime\prime}f^{k\prime\prime\prime},$$

and when these expressions for a_i are substituted in the first of equations (1.14) we obtain

(1.21) $$e_{ijk}f^{i\prime}f^{j\prime\prime}f^{k\prime\prime\prime} = 0,$$

which is equation (1.15), as one verifies by forming the sum indicated by the summation convention as each of the indices independently takes the values 1, 2, 3.

EXERCISES

1. Parametric equations of a line normal to the plane (1.4) and passing through a point of the curve (1.10) are

(i) $X^i = f^i(t) + a_i u,$

where u is a parameter; as t and u take all values equations (i) give the coordinates of points on the cylinder whose generators pass through points of the curve and are normal to the plane (1.4); when in equations (i) we put

$$u = -\frac{1}{\sum_j (a_j)^2} \left(a + \sum_j a_j f^j\right),$$

the resulting equations are parametric equations of the projection of the curve (1.10) on the plane (1.4).

2. The equation

$$c_2 c_3(t_1 t_2 + t_2 t_3 + t_3 t_1)x^1 - c_1 c_3(t_1 + t_2 + t_3)x^2 + c_1 c_2 x^3 - c_1 c_2 c_3 t_1 t_2 t_3 = 0$$

is an equation of the plane through three points of the twisted cubic (1.17) with parameters t_1, t_2 and t_3.

3. The plane

$$3c_2 c_3 a^2 x^1 - 3c_1 c_3 a x^2 + c_1 c_2 x^3 - c_1 c_2 c_3 a^3 = 0$$

for each value of the constant a meets the curve (1.17) in three coincident points at the point $t = a$.

4. There pass through a given point x_1^i in space three planes with equations of the form of the equation of Ex. 3, and if a_1, a_2, a_3 denote the corresponding values of a, we have

$$a_1 + a_2 + a_3 = \frac{3x_1^1}{c_1}, \qquad a_1 a_2 + a_2 a_3 + a_3 a_1 = \frac{3x_1^2}{c_2}, \qquad a_1 a_2 a_3 = \frac{x_1^3}{c_3};$$

from this result and Ex. 2 it follows that an equation of the plane through the three points in which these three planes meet the cubic (each in three coincident points) is

$$3c_3(x_1^2 x^1 - x_1^1 x^2) + c_1 c_2(x^3 - x_1^3) = 0,$$

which plane passes through the point x_1^i.

5. Four planes determined by a variable chord of the cubic (1.17) and four fixed points of the cubic are in constant cross-ratio.

6. The curve

$$x^1 = a \cos t, \qquad x^2 = a \sin t, \qquad x^3 = b \sin 2t$$

is the intersection of a circular cylinder and a hyperbolic paraboloid.

7. Determine $f(t)$ so that the curve

$$x^1 = a \cos t, \qquad x^2 = a \sin t, \qquad x^3 = f(t)$$

shall be plane; what is the form of the curve?

8. By means of the quantities e_{ijk} and e^{ijk} one has

$$\begin{vmatrix} a_{11} & a_{12} & a_{13} \\ a_{21} & a_{22} & a_{23} \\ a_{31} & a_{32} & a_{33} \end{vmatrix} = e^{ijk} a_{i1} a_{j2} a_{k3} = e^{ijk} a_{1i} a_{2j} a_{3k},$$

and, if the determinant is denoted by a, then

$$e_{ijk} a = e^{lmn} a_{li} a_{mj} a_{nk}.$$

9. Show that

$$\sum_h e_{ijh} e_{klh} = \delta_{ik} \delta_{jl} - \delta_{il} \delta_{jk},$$

where

$$\delta_{ij} = 1 \text{ or } 0 \text{ according as } i = j \text{ or } i \neq j;$$

from this result it follows that

$$\sum_h (e_{ijh} e_{klh} + e_{jkh} e_{ilh} + e_{kih} e_{jlh}) = 0.$$

2. LENGTH OF A CURVE. LINEAR ELEMENT

Consider a curve with the equations

$$(2.1) \qquad\qquad x^i = f^i(t),$$

and the arc of the curve between the points P_0 and P_α for which the parametric values are t_0 and t_α respectively. Consider also intermediate points P_1, P_2, $\cdots$ for which the values of the parameter are t_1, t_2, $\cdots$. The length l_k of the chord $P_k P_{k+1}$ is given by

$$l_k = \sqrt{\sum_i [f^i(t_{k+1}) - f^i(t_k)]^2}$$

$$= \sqrt{\sum_i [f^{i\prime}(\xi_i)]^2}\,(t_{k+1} - t_k)$$

where $\qquad \xi_i = t_k + \theta_i(t_{k+1} - t_k) \qquad 0 < \theta_i < 1,$

the second expression for l_k following from the mean value theorem of the differential calculus, where the prime denotes the derivative. As the number of intermediate points P_k increases indefinitely and each l_k approaches zero, the limit of the sum of the l_k's is the definite integral

$$\int_{t_0}^{t_\alpha} \sqrt{\sum_i \frac{df^i}{dt}\frac{df^i}{dt}}\, dt.$$

By definition this is the *length* of the arc $P_0 P_\alpha$. If then s denotes the length of the arc from the point of parameter t_0 to a representative point of parameter t, we have

$$(2.2) \qquad\qquad s = \int_{t_0}^{t} \sqrt{\sum_i \frac{df^i}{dt}\frac{df^i}{dt}}\, dt.$$

This gives s as a function of t; we denote it by

$$(2.3) \qquad\qquad s = \varphi(t),$$

where φ involves t_0 also.

From (2.2) we have

$$(2.4) \qquad ds^2 = (dx^1)^2 + (dx^2)^2 + (dx^3)^2 \equiv \sum_i dx^i\, dx^i,$$

where $dx^i = \dfrac{df^i}{dt}\, dt$. As thus expressed ds is called the *element of length*, or *linear element*, of the curve.

As remarked in §1 there is a high degree of arbitrariness in the choice of a parameter for a curve. In what follows we shall often find that it adds to the simplicity of a result, if the arc s is taken as parameter. From (2.2) we have

[2.1] *For a curve with equations* (2.1) *the parameter t is the length of the curve measured from a given point, if and only if*

$$(2.5) \qquad\qquad \sum_i \frac{df^i}{dt}\frac{df^i}{dt} = 1.$$

It is evident that s is defined by (2.2) except when

(2.6) $$\sum_i \frac{df^i}{dt} \frac{df^i}{dt} = 0.$$

Since it is understood now, and in what follows, that we are considering only real functions f^i of a real parameter, that is, one assuming only real values, there is no real solution of (2.6) other than f^i constant, that is, the locus is a point. If we admit complex functions of a complex parameter, the curves for which (2.6) hold are called *curves of length zero*, or *minimal curves*. There are cases in which it is advisable to consider such curves, but unless otherwise stated they are not involved in what follows.

Consider a curve defined in terms of the arc s as parameter. Let P and $\bar{P}$ of coordinates x^i and $\bar{x}^i$ be points for which the parameter has the values s and $s + \epsilon$. By Taylor's theorem we have

(2.7) $$\bar{x}^i = x^i + x^{i\prime}\epsilon + \frac{1}{2}x^{i\prime\prime}\epsilon^2 + \frac{1}{\underline{3}}x^{i\prime\prime\prime}\epsilon^3 + \cdots \qquad (i = 1, 2, 3).$$

Here and in what follows an x with one or more primes means that the arc s is the parameter and the primes indicate derivatives with respect to s; if the parameter is other than s, we write $\dfrac{dx^i}{dt}$ and similarly for higher derivatives.

In this notation (2.5) is

(2.8) $$\sum_i x^{i\prime} x^{i\prime} = 1.$$

Differentiating this equation with respect to s, we have

(2.9) $$\sum_i x^{i\prime} x^{i\prime\prime} = 0.$$

If we denote by l the length of the chord $P\bar{P}$, it follows from (2.7), (2.8) and (2.9) that

(2.10) $$\frac{l^2}{\epsilon^2} = \frac{1}{\epsilon^2}\sum_i (\bar{x}^i - x^i)^2 = 1 + \sum_i (\tfrac{1}{4}x^{i\prime\prime} x^{i\prime\prime} + \tfrac{1}{3}x^{i\prime} x^{i\prime\prime\prime})\epsilon^2 + \cdots.$$

From this result it follows that as $\bar{P}$ approaches P along the curve the ratio of the chord to the arc ϵ approaches unity as limit.

3. TANGENT TO A CURVE. ORDER OF CONTACT. OSCULATING PLANE

The quantities $\bar{x}^i - x^i$ in (2.10) are direction numbers of the line through P and $\bar{P}$, and $(\bar{x}^i - x^i)/l$ are its direction cosines. From (2.7) and (2.10) it follows that

$$\lim_{\bar{P} \to P} \frac{\bar{x}^i - x^i}{l} = \lim \frac{\bar{x}^i - x^i}{\epsilon} \frac{\epsilon}{l} = x^{i'}.$$

Since by definition the limiting position of the line through P and $\bar{P}$ as $\bar{P}$ approaches P along the curve is the *tangent to the curve* at P, we have

[3.1] *When for a curve x^i are expressed in terms of the length of the arc from a given point as parameter, the quantities $x^{i'}$ are direction cosines of the tangent at a point x^i.*

If x^i for a curve are expressed in terms of a parameter t, the quantities $\dfrac{dx^i}{dt}$ are direction numbers of the tangent. Thus the tangent is not defined if all of these quantities are zero, which possibility was excluded in §1.

As a result of this theorem we have as parametric equations of the tangent to a curve at a point x^i

$$(3.1) \qquad\qquad X^i = x^i + x^{i'}\, d,$$

where X^i are coordinates of a representative point on the tangent and d is the distance from the point x^i to the point X^i.* We define *positive* sense along the tangent as that for which d in (3.1) is positive, this means that a half line drawn from the origin parallel to the tangent makes with the coordinate axes angles whose cosines are $x^{i'}$. This same convention applies to any line associated with a curve when direction cosines of such a line are given in terms of quantities defining the curve.†

---

* C. G., p. 85.

† Here we define sense by means of direction cosines, which means that a line has two sets of direction cosines, differing in sign. This is not the convention adopted in C. G., pp. 77, 78.

If we denote by α^i the direction cosines of the tangent, we have from the above result and (2.2), when the parameter t is any whatever,

$$(3.2) \qquad \alpha^i = \frac{dx^i}{ds} = \frac{\dfrac{dx^i}{dt}}{\sqrt{\sum_i \dfrac{dx^i}{dt}\dfrac{dx^i}{dt}}}.$$

The plane through a point of a curve and normal to the tangent at the point is called the *normal plane* at the point; an equation of this plane is

$$(3.3) \qquad \sum_i (X^i - x^i)\alpha^i = 0,$$

where α^i are given by (3.2).*

Parametric equations of any line through the point x^i of a curve are

$$(3.4) \qquad X^i = x^i + u^i t,$$

where u^i are direction cosines of the line. The square of the distance of the point $\bar{P}(\bar{x}^i)$ from this line is given by†

$$(3.5) \qquad d^2 = \begin{vmatrix} \bar{x}^1 - x^1 & \bar{x}^2 - x^2 \\ u^1 & u^2 \end{vmatrix}^2 + \begin{vmatrix} \bar{x}^2 - x^2 & \bar{x}^3 - x^3 \\ u^2 & u^3 \end{vmatrix}^2 + \begin{vmatrix} \bar{x}^3 - x^3 & \bar{x}^1 - x^1 \\ u^3 & u^1 \end{vmatrix}^2.$$

When the point $\bar{P}$ is a point of the curve, its coordinates being given by (2.7), the expression (3.5) becomes

$$(3.6) \qquad d^2 = \left[(x^{1'} u^2 - x^{2'} u^1)\epsilon + \frac{1}{2}(x^{1''} u^2 - x^{2''} u^1)\epsilon^2 \right.$$
$$\left. + \frac{1}{|3}(x^{1'''} u^2 - x^{2'''} u^1)\epsilon^3 + \cdots \right]^2$$
$$+ [(x^{2'} u^3 - x^{3'} u^2)\epsilon + \cdots]^2 + [(x^{3'} u^1 - x^{1'} u^3)\epsilon + \cdots]^2.$$

Thus d is of the order of ϵ unless the u^i are proportional to $x^{i'}$; since in the latter case both of these sets of quantities are by theorem [3.1] direction cosines, it follows that $u^i = ex^{i'}$, where e is $+1$ or -1, and that the distance of a point on the curve nearby x^i is of the second, or

* C. G., p. 92.
† C. G., p. 96.

higher, order. The distance is of the third order if also $x^{i'''}$ are proportional to $x^{i'}$, and of the $(n + 1)^{\text{th}}$ order if $x^{i''}, x^{i'''}, \cdots, x^{i(n)}$ are proportional to $x^{i'}$. In general, $n = 1$ and the contact of the tangent with the curve is said to be of the *first order*; for $n > 1$, the *contact is of the nth order*.

If the equations of the curve are in terms of a general parameter t, as (2.1), we have since t is a function of s

(3.7) $\qquad \dfrac{dx^i}{ds} = \dfrac{dx^i}{dt}\dfrac{dt}{ds}, \qquad \dfrac{d^2 x^i}{ds^2} = \dfrac{d^2 x^i}{dt^2}\left(\dfrac{dt}{ds}\right)^2 + \dfrac{dx^i}{dt}\dfrac{d^2 t}{ds^2}, \cdots.$

Hence the tangent at a point t for which t satisfies for some value of n (> 1), if any, the equations

(3.8) $\qquad \dfrac{\dfrac{d^k x^1}{dt^k}}{\dfrac{dx^1}{dt}} = \dfrac{\dfrac{d^k x^2}{dt^k}}{\dfrac{dx^2}{dt}} = \dfrac{\dfrac{d^k x^3}{dt^k}}{\dfrac{dx^3}{dt}} \qquad\qquad (k = 2, \cdots, n)$

has contact of the nth order.

By definition the *osculating plane* of a curve at a point $P(x^i)$ is the limiting position of the plane determined by the tangent at P and a point $\bar{P}$ of the curve as $\bar{P}$ approaches P along the curve. Since the plane passes through P its equation is of the form

(3.9) $\qquad\qquad a_i(X^i - x^i) = 0,$

where a_i being direction numbers of the normal to the plane must be such that

(3.10) $\qquad\qquad a_i x^{i'} = 0.$

Equations (3.9) and (3.10) express the condition that the tangent at P lies in the plane.* Substituting in (3.9) for X^i the expressions (2.7) for $\bar{x}^i$ and making use of (3.10) we obtain

(3.11) $\qquad\qquad a_i\left(\dfrac{1}{2}x^{i''} + \dfrac{1}{|3}x^{i'''}\epsilon + \cdots\right) = 0.$

As $\bar{P}$ approaches P, that is, as ϵ approaches zero, we have in the limit

(3.12) $\qquad\qquad a_i x^{i''} = 0.$

* C. G., p. 120.

In order that equations (3.9), (3.10) and (3.12) be satisfied by a's not all zero, we must have*

$$(3.13) \quad \begin{vmatrix} X^1 - x^1 & X^2 - x^2 & X^3 - x^3 \\ x^{1'} & x^{2'} & x^{3'} \\ x^{1''} & x^{2''} & x^{3''} \end{vmatrix} \equiv e_{ijk}(X^i - x^i)x^{j'}x^{k''} = 0,$$

which is an equation of the osculating plane at the point x'.

When the curve is defined by (2.1) in terms of a general parameter t, it follows from (3.7) that an equation of the osculating plane is

$$(3.14) \quad \begin{vmatrix} X^1 - x^1 & X^2 - x^2 & X^3 - x^3 \\ \dfrac{dx^1}{dt} & \dfrac{dx^2}{dt} & \dfrac{dx^3}{dt} \\ \dfrac{d^2x^1}{dt^2} & \dfrac{d^2x^2}{dt^2} & \dfrac{d^2x^3}{dt^2} \end{vmatrix} = 0.$$

If the tangent at a point has contact of higher order than the first, equation (3.14) is satisfied identically. In this case from (3.11) and (3.7) it follows that an equation of the osculating plane at a point for which the tangent has contact of order $n - 1$ is

$$\begin{vmatrix} X^1 - x^1 & X^2 - x^2 & X^3 - x^3 \\ \dfrac{dx^1}{dt} & \dfrac{dx^2}{dt} & \dfrac{dx^3}{dt} \\ \dfrac{d^n x^1}{dt^n} & \dfrac{d^n x^2}{dt^n} & \dfrac{d^n x^3}{dt^n} \end{vmatrix} = 0.$$

When a curve is plane and its plane is taken for the plane $x^3 = 0$, equation (3.13) is equivalent to $x^3 = 0$, that is, the osculating plane of a plane curve is the plane of the curve. Conversely, when all the osculating planes of a curve coincide, the curve is a plane curve since all of the points of the curve lie in this plane.

EXERCISES

1. The curve

$$x^1 = a \cos t, \qquad x^2 = a \sin t, \qquad x^3 = bt$$

lies on a circular cylinder (see Fig. 2); find the direction cosines of the tangents to the curve and show that the tangent makes a constant angle with the generators of the cylinder; the curve is called a *circular helix*.

---

* C. G., pp. 115, 121.

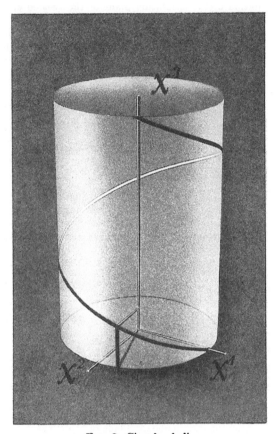

FIG. 2. Circular helix

2. By definition a *cylindrical helix* is a curve lying on a cylinder and which meets all the generators under the same angle; if this constant angle is denoted by θ,

$$x^1 = f^1(t), \qquad x^2 = f^2(t), \qquad x^3 = \cot\theta \int \sqrt{(f^{1\prime})^2 + (f^{2\prime})^2}\, dt,$$

are parametric equations of a cylindrical helix; is any cylindrical helix so defined?

3. By definition a *conical helix* is a curve lying on a cone which meets all the generators of the cone under the same angle; if this angle is denoted by θ, $x^i = f^i(t)$ are equations of a conical helix, if the functions f^i satisfy the conditions

$$a_1(f^1)^2 + a_2(f^2)^2 + a_3(f^3)^2 = 0,$$

$$\sum_i f^i f^{i\prime} = \cos\theta \, \sqrt{\sum_i f^i f^i} \, \sqrt{\sum_i f^{i\prime} f^{i\prime}},$$

where the a's are constants not all of the same sign; is any conical helix so defined?

4. Find an equation of the osculating plane of the twisted cubic (1.17) and compare the result with Ex. 3 of §1.

5. The distance of a point $\bar{P}$ on a curve from the osculating plane at a nearby point P is of the third order at least in the arc $P\bar{P}$; and for any other plane through P not containing the tangent at P the distance is of the first order; discuss the case of planes containing the tangent at P.

6. The curve with the equations

$$x^1 = \frac{1}{2}(1 - t^2)f''(t) + tf'(t) - f(t),$$

$$x^2 = \frac{i}{2}(1 + t^2)f''(t) - itf'(t) + if(t),$$

$$x^3 = tf''(t) - f'(t),$$

where $i = \sqrt{-1}$, $f(t)$ is any function of t, and primes indicate differentiation with respect to the parameter t, is a *minimal* curve, and any minimal curve is so defined; discuss the case when $f(t) = c_1t^2 + c_2t + c_3$, where the c's are constants.

7. If at every point of a curve the tangent has contact of the second order with the curve, the latter is a straight line.

8. In terms of the arc s as parameter equations of the circular helix (Ex. 1) are

$$x^1 = a \cos \frac{s}{\sqrt{a^2 + b^2}}, \qquad x^2 = a \sin \frac{s}{\sqrt{a^2 + b^2}}, \qquad x^3 = \frac{bs}{\sqrt{a^2 + b^2}};$$

each osculating plane of the helix meets the circular cylinder on which it lies in an ellipse.

9. The curve

$$x^1 = a \sin^2 t, \qquad x^2 = a \sin t \cos t, \qquad x^3 = a \cos t$$

is a *spherical curve*, that is, lies on a sphere; its normal planes pass through the center of the sphere; the curve has a double point at $(a, 0, 0)$, and the tangents to the curve at this point are perpendicular to one another.

10. Find an equation of the osculating plane of the curve

$$x^1 = a \cos t + b \sin t, \qquad x^2 = a \sin t + b \cos t, \qquad x^3 = c \sin 2t;$$

find also two equations of the form (1.9) as equations of the curve.

4. CURVATURE. PRINCIPAL NORMAL. CIRCLE OF CURVATURE

Let P and $\bar{P}$ be two points on a curve C, Δs the length of the arc between these points, and $\Delta\theta$ the angle of the tangents at P and $\bar{P}$, that is, the angle between two half-lines through any point and having the positive senses of the two tangents. The limit of $\left|\dfrac{\Delta\theta}{\Delta s}\right|$ as Δs approaches zero, measures the rate of change of the direction of the tangent at P. This limiting value, denoted by κ, is called the *curvature* of C at P, and its reciprocal, denoted by ρ, the *radius of curvature*; from their definition it follows that κ and ρ are non-negative. When C is a plane curve, the definition of curvature here given is that usually given in the differential calculus.

In order to derive in terms of s an expression for κ in terms of the quantities defining a curve C, we consider the auxiliary curve Γ with the equations

$$(4.1) \qquad\qquad X^i = \frac{dx^i}{ds},$$

which in consequence of (2.8) is a curve upon the sphere of unit radius with center at the origin. The radius of the sphere at any point of Γ is parallel to the tangent to C at the corresponding point, that is, the point with this tangent. The curve Γ is called the *spherical indicatrix of the tangents to* C. If we denote by σ the arc of Γ, it follows from (4.1) and (2.4) that

$$(4.2) \qquad\qquad d\sigma^2 = \sum_i \frac{d^2x^i}{ds^2}\frac{d^2x^i}{ds^2}\,ds^2.$$

From the definition of κ we have

$$\kappa = \lim_{\Delta s = 0}\left|\frac{\Delta\theta}{\Delta s}\right| = \lim\left|\frac{\Delta\theta}{\Delta\sigma}\cdot\frac{\Delta\sigma}{\Delta s}\right|.$$

Since $\Delta\theta$ is the length of the arc of the great circle between the points on the unit sphere corresponding to P and $\overline{P}$ on C, the limit of $\dfrac{\Delta\theta}{\Delta\sigma}$ is unity in consequence of the result at the close of §2 and the fact that $\Delta\theta$ and $\Delta\sigma$ have a common chord. Consequently as follows from (4.2)

$$(4.3) \qquad\qquad \kappa = \sqrt{\sum_i \frac{d^2x^i}{ds^2}\frac{d^2x^i}{ds^2}}.$$

For a straight line, with equations (1.1) in which t is the arc, one has $\kappa = 0$, which is evident also geometrically from the fact that the tangent to a straight line at each point is the line itself. In order to obtain the expression for κ when the equations of the curve are in terms of a general parameter t, we observe that from (2.3) we have

$$(4.4) \qquad\qquad \frac{dt}{ds} = \frac{1}{\varphi'}, \qquad \frac{d^2t}{ds^2} = -\frac{\varphi''}{\varphi'^3},$$

where primes denote differentiation with respect to t, and from (2.2)

$$(4.5) \qquad\qquad \varphi'^2 = \sum_i f^{i\prime}f^{i\prime}, \qquad \varphi'\varphi'' = \sum_i f^{i\prime}f^{i\prime\prime},$$

the second following from the differentiation of the first. Substituting in (4.3) from (3.7) and making use of (4.4) and (4.5), we obtain

$$(4.6) \qquad \kappa = \frac{\sqrt{\sum_i \left(\frac{d^2 x^i}{dt^2}\right)^2 - \varphi''^2}}{\varphi'^2}.$$

From (2.9) it follows that the line through the point x^i of a curve and with direction numbers $\frac{d^2 x^i}{ds^2}$ is perpendicular to the tangent at the point, and thus is one of an endless number of normals to the curve at the point. If we define quantities β^i by

$$(4.7) \qquad \frac{d^2 x^i}{ds^2} = \kappa \beta^i,$$

it follows from (4.3) that β^i are direction cosines of the positive sense of this normal.* Its equations are

$$(4.8) \qquad X^i = x^i + \beta^i d,$$

where d denotes the distance of the point X^i from the point x^i of the curve. This normal is called the *principal normal* of the curve at the point. When the expressions (4.8) are substituted in the equation (3.13) of the osculating plane, the equation is satisfied for all values of d, in consequence of (4.7), that is, the principal normal at a point lies in the osculating plane at the point. Hence the osculating plane at a point of a skew curve is the plane determined by the tangent and principal normal of the curve at the point.

The circle in the osculating plane with center at the point

$$(4.9) \qquad X^i = x^i + \rho \beta^i = x^i + \frac{1}{\kappa} \beta^i$$

and of radius ρ is called the *circle of curvature* of the curve at the point x^i and its center the *center of curvature* of the curve for the point x^i. Evidently this circle and the curve have a common tangent at x^i.

EXERCISE

1. When a curve is defined in terms of a general parameter t by (1.10), the direction cosines β^i of the principal normal are given by

$$\beta^i = \frac{\rho}{\varphi'^3} (\varphi' f^{i''} - \varphi'' f^{i'}),$$

where $\varphi(t)$ is defined by (2.3), and primes denote differentiation with respect to t.

---

* See the statement about positive sense after equations (3.1).

2. For a cylindrical helix, as defined in §3, Ex. 2,

$$\varphi'(t) = \csc \theta \sqrt{(f^{1'})^2 + (f^{2'})^2}, \qquad \beta^3 = 0,$$

$$\kappa = \frac{1}{\rho} = e \frac{\csc \theta}{\varphi'^3} (f^{1'}f^{2''} - f^{1''}f^{2'}),$$

where e is $+1$ or -1 so that κ is positive.

3. Let P be a point of a curve; a circle C with a common tangent to the curve at P is determined by requiring that it pass through another point Q of the curve; the limiting circle as Q approaches P along the curve is the circle of curvature of the given curve at P.

4. Find the function $\varphi(t)$ so that the curve

$$x^1 = \int \varphi(t) \sin t \, dt, \qquad x^2 = \int \varphi(t) \cos t \, dt, \qquad x^3 = \int \varphi(t) \tan t \, dt$$

shall be a curve of constant curvature.

5. Determine the form of the function $\varphi(t)$ so that the principal normals to the curve

$$x^1 = t, \qquad x^2 = \sin t, \qquad x^3 = \varphi(t)$$

are parallel to the x^2x^3-plane.

6. The circle of curvature of a curve at a point of the curve has contact of the second order with the curve; every other circle which lies in the osculating plane and is tangent to the curve has contact of the first order; accordingly a circle of curvature is called an *osculating circle* of the curve.

7. Find equations of the surface consisting of the principal normals of a circular helix (see §3, Ex. 1), and show that the locus of the center of curvature is a circular helix.

8. Find the coordinates of the center of curvature of the curve

$$x^1 = a \cos t, \qquad x^2 = a \sin t, \qquad x^3 = a \cos 2t.$$

5. BINORMAL. TORSION

The normal to a curve at a point P which is normal to the osculating plane at P is called the *binormal* at P. Evidently it is perpendicular to the tangent and to the principal normal at P. From (3.13) it follows that

$$\begin{vmatrix} x^{2'} & x^{3'} \\ x^{2''} & x^{3''} \end{vmatrix}, \qquad \begin{vmatrix} x^{3'} & x^{1'} \\ x^{3''} & x^{1''} \end{vmatrix}, \qquad \begin{vmatrix} x^{1'} & x^{2'} \\ x^{1''} & x^{2''} \end{vmatrix}$$

are direction numbers of the binormal. In order to find direction cosines of the binormal we make use of the identity

$$(5.1) \quad \begin{vmatrix} \sum_i x^{i'} x^{i'} & \sum_i x^{i'} x^{i''} \\ \sum_i x^{i'} x^{i''} & \sum_i x^{i''} x^{i''} \end{vmatrix}$$

$$= \begin{vmatrix} x^{2'} & x^{3'} \\ x^{2''} & x^{3''} \end{vmatrix}^2 + \begin{vmatrix} x^{3'} & x^{1'} \\ x^{3''} & x^{1''} \end{vmatrix}^2 + \begin{vmatrix} x^{1'} & x^{2'} \\ x^{1''} & x^{2''} \end{vmatrix}^2.$$

The reader should verify that this is an identity whatever be the quantities involved without any use of the primes as indicating derivatives. When in particular the quantities $x^{i'}$ and $x^{i''}$ have the meaning ascribed to them in §2, it follows from (2.8), (2.9) and (4.3) that the left-hand member of (5.1) is equal to κ^2. Hence direction cosines γ^i of the binormal and the positive sense along the latter are defined by

$$(5.2) \quad \gamma^1 = \rho(x^{2'} x^{3''} - x^{2''} x^{3'}), \qquad \gamma^2 = \rho(x^{3'} x^{1''} - x^{3''} x^{1'}),$$
$$\gamma^3 = \rho(x^{1'} x^{2''} - x^{1''} x^{2'}).$$

These expressions may be written in the form

$$(5.3) \qquad \gamma^i = \rho(x^{j'} x^{k''} - x^{j''} x^{k'}),$$

with the understanding that i, j, k take the values 1, 2, 3 cyclically. Hence equations of the binormal are

$$(5.4) \qquad X^i = x^i + \gamma^i d.$$

The significance of the choice of sign in (5.2) is seen when we observe that the expressions (3.2), (4.7) and (5.2) for α^i, β^i, and γ^i respectively are such that

$$(5.5) \qquad \begin{vmatrix} \alpha^1 & \alpha^2 & \alpha^3 \\ \beta^1 & \beta^2 & \beta^3 \\ \gamma^1 & \gamma^2 & \gamma^3 \end{vmatrix} = +1,$$

as is readily verified since the right-hand member of (5.1) is equal to κ^2. The result (5.5) means that the positive directions of the tangent, principal normal and binormal of a curve at each point of a curve have the mutual orientation of the x^1-, x^2- and x^3-axes respectively (Fig. 3).*

* C. G., p. 162.

From the equations

(5.6)
$$\sum_i \alpha^i \beta^i = 0, \qquad \sum_i \alpha^i \gamma^i = 0, \qquad \sum_i \beta^i \gamma^i = 0,$$
$$\sum_i \alpha^{i^2} = 1, \qquad \sum_i \beta^{i^2} = 1, \qquad \sum_i \gamma^{i^2} = 1$$

and from (5.5) it follows that each element in the determinant equation (5.5) is equal to its cofactor.* This result may be written

(5.7) $\alpha^i = \beta^j \gamma^k - \beta^k \gamma^j, \quad \beta^i = \gamma^j \alpha^k - \gamma^k \alpha^j, \quad \gamma^i = \alpha^j \beta^k - \alpha^k \beta^j,$

as i, j, k take the values 1, 2, 3 cyclically.

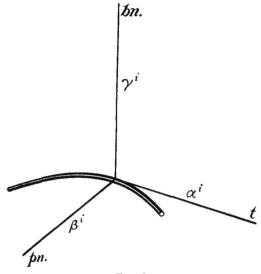

FIG. 3

From the definition of the binormal it follows that the binormals of a plane curve are the normals to the plane at points of the curve, and consequently have the same direction at all points of the curve. For a skew curve the direction of the binormal changes. If $\Delta\theta$ is the angle of the positive directions of the binormals at two points of parameters s and $s + \Delta s$, the limit of $\dfrac{\Delta\theta}{\Delta s}$ as Δs approaches zero measures the rate of change of the direction of the binormal at the point of parameter s, and consequently the rate of change of the orientation of the osculating

* C. G., p. 161.

plane. This limit is called the *torsion* of the curve and is denoted by τ. Sometimes the curvature, as defined in §4, and the torsion are called the *first* and *second* curvatures respectively of the curve.

In order to obtain an expression for τ, we introduce *the spherical indicatrix of the binormal*, that is, the curve defined by

$$(5.8) \qquad X^i = \gamma^i.$$

Evidently this is a curve upon the unit sphere with center at the origin, and such that the radius of the sphere to any point of the curve is parallel to the positive binormal to the given curve at the point with the same value of the parameter s. The linear element of the indicatrix is given by

$$d\sigma^2 = \sum_i \frac{d\gamma^i}{ds} \frac{d\gamma^i}{ds} ds^2.$$

By an argument similar to that used in §4 we have

$$(5.9) \qquad \tau^2 = \sum_i \frac{d\gamma^i}{ds} \frac{d\gamma^i}{ds}.$$

In order to find expressions for $\dfrac{d\gamma^i}{ds}$, we differentiate with respect to s the equations

$$\sum_i \gamma^i \gamma^i = 1, \qquad \sum_i \alpha^i \gamma^i = 0,$$

and obtain

$$(5.10) \qquad \sum_i \gamma^i \frac{d\gamma^i}{ds} = 0, \qquad \sum_i \left(\alpha^i \frac{d\gamma^i}{ds} + \gamma^i \frac{d\alpha^i}{ds} \right) = 0.$$

From (4.7) and (3.2) we have

$$(5.11) \qquad \frac{d\alpha^i}{ds} = \kappa \beta^i,$$

from which and the third of (5.6) it follows that the second of (5.10) reduces to $\sum_i \alpha^i \dfrac{d\gamma^i}{ds} = 0$. From this equation and the first of (5.10) we have* in consequence of the second set of equations (5.7) that $\dfrac{d\gamma^i}{ds}$ is proportional to β^i, and from (5.9) that the factor of proportionality

---

* C. G., p. 104.

is τ or $-\tau$. Thus far τ is defined by (5.9) to within sign; we choose the sign so that we have

$$(5.12) \qquad\qquad \frac{d\gamma^i}{ds} = \tau\beta^i.$$

We are now in position to obtain an expression for τ in terms of the derivatives of x^i. In fact, if we differentiate (5.3) with respect to s, the result may be written in consequence of (5.12)

$$\tau\beta^i = \frac{1}{\rho}\frac{d\rho}{ds}\,\gamma^i + \rho(x^{j'}\,x^{k'''} - x^{j'''}\,x^{k'}).$$

If this equation is multiplied by β^i and summed with respect to i, the result becomes in consequence of the third of (5.6) and (4.7)

$$(5.13) \qquad\qquad \tau = -\frac{1}{\kappa^2}\begin{vmatrix} x^{1'} & x^{2'} & x^{3'} \\ x^{1''} & x^{2''} & x^{3''} \\ x^{1'''} & x^{2'''} & x^{3'''} \end{vmatrix}.$$

From the definition of torsion it follows that τ is zero for a plane curve. Conversely, if τ is zero at every point of a curve, the latter is plane in accordance with theorem [1.1] and equation (5.13). At points of a curve, if any, for which the determinant in (5.13) is zero the osculating plane is said to be *stationary*.

EXERCISES

1. When a curve is defined in terms of a general parameter t by (1.10), the direction cosines γ^i of the binormal are given by

$$\gamma^i = \frac{\rho}{\varphi'^3}\,(f^{j'}f^{k''} - f^{j''}f^{k'}),$$

as i, j, k take the values 1, 2, 3 cyclically, where $\varphi(t)$ is defined by (2.3) and primes denote differentiation with respect to t; also the torsion of the curve is given by

$$\tau = -\frac{1}{\kappa^2\,\varphi'^6}\begin{vmatrix} f^{1'} & f^{2'} & f^{3'} \\ f^{1''} & f^{2''} & f^{3''} \\ f^{1'''} & f^{2'''} & f^{3'''} \end{vmatrix}.$$

2. The curvature and torsion of a circular helix, as defined in §3, Ex. 1, are constants, namely

$$\kappa = \frac{a}{a^2 + b^2}, \qquad \tau = \frac{-b}{a^2 + b^2}.$$

3. For a cylindrical helix, as defined in §3, Ex. 2,

$$\tau = -\frac{\cot\theta}{\kappa^2\varphi'^6}\frac{(f_1'f_2'' - f_1''f_2')^3}{[(f_1')^2 + (f_2')^2]^{3/2}},$$

from which result and that of §4, Ex. 2 it follows that

$$\tau = -e\kappa\cot\theta,$$

that is, κ/τ is a constant.

4. Find the curvature and torsion of the curve

$$x^1 = e^t, \qquad x^2 = e^{-t}, \qquad x^3 = \sqrt{2}\,t.$$

5. The curvature and torsion of the curve

$$x^1 = a(3t - t^3), \qquad x^2 = 3at^2, \qquad x^3 = a(3t + t^3)$$

are given by

$$\kappa = -\tau = \frac{1}{3a(1 + t^2)^2}.$$

6. Find the points of the curve of §3, Ex. 9 at which the torsion is equal to zero.

7. When two curves are symmetric with respect to a point, or a plane, their curvatures at corresponding points are equal and their torsions differ in sign.

8. A necessary and sufficient condition that the circle of curvature have contact of the third order with the curve at a point is that at the point $\tau = 0$, $\dfrac{d\kappa}{ds} = 0$ (see §4, Ex. 6); at such a point the circle is said to *superosculate* the curve.

9. If θ and φ are the angles made with a fixed line in space by the tangent and binormal respectively of a curve,

$$\frac{\sin\theta\,d\theta}{\sin\varphi\,d\varphi} = \frac{\kappa}{\tau}.$$

10. A necessary and sufficient condition that the principal normals of a curve are parallel to a fixed plane is that the curve be a cylindrical helix.

11. If the curve $x^i(s)$ is a cylindrical helix so also is the curve with the equations

$$X^i = \rho\alpha^i - \int \beta^i\,ds.$$

12. The curvature and torsion of the curve

$$x^1 = \int f(t)\sin t\,dt, \qquad x^2 = \int f(t)\cos t\,dt, \qquad x^3 = \int f(t)\psi(t)\,dt$$

are given by

$$\kappa = \frac{1}{f}\sqrt{\frac{1 + \psi^2 + \psi'^2}{(1 + \psi^2)^3}}, \qquad \tau = \frac{1}{f}\frac{\psi'' + \psi}{1 + \psi^2 + \psi'^2};$$

from this result it follows that the curves for which κ or τ is a constant can be found by quadratures.

6. THE FRENET.FORMULAS. THE FORM OF A CURVE IN THE NEIGHBORHOOD OF A POINT

If the second set of equations (5.7), that is,

$$\beta^i = \gamma^j \alpha^k - \gamma^k \alpha^j$$

as i, j, k take the values 1, 2, 3 cyclically, are differentiated with respect to s, the result is reducible by (5.11), (5.12) and (5.7) to

$$\frac{d\beta^i}{ds} = \kappa(\gamma^j \beta^k - \gamma^k \beta^j) + \tau(\beta^j \alpha^k - \beta^k \alpha^j) = -(\kappa \alpha^i + \tau \gamma^i).$$

Gathering together this result, (5.11), and (5.12), we have the following set of equations fundamental in the theory of skew curves and called the *Frenet formulas*:

$$(6.1) \qquad \frac{d\alpha^i}{ds} = \kappa \beta^i, \qquad \frac{d\beta^i}{ds} = -(\kappa \alpha^i + \tau \gamma^i), \qquad \frac{d\gamma^i}{ds} = \tau \beta^i.$$

On replacing β^i in the second set of equations (6.1) by $\frac{1}{\kappa} x^{i''}$ (see equations (4.7)), the resulting equations are reducible to

$$(6.2) \qquad x^{i'''} = -\kappa^2 \alpha^i + \frac{d\kappa}{ds} \beta^i - \kappa \tau \gamma^i.$$

Differentiating with respect to s and making use of (6.1), we obtain

$$(6.3) \quad x^{i''''} = -\frac{3}{2} \frac{d\kappa^2}{ds} \alpha^i + \left(\frac{d^2\kappa}{ds^2} - \kappa^3 - \kappa \tau^2\right) \beta^i - \left[\frac{d}{ds}(\kappa \tau) + \tau \frac{d\kappa}{ds}\right] \gamma^i.$$

We observed following equation (5.5) that the positive directions of the tangent, principal normal and binormal at each point of a curve have the mutual orientation of the coordinate axes. If then we take for coordinate axes these lines at a point P_0 of a curve and measure the arc from the point, we have at P_0

$$(6.4) \qquad \alpha^i = \delta_1^i, \qquad \beta^i = \delta_2^i, \qquad \gamma^i = \delta_3^i,$$

where δ_j^i are Kronecker deltas defined by

$$(6.5) \qquad \delta_j^i = 1 \text{ or } 0 \text{ according as } i = j \text{ or } i \neq j.$$

By Maclaurin's theorem we have that the coordinates x^i of any point on the curve are given by

$$(6.6) \quad x^i = (x^{i'})_0 s + \frac{1}{2}(x^{i''})_0 s^2 + \frac{1}{\lfloor 3}(x^{i'''})_0 s^3 + \frac{1}{\lfloor 4}(x^{i''''})_0 s^4 + \cdots,$$

where a subscript zero indicates the value of the quantity at P_0. From (3.2), (4.7) and (6.4) we have

$$(x^{i'})_0 = \delta_1^i, \qquad (x^{i''})_0 = \kappa_0 \delta_2^i.$$

From these results, (6.2), and (6.3) we have from (6.6) for this choice of coordinate axes

$$x^1 = s - \frac{\kappa_0^2}{6} s^3 - \frac{1}{16} \left(\frac{d\kappa^2}{ds} \right)_0 s^4 + \cdots,$$

(6.7) $$x^2 = \frac{\kappa_0}{2} s^2 + \frac{1}{\lfloor 3} \left(\frac{d\kappa}{ds} \right)_0 s^3 + \frac{1}{\lfloor 4} \left(\frac{d^2\kappa}{ds^2} - \kappa^3 - \kappa\tau^2 \right)_0 s^4 + \cdots,$$

$$x^3 = -\frac{\kappa_0 \tau_0}{6} s^3 - \frac{1}{\lfloor 4} \left[\frac{d}{ds}(\kappa\tau) + \tau \frac{d\kappa}{ds} \right] s^4 + \cdots.$$

It follows from these equations that in the neighborhood of a point, if any, at which $\kappa = 0$ the curve approximates a straight line. Also if $\kappa \neq 0$ the curve with s increasing crosses the osculating plane at the point, from the positive to the negative side when $\tau > 0$ and vice-versa when $\tau < 0$; in the former case the curve is said to be *left-handed* and in the latter *right-handed* at the point. At a stationary point, that is, when $\tau = 0$, the curve remains on the same side of the osculating plane in the neighborhood of the point (provided $d\tau/ds \neq 0$), since in this case the sign of x^3 does not change with s for sufficiently small values of s.

These results follow in fact when we consider only the first terms in each of equations (6.7), that is, the approximate curve

(6.8) $$x^1 = s, \qquad x^2 = \frac{\kappa_0}{2} s^2, \qquad x^3 = -\frac{\kappa_0 \tau_0}{6} s^3,$$

in which κ_0 and τ_0 are constants. This curve is a twisted cubic, whose projections upon the coordinate planes are shown in Fig. 1, the x^1-, x^2-, and x^3-axes being respectively the tangent, principal normal, and binormal of the curve at the point of the given curve which is the origin of the coordinate system used.

The coordinates X^i of a point in the osculating plane to a twisted curve are given by

(6.9) $$X^i = x^i + u\alpha^i + v\beta^i$$

for suitable values of u and v. We raise the question of determining u and v as functions of s so that the locus of the points of coordinates

X^i given by (6.9) shall be an orthogonal trajectory of the osculating planes. Differentiating (6.9) with respect to s and making use of (6.1), we obtain

$$\frac{dX^i}{ds} = \left(1 + \frac{du}{ds} - v\kappa\right)\alpha^i + \left(u\kappa + \frac{dv}{ds}\right)\beta^i - v\tau\gamma^i.$$

Since $\dfrac{dX^i}{ds}$ are direction numbers of the tangent to the desired locus, u and v must be such that $\dfrac{dX^i}{ds}$ are proportional to γ^i, that is, the co-efficients of α^i and β^i must be zero. If we introduce the parameter σ, defined by

$$\sigma = \int \kappa \, ds,$$

this gives the following conditions to be satisfied:

(6.10) $$\frac{du}{d\sigma} + \frac{1}{\kappa} - v = 0, \qquad \frac{dv}{d\sigma} + u = 0.$$

Differentiating the second of equations (6.10) with respect to σ and substituting from the first of (6.10), we obtain

(6.11) $$\frac{d^2v}{d\sigma^2} + v = \frac{1}{\kappa}.$$

From the theory of linear ordinary differential equations it follows that the general solution of equation (6.11) may be obtained by quadratures. When such a solution has been obtained and substituted in the second of equations (6.10), u is given directly. Hence we have

[6.1] *The orthogonal trajectories of the osculating planes of a skew curve can be obtained by quadratures.*

EXERCISES

1. For a plane curve the Frenet formulas are

$$\frac{d\alpha^i}{ds} = \kappa\beta^i, \qquad \frac{d\beta^i}{ds} = -\kappa\alpha^i \qquad (i = 1, 2),$$

and equations of the curve are

(i) $$x^1 = \int \cos \sigma \, ds, \qquad x^2 = \int \sin \sigma \, ds,$$

where

$$\sigma = \int_0^s \kappa \, ds;$$

from equations (i) it follows that σ is the angle which the tangent to the curve makes with the x^1-axis.

2. When all the osculating planes of a curve have a point in common, the curve is plane.

3. The locus of the centers of curvature of a twisted curve of constant curvature is an orthogonal trajectory of the osculating planes of the curve, and is a curve of constant curvature.

4. A tangent to the locus of the centers of curvature of a twisted curve C is perpendicular to the corresponding tangent to C; it coincides with, or is perpendicular to, the principal normal to C only at points for which $\tau = 0$, or $\dfrac{d\kappa}{ds} = 0$.

5. When for a cylindrical helix (see §3, Ex. 2) the generators of the cylinder are parallel to the x^3-axis, then $\alpha^3 = \cos \theta$, and from the Frenet formulas it follows that

(i) $$\beta^3 = 0, \qquad \frac{d\gamma^3}{ds} = 0, \qquad \kappa\alpha^3 + \tau\gamma^3 = 0.$$

If a function σ is defined by

$$\alpha^1 = \sin \theta \cos \sigma, \qquad \alpha^2 = \sin \theta \sin \sigma,$$

then

$$\beta^1 = -e \sin \sigma \qquad \beta^2 = e \cos \sigma, \qquad \kappa = e \sin \theta \frac{d\sigma}{ds},$$

where e is $+1$ or -1 so that κ is positive, and

$$\gamma^3 = e \sin \theta, \qquad \tau = -e\kappa \cot \theta$$

(see §5, Ex. 3).

6. For a curve for which $\tau/\kappa = c$, where c is a constant, it follows from the Frenet formulas that

$$\gamma^i = c\alpha^i + b^i,$$

where b^i are constants, from which it follows that $\sum_i \alpha^i b^i = $ const., that is, the curve makes a constant angle with the lines of direction numbers b^i, and hence is a cylindrical helix.

7. The equations

$$x^i = -a \int (\gamma^j \, d\gamma^k - \gamma^k \, d\gamma^j)$$

as i, j, k take the values 1, 2, 3 cyclically, where a is a constant, and γ^i are functions of a single parameter such that $\sum_i \gamma^i \gamma^i = 1$, are equations of a curve of torsion $1/a$ and γ^i are direction cosines of the binormal. Does it follow from this result that any curve on the unit sphere can serve as the spherical indicatrix of the binormals of a curve of constant torsion?

.8. If C is a curve of constant torsion, for the associated curve with the equations

$$x^i = \frac{\beta^i}{\tau} + \int \gamma^i \, ds$$

the curvature is constant.

9. When two twisted curves are in one-to-one correspondence with tangents at corresponding points parallel, the principal normals at corresponding points are parallel, and also the binormals; two curves so related are said to be deducible from one another by a *transformation of Combescure*.

10. The equations

(i) $\bar{x}^i = x^i + a^i a,$

where a is a constant, are equations of a curve $\bar{C}$ whose points are on the tangents to the curve $x^i(s)$ and at the constant distance a from the corresponding points of contact; the arc $\bar{s}$, the direction cosines $\bar{\alpha}^i$ of the tangent, and the curvature $\bar{\kappa}$ of $\bar{C}$ are given by

$$\bar{s} = \int_0^s \sqrt{1 + a^2 \kappa^2} \, ds, \qquad \bar{\alpha}^i = \frac{\alpha^i + a\kappa\beta^i}{\sqrt{1 + a^2\kappa^2}},$$

$$\bar{\kappa}^2 = \frac{\kappa^2 \left(1 + a^2\kappa^2 + \frac{a}{\kappa}\frac{d\kappa}{ds}\right)^2}{(1 + a^2\kappa^2)^3} + \frac{a^2\kappa^2\tau^2}{(1 + a^2\kappa^2)^2};$$

the tangents to $\bar{C}$ are parallel to the corresponding osculating planes of the given curve.

11. In order that the curve $\bar{C}$ in Ex. 10 be a straight line it is necessary and sufficient that $\tau = 0$ and $a^2 + \frac{1}{\kappa^2} = ce^{2s/a}$, where c is an arbitrary constant. If we put $c = a^2$, we have

$$\sigma = \int \kappa ds = \frac{1}{a} \int \frac{ds}{\sqrt{e^{2s/a} - 1}} = \cos^{-1} e^{-s/a}.$$

From this result and Ex. 1 we have

$$x^1 = -ae^{-s/a}, \qquad x^2 = \int \sqrt{1 - e^{-2s/a}} \, ds,$$

from which and equation (i) of Ex. 10 for $i = 1$ we have that the locus $\bar{C}$ is the x^2-axis. The curve is called the *tractrix*. In terms of σ, the angle which the tangent makes with the x^1-axis, equations of the curve are

$$x^1 = -a \cos \sigma, \qquad x^2 = a \int \frac{\sin^2 \sigma}{\cos \sigma} \, d\sigma = a \, [\log (\sec \sigma + \tan \sigma) - \sin \sigma].$$

12. A curve whose principal normals are the principal normals of another curve is called a *Bertrand curve*; if x^i and $\bar{x}^i$ are the coordinates of corresponding points on the respective curves C and $\bar{C}$ and s and $\bar{s}$ corresponding arcs, one has

$$\bar{x}^i = x^i + h\beta^i,$$

$$\bar{\alpha}^i \frac{d\bar{s}}{ds} = (1 - \kappa h)\,\alpha^i + \frac{dh}{ds}\,\beta^i - \tau h\gamma^i.$$

Since $\bar{\beta}^i = e\beta^i$ by hypothesis, where e is $+1$ or -1, it follows that h is a constant; denoting by ω the angle between the osculating planes of C and its *conjugate* $\bar{C}$, one has $\bar{\alpha}^i = \cos\omega\,\alpha^i + \sin\omega\,\gamma^i$; $\bar{\gamma}^i = e\,(-\sin\omega\,\alpha^i + \cos\omega\,\gamma^i)$; from the Frenet formulas for $\bar{C}$ it follows that ω is a constant and

$$\kappa \sin\omega - \tau\cos\omega = \frac{\sin\omega}{h}, \qquad \frac{d\bar{s}}{ds} = -\frac{h\tau}{\sin\omega}.$$

Also $\bar{\kappa}$ and $\bar{\tau}$ for $\bar{C}$ are given by

$$\bar{\kappa} + e\bar{\tau}\cot\omega + \frac{e}{h} = 0, \qquad \tau\bar{\tau} = \frac{\sin^2\omega}{h^2};$$

thus e is to be chosen so that $\bar{\kappa}$ is non-negative.

13. A curve for which

$$a\kappa + b\tau = 1,$$

where a and b are constants different from zero, is a Bertrand curve; the equations (see §5, Ex. 12)

$$x^1 = \int f(t)\sin t\,dt, \qquad x^2 = \int f(t)\cos t\,dt, \qquad x^3 = \int f(t)\psi(t)\,dt,$$

where $\psi(t)$ is any function of t and

$$f(t) = a\sqrt{\frac{1 + \psi^2 + \psi'^2}{(1 + \psi^2)^3}} + b\,\frac{\psi'' + \psi}{1 + \psi^2 + \psi'^2},$$

a and b being constants different from zero, are equations of a Bertrand curve.

14. A circular helix is a Bertrand curve; it has an infinite number of conjugates, each lying on a circular cylinder with the same axis as that of the given helix.

15. A necessary and sufficient condition that the osculating planes of a Bertrand curve and of its conjugate coincide is that the curve be a plane curve; any curve parallel to the given curve is a conjugate curve.

16. If C is a curve of constant torsion, the curve with equations

$$\bar{x}^i = ax^i + b\left(\frac{\beta^i}{\tau} + \int \gamma^i\,ds\right),$$

where a and b are constants, is a Bertrand curve.

17. The binormals of a curve are the binormals of another curve, if and only if the given curve is plane.

18. In order that the principal normals of a curve C be the binormals of a curve $\bar{C}$, it is necessary and sufficient that

(i) $$\kappa = a(\kappa^2 + \tau^2),$$

where a is a constant; then equations of $\bar{C}$ are

$$\bar{x}^i = x^i + a\beta^i;$$

curves C satisfying the condition (i) can be found by quadratures (see §5, Ex. 12).

7. INTRINSIC EQUATIONS OF A CURVE

When equations (6.3) are differentiated successively with respect to s and in each case the derivatives of α^i, β^i, and γ^i are replaced by their expressions from the Frenet formulas (6.1), we find that each derivative of x^i is expressible linearly and homogeneously in α^i, β^i, and γ^i, the coefficients being functions of κ, τ, and their derivatives of various orders with respect to s. Consequently the coefficients of further terms in (6.7) as derived from (6.6) involve only the values of κ, τ, and their derivatives for $s = 0$, because of the particular values (6.4) at the origin in the coordinate system used. Hence, if for two curves the functions κ and τ of s are the same functions, the expressions for x^i for each curve relative to the axes consisting of the tangent, principal normal, and binormal of each curve at the point $s = 0$ are the same. Since either set of axes can be brought into coincidence with the other by a rigid motion, we have:

[7.1] *Two curves whose curvature and torsion are the same functions respectively of the arc are congruent.*

From this it follows that a curve is determined to within its position in space by the expressions for κ and τ in terms of s. Consequently

(7.1) $$\kappa = f_1(s), \qquad \tau = f_2(s)$$

are equations of the curve. Since they are independent of the coordinate system used, they are called *intrinsic equations* of the curve.

From the manner in which equations (6.7) were obtained, it follows that κ and τ derived from these equations by means of (4.3) and (5.13) are power series in s, the coefficients being values of κ, τ, and their derivatives evaluated for $s = 0$. Consequently if we have any two equations (7.1) in which $f_1(s)$ is a non-negative function of s, the corresponding equations (6.7) are equations of the curve for which (7.1) are the intrinsic equations. Although this method of obtaining x^i as functions of s gives the equations of a curve for given functions f_1 and

f_2 in equations (7.1), it gives these equations as infinite series. We shall consider another approach to this problem which may in certain cases lead to finite expressions for x^i.

The three sets of quantities α^i, β^i, and γ^i as i takes the values 1, 2, 3 are seen from (6.1) to be solutions of the following system of ordinary differential equations where κ and τ are given functions of s:

$$(7.2) \qquad \frac{du}{ds} = \kappa v, \qquad \frac{dv}{ds} = -(\kappa u + \tau w), \qquad \frac{dw}{ds} = \tau v.$$

If we have three sets of solutions, u^1, v^1, w^1; u^2, v^2, w^2; u^3, v^3, w^3 of these equations, which may be denoted by u^i, v^i, w^i for $i = 1, 2, 3$, and put

$$(7.3) \qquad u^i u^j + v^i v^j + w^i w^j = c^{ij} \qquad (i, j = 1, 2, 3),$$

we find by differentiation that in consequence of (7.2) the c's are constants. We define quantities by

$$(7.4) \qquad \alpha^i = a_k^i u^k, \qquad \beta^i = a_k^i v^k, \qquad \gamma^i = a_k^i w^k,$$

where the a's are constants, and seek under what conditions these a's can be chosen so that

$$(7.5) \qquad \alpha^i \alpha^j + \beta^i \beta^j + \gamma^i \gamma^j = \delta^{ij} \qquad (i, j = 1, 2, 3),$$

where

$$(7.6) \qquad \delta^{ij} = 1 \text{ or } 0 \text{ according as } i = j \text{ or } i \neq j.$$

This choice is made in order that α^i, β^i, and γ^i shall be direction cosines of three mutualy perpendicular vectors. Substituting from (7.4) in (7.5) and making use of (7.3) we have

$$(7.7) \qquad a_k^i a_l^j c^{kl} = \delta^{ij}.$$

If the determinant $|\, c^{kl} \,|$ is different from zero,

$$c^{kl} x_k x_l = 0$$

is an equation in homogeneous coordinates x_1, x_2, x_3 of a non-degenerate conic in the plane. With respect to this conic a_1^1, a_2^1, a_3^1; a_1^2, a_2^2, a_3^2; a_1^3, a_2^3, a_3^3 satisfying (7.7) are the coordinates of the vertices of a self-polar triangle, and consequently an endless number of sets of a_j^i satisfying (7.7) can be found.* For such a set of a_j^i the quantities α^i, β^i, γ^i defined by (7.4) are solutions of equations (7.2), that is, we have equations (6.1), the signs of a_j^i having been chosen so that equation (5.5) holds.

---

* Veblen and Young, 1910, 1, p. 282.

If then we define x^i by

$$(7.8) \qquad\qquad x^i = \int \alpha^i \, ds,$$

for the curve so defined s is the arc and α^i direction cosines of the tangent. Also from (7.8) and (6.1) we obtain by differentiation equations (4.7) and (6.2), from which it follows that κ and τ are the curvature and torsion of the curve, and β^i and γ^i direction cosines of the principal normal and binormal respectively.

Thus we have shown that three sets of solutions of equations (7.2) for given functions κ and τ of s lead by quadratures (7.8) to a curve for which κ and τ are the curvature and torsion, provided that the determinant $|c^{kl}|$ is not equal to zero. That there are sets of solutions of equations (7.2) satisfying this condition follows from the theory of such sets of equations, namely that there exists a unique solution for a given set of initial values of u, v and w.* Since c^{kl} are constants, it follows that one has only to choose the initial values of the three sets of solutions so that it shall follow from (7.3) that the determinant of the c's is not equal to zero.

EXERCISES

1. A solution of the equation

$$\frac{d^3u}{ds^3} - \frac{d}{ds} \log \tau \kappa^2 \frac{d^2u}{ds^2} + \left[\kappa \frac{d^2}{ds^2}\left(\frac{1}{\kappa}\right) + \frac{1}{\kappa\tau} \frac{d\kappa}{ds}\frac{d\tau}{ds} + \kappa^2 + \tau^2 \right] \frac{du}{ds} + \kappa\tau \frac{d}{ds}\left(\frac{\kappa}{\tau}\right) u = 0,$$

and v and w given by

$$v = \frac{1}{\kappa} \frac{du}{ds}, \qquad w = -\frac{1}{\kappa\tau}\left(\frac{d^2u}{ds^2} - \frac{1}{\kappa}\frac{d\kappa}{ds}\frac{du}{ds} + \kappa^2 u \right)$$

constitute a solution of equations (7.2); and any solution of (7.2) is expressed in terms of three sets of solutions by (7.4) for suitable values of the constants a_k^i.

2. If u, v, w are solutions of equations (7.2) such that $u^2 + v^2 + w^2 = 1$, the quantities σ and ω defined by

$$\frac{u+iv}{1-w} = \frac{1+w}{u-iv} = \sigma, \qquad \frac{u+iv}{1+w} = \frac{1-w}{u-iv} = -\omega$$

are solutions of the Riccati equation

$$\frac{d\theta}{ds} = \frac{i\tau}{2}\left(1 - \frac{2\kappa}{\tau}\theta - \theta^2 \right).$$

---

* Goursat, 1924, 1, vol. 2, p. 368.

3. The general integral of an equation of Riccati

$$\frac{d\theta}{ds} = L + 2M\theta + N\theta^2,$$

where L, M, N are functions of s, is of the form

$$\theta = \frac{aP + Q}{aR + S},$$

where a is an arbitrary constant, and P, Q, R, S are functions of s.

4. From theorem [7.1] and §5 Ex. 2 it follows that a necessary and sufficient condition that a curve be a circular helix is that its curvature and torsion be constant; show also by means of Ex. 1 that this condition is sufficient.

5. Establish the statement made about the number of solutions of equations (7.7) by purely algebraic methods.

8. INVOLUTES AND EVOLUTES OF A CURVE

As shown in §3 the equations

(8.1) $X^i = x^i + u\alpha^i$

are parametric equations of the tangents to the curve C defined by x^i as functions of the arc s. For a particular tangent u is the distance between the points x^i and X^i. If u is replaced in equations (8.1) by a function of s, the resulting equations are equations of a curve Γ whose points lie on the tangents to the given curve. Differentiating equation (8.1) with respect to s, one has in consequence of (3.2) and (6.1)

(8.2) $\frac{dX^i}{ds} = \left(1 + \frac{du}{ds}\right)\alpha^i + \kappa u\beta^i.$

Since $\dfrac{dX^i}{ds}$ are direction numbers of the tangent to Γ, if the latter curve is to be such that its tangent at each point is perpendicular to the tangent to C through the point, it is necessary and sufficient that

$$\sum_i \alpha^i \frac{dX^i}{ds} = 0.$$

Since $\sum_i \alpha^i\beta^i = 0$, this condition is that

(8.3) $\frac{du}{ds} + 1 = 0$

from which it follows that

(8.4) $u = c - s,$

where c is an arbitrary constant. Hence there is an infinity of such curves Γ, each defined by

$$(8.5) \qquad\qquad X^i = x^i + (c - s)\alpha^i$$

for a particular value of c. They are called the *involutes* of the given curve. From (8.4) it follows that the length of the segment of any tangent to the curve determined by two involutes has the same value, the difference of the c's of the two involutes.

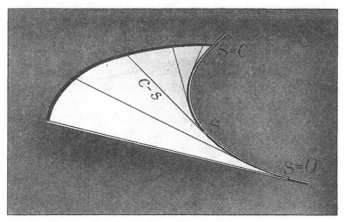

Fig. 4. Involute of the circular helix of Fig. 2

When a curve C is defined in terms of a general parameter t, the determination of s requires a quadrature (2.2), and then the involutes are given directly by (8.5).

An involute when $c - s$ is positive may be described mechanically as follows: Take a string of length c, fasten one end at the point of the curve for which $s = 0$ and bring the string into coincidence with the curve; when the string is unwound from the curve and is kept taut, the other end point describes the involute as is seen from (8.5).

From (8.2) and (8.4) we have

$$(8.6) \qquad\qquad \frac{dX^i}{ds} = (c - s)\kappa\beta^i .$$

Hence we have

[8.1] *A curve has an infinity of involutes; a tangent to an involute at a point X^i is parallel to the principal normal to the curve at the corresponding point x^i.*

If Γ is an involute of a curve C, we say that C is an *evolute* of Γ, that is, it is a curve whose tangents are normal to Γ. Suppose then that we start with a curve of coordinates x^i, and seek its evolutes. Since the points of an evolute lie in the normal planes to the curve, its equations are of the form

$$(8.7) \qquad X^i = x^i + u\beta^i + v\gamma^i,$$

where u and v are functions of s to be determined. From these equations in consequence of (3.2) and (6.1) we have by differentiation with respect to s

$$(8.8) \qquad \frac{dX^i}{ds} = (1 - u\kappa)\alpha^i + \left(\frac{du}{ds} + v\tau\right)\beta^i + \left(\frac{dv}{ds} - u\tau\right)\gamma^i.$$

Since these quantities are direction numbers of the tangent to an evolute, they must be proportional to $X^i - x^i$, that is, to $u\beta^i + v\gamma^i$, which as follows from (8.7) are direction numbers of the line joining the points X^i and x^i. Consequently we must have $u = 1/\kappa$ and

$$v\left(\frac{du}{ds} + v\tau\right) = u\left(\frac{dv}{ds} - u\tau\right).$$

When this equation is written in the form

$$u\frac{dv}{ds} - v\frac{du}{ds} = (u^2 + v^2)\tau,$$

we see that its integral is

$$(8.9) \qquad \frac{v}{u} = \tan(\omega + c),$$

where by definition

$$(8.10) \qquad \omega = \int \tau\, ds,$$

and c is an arbitrary constant. Substituting these results in (8.7), we obtain

$$(8.11) \qquad X^i = x^i + \frac{1}{\kappa}(\beta^i + \tan(\omega + c)\,\gamma^i).$$

For each value of c these are equations of an evolute. Consequently a curve has an infinity of evolutes. Since the curve is an involute of each evolute, we have

[8.2] *A curve C has an infinity of evolutes; the principal normal to an evolute is parallel to the tangent to C at the corresponding point.*

From equations (8.11) and (4.9) it follows that the points of all of these evolutes corresponding to a given point on the given curve lie on the line, called the *polar line*, parallel to the binormal and through the center of curvature for the given point on the curve. Moreover from (8.9) it follows that $\omega + c$ is the angle which the line joining a point on the curve to the corresponding point on an evolute makes with the osculating plane of the curve at this point. Hence we have

[8.3] *When each of the normals to a curve C which are tangent to an evolute s turned through the same angle about the corresponding tangent to C, the inormals in their new position are tangent to another evolute of C.*

From (8.11) and (8.10) it follows that it is possible to choose c so that the points of the corresponding evolute lie in the osculating planes only in case ω is a constant, in which case $\tau = 0$, that is, when the curve is plane. In this exceptional case this evolute is the locus of the centers of the circle of curvature and is a plane curve, except when the given curve is a circle, in which case the locus is the center of the circle. This evolute is the one which in the differential calculus is called *the* evolute of the plane curve. However, a plane curve has an infinity of evolutes, as c in equations (8.11) takes all possible values. From the form of these equations it follows that these evolutes lie on the cylinder whose generators are the normals to the plane of the given curve at points of the evolute in the plane, that is, the evolute for which $c = -\omega$. Moreover, from the remark preceding theorem [8.3] it follows that the tangents to each evolute make constant angles with the generators of this cylinder and consequently these evolutes are cylindrical helices (see §3 Ex. 2). Hence we have

[8.4] *A plane curve other than a circle has an infinity of evolutes, each of which is a helix of the cylinder whose right section by the plane of the curve is the plane evolute of the curve.*

EXERCISES

1. The involutes of a circular helix (§3, Exs. 1, 8) are plane curves, which also are involutes of circular sections of the circular cylinder upon which the helix lies.

2. For an involute (8.5) for which $c - s$ is positive the arc, direction cosines of the tangent, principal normal, binormal, and the curvature and torsion are

given by

$$8 = \int_0^8 (c - 8)\kappa \, ds, \qquad \bar{\alpha}^i = \beta^i,$$

$$\bar{\beta}^i = - \frac{\kappa\alpha^i + \tau\gamma^i}{\sqrt{\kappa^2 + \tau^2}}, \qquad \bar{\gamma}^i = - \frac{\tau\alpha^i - \kappa\gamma^i}{\sqrt{\kappa^2 + \tau^2}},$$

$$\bar{\kappa} = \frac{\sqrt{1 + \left(\dfrac{\tau}{\kappa}\right)^2}}{c - 8}, \qquad \bar{\tau} = \frac{\kappa \dfrac{d}{ds}\left(\dfrac{\tau}{\kappa}\right)}{(c - 8)(\kappa^2 + \tau^2)}.$$

3. A necessary and sufficient condition that the involutes of a twisted curve be plane curves is that the curve be a cylindrical helix (see §6, Ex. 6).

4. Find the evolutes of the curves of §5, Exs. 4 and 5.

5. For an evolute (8.11) the arc, direction cosines of the tangent, principal normal, and binormal, and the curvature and torsion are given by

$$8 = \int_0^8 \left[\frac{d}{ds}\left(\frac{1}{\kappa}\right) + \frac{\tau}{\kappa} \tan(\omega + c)\right] \sec(\omega + c) \, ds,$$

$$\bar{\alpha}^i = \cos(\omega + c)\beta^i + \sin(\omega + c)\gamma^i, \qquad \bar{\beta}^i = - e\alpha^i,$$

$$\bar{\gamma}^i = e[-\sin(\omega + c)\beta^i + \cos(\omega + c)\gamma^i],$$

$$\bar{\kappa}, \bar{\tau} = \frac{e\kappa \cos(\omega + c), \ - \kappa \sin(\omega + c)}{\sec(\omega + c)\left[\dfrac{d}{ds}\left(\dfrac{1}{\kappa}\right) + \dfrac{\tau}{\kappa} \tan(\omega + c)\right]},$$

where e is +1 or −1 so that $\bar{\kappa}$ is positive.

9. THE TANGENT SURFACE OF A CURVE.   THE POLAR SURFACE.   OSCULATING SPHERE

When for any curve the two parameters s and u are eliminated from the three parametric equations of its tangents, namely

(9.1) $$X^i = x^i(s) + u\alpha^i(s),$$

we obtain a single equation in the X^i.   Consequently (see §1) the locus of points on the tangents to a curve is a surface, called the *tangent surface* of the curve, and each of the tangent lines is called a *generator* of the surface.   When the curve is defined in terms of a general parameter t equations of the tangent surface are

(9.2) $$X^i = x^i(t) + u \frac{dx^i}{dt},$$

where now u is not the distance of the point X^i from the point x^i as it is in (9.1).

When s and u in (9.1) are given particular values, equations (9.1) give the coordinates of a point on the surface. Thus the locus is two dimensional, which is another proof that the locus is a surface. If in (9.1) we replace u by a function of s, say $\varphi(s)$, the resulting equations are parametric equations of a curve on the surface. In particular, if we put $u = c - s$, where c is a constant, then, as follows from §8, the curve for each value of c is an involute of the given curve. Consequently

Fig. 5. Tangent surface of the circular helix of Fig. 2

all the involutes of a curve lie on its tangent surface. They are the curves which intersect the generators of the surface at right angles, that is, the involutes are the *orthogonal trajectories* of the generators. Since there is only one set of orthogonal trajectories of a set of lines, we have

[9.1] *The orthogonal trajectories of the generators of the tangent surface of a curve are the involutes of the curve.*

According as u in equation (9.1) has a positive or negative value the point lies on the portion of the tangent drawn in the positive direction from a point on the curve or in the opposite direction. Hence the surface consists of two parts, or sheets, one part consisting of all the points for which $u \geq 0$, the other of the points for which $u \leq 0$. Thus the curve forms a common boundary of the two sheets.

In order to get an idea of the form of the surface in the neighborhood of the curve, we recall from §6 that in the neighborhood of the point P_0 ($s = 0$) the curve approximates the twisted cubic

$$(9.3) \qquad x^1 = s, \qquad x^2 = \frac{\kappa_0 s^2}{2} \qquad x^3 = -\frac{\kappa_0 \tau_0 s^3}{6},$$

where κ_0 and τ_0 are the curvature and torsion of the given curve at the point P_0, the tangent, principal normal, and binormal at P_0 being the coordinate axes. Noting that s is the arc of the given curve but not of the cubic, we have that equations of the tangent surface to the curve (9.3) are

$$(9.4) \quad X^1 = s + u, \qquad X^2 = \kappa_0 \left(\frac{s^2}{2} + su \right), \qquad X^3 = -\frac{\kappa_0 \tau_0}{2} \left(\frac{s^3}{3} + s^2 u \right).$$

In this coordinate system the plane $x^1 = 0$ is the plane normal to the given curve and to the cubic at P_0, and cuts this tangent surface in the curve Γ for which $u = -s$. From the second and third of equations (9.4) it follows that equations of this plane section are

$$X^1 = 0, \qquad X^2 = \frac{-\kappa_0 s^2}{2}, \qquad X^3 = \frac{\kappa_0 \tau_0 s^3}{3}.$$

On eliminating s from the second and third equations we see that the curve is a semi-cubical parabola with the negative half of the principal normal for cuspidal tangent. Since this is the case at every ordinary point of the curve, that is, every point at which neither κ nor τ is zero, we have

[9.2] *The tangent surface of a curve consists of two sheets which are tangent to one another along the curve, and thus form a sharp edge, namely the curve.*

The curve is called the *edge of regression* of the surface. An idea of the form of the surface in the neighborhood of the curve may be had from Figs. 5 and 6.

The line with the equations

$$(9.5) \qquad X^i = x^i + \rho \beta^i + u \gamma^i$$

for each value of s is the polar line, as defined in §8, corresponding to the point x^i on the curve for this value of s, and the parameter u is the distance of a point on the line from the corresponding center of curvature, namely the point $x^i + \rho\beta^i$. Since equations (9.5) involve two parameters, s and u, it follows that the totality of the polar lines of a curve constitute a surface; it is called the *polar surface* of the curve. We shall show that this surface is the tangent surface of another curve defined by (9.5) when u is replaced by a suitable function of s.

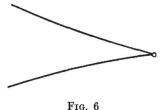

Fɪɢ. 6

If u is any function of s, we have from (9.5) in consequence of (3.2) and (6.1)

$$(9.6) \qquad \frac{dX^i}{ds} = \left(\frac{d\rho}{ds} + \tau u\right)\beta^i + \left(\frac{du}{ds} - \rho\tau\right)\gamma^i.$$

These are direction numbers of the tangent to the curve (9.5) for u a given function of s. If the polar line with respect to the given curve is to be the tangent to the curve (9.5) for u equal to some function of s, the direction numbers must be proportional to γ^i. This requirement is satisfied, if and only if $u = -\dfrac{1}{\tau}\dfrac{d\rho}{ds}$. Hence we have

[9.3] *The polar surface of a curve is the tangent surface of the curve with the equations*

$$(9.7) \qquad X^i = x^i + \rho\beta^i - \frac{1}{\tau}\frac{d\rho}{ds}\gamma^i.$$

Consider the sphere Σ with center at the point of coordinates (9.7) for a given value of s and passing through the corresponding point x^i on the curve, that is, the point for this value of s. Evidently the circle of curvature for the point x^i lies on Σ, and the square of radius R of Σ is $\rho^2 + \left(\dfrac{1}{\tau}\dfrac{d\rho}{ds}\right)^2$. We shall show that Σ has contact of the third order with the curve at P, that is, if $R + \delta$ denotes the distance between the center of Σ and a point $\bar{P}$ on the curve such that the arc $P\bar{P}$ is s, then δ is of the order of s^4. In order to establish this result we make use of equations (6.7). The center of the sphere corresponding to the point

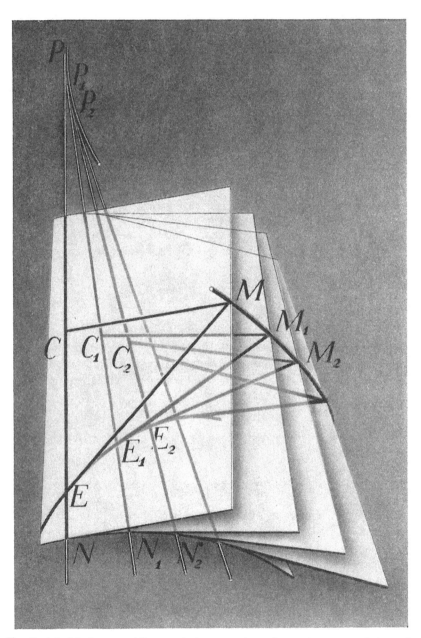

FIG. 7. A twisted curve with normal planes, centers of curvature, an evolute, and the polar developable

of the curve at the origin of the coordinate system used in (6.7) is at the point $\left(0, \rho, -\dfrac{1}{\tau}\dfrac{d\rho}{ds}\right)$. Hence we have

$$(R + \delta)^2 = \bar{x}^{1^2} + (\rho - \bar{x}^2)^2 + \left(-\frac{1}{\tau}\frac{d\rho}{ds} - \bar{x}^3\right)^2.$$

On substituting for $\bar{x}^i$ from (6.7) we find that $2R\delta + \delta^2$ is equal to an expression in the fourth and higher powers in s as was to be proved. Hence we have

[9.4] *The sphere through an arbitrary point P of a skew curve and with center given by (9.7) has contact of the third order with the curve at P.*

This sphere is called the *osculating sphere* of the curve at the point P. Hence we have

[9.5] *The polar line for a point of a curve is tangent to the locus of the center of the osculating sphere of the curve at the corresponding point.*

The previous results are represented in Fig. 7 in which the curve is the locus of the points M, M_1, M_2, $\cdots$; the points C, C_1, C_2, $\cdots$ are the corresponding centers of curvature; the planes MCN, $M_1C_1N_1$, $\cdots$ are normal to the curve; the lines CP, C_1P_1, $\cdots$ are the polar lines; and the locus of the points P, P_1, $\cdots$ is the edge of regression of the polar surface.

A curve all of whose points lie on a sphere is called a *spherical* curve. The normal planes to such a curve pass through the center of the sphere. The coordinates of the center are given by equations (8.7), where u and v must be such functions of s that $\dfrac{dX^i}{ds} = 0$. From (8.8) it follows that $u = \dfrac{1}{\kappa} = \rho$ and that

(9.8) $$\frac{d\rho}{ds} + \tau v = 0, \qquad \frac{dv}{ds} - \rho\tau = 0.$$

From the first of these equations we have $v = -\dfrac{1}{\tau}\dfrac{d\rho}{ds}$ and hence the coordinates of the point are given by (9.7). When this value of v is substituted in the second of (9.8), we have the following condition to be satisfied by ρ and τ:

(9.9) $$\rho\tau + \frac{d}{ds}\left(\frac{1}{\tau}\frac{d\rho}{ds}\right) = 0.$$

From (9.7) it follows that the square of the distance between the points X^i and x^i is $\rho^2 + \left(\dfrac{1}{\tau} \dfrac{d\rho}{ds} \right)^2$, which is a constant when the condition (9.9) is satisfied.

Conversely, when the condition (9.9) is satisfied it follows from (9.7) that $\dfrac{dX^i}{ds} = 0$. Thus, when equation (9.9) is satisfied, all the normal planes to the curve pass though a fixed point, which is at the same distance from the points of the curve. Hence we have:

[9.6] *A necessary and sufficient condition that a curve be spherical is that its intrinsic equations satisfy the condition* (9.9).

Also we have proved incidentally that

[9.7] *When all the normal planes to a curve have a point in common, the curve is spherical.*

EXERCISES

1. The tangent surface of the cubic (1.17) is an algebraic surface of the fourth order.

2. The osculating plane of the curve (9.3) at the point $s = 0$ is the plane $x^3 = 0$, which meets the tangent surface (9.4) in the generator and in the parabola $x^1 = 2s/3$, $x^2 = \kappa_0 s^2/6$, whose curvature is $\frac{3}{4}\kappa_0$; thus the osculating plane at a point P of a curve meets the tangent surface in a generator and in a curve whose curvature at P is three-fourths of the curvature of the curve at P.

3. The polar surface of a plane curve is a cylinder, whose right section is the plane evolute of the curve.

4. Any sphere which contains the circle of curvature for a point P of a non-spherical twisted curve and which is not the osculating sphere at P has contact of the second order with the curve at P.

5. The angle between the radius of the osculating sphere of a twisted curve at a point P and the locus of the center of the sphere is equal to the angle between the radius of the circle of curvature and the locus of its center.

6. When a twisted curve is spherical, the center of curvature for a point is the orthogonal projection of the center of the sphere upon the osculating plane.

7. The only spherical curves of constant curvature are circles.

10. PARAMETRIC EQUATIONS OF A SURFACE.
COORDINATES AND COORDINATE CURVES
IN A SURFACE

Equations (9.2) and (9.5) of the tangent surface and polar surface of a curve respectively are particular cases of three equations of the form

(10.1) $$ x^i = f^i(u^1, u^2) \qquad (i = 1, 2, 3). $$

Suppose now that one has three such equations where f^i are one-valued functions of two variables u^1 and u^2. If the functions f^i are such that it is possible to eliminate u^1 and u^2 from these equations and obtain a single equation

(10.2) $$F(x^1, x^2, x^3) = 0,$$

then in accordance with the definition of a surface in §1 equations (10.1) are equations of a surface. This idea was used in §9 in establishing that equations (9.2) are equations of a surface.

In order to determine whether three equations (10.1) in which the f's are functions of the u's are equations of a surface, we define quantities A^{ij} thus

(10.3) $$A^{ij} \equiv \frac{\partial(f^i, f^j)}{\partial(u^1, u^2)} \equiv \begin{vmatrix} \dfrac{\partial f^i}{\partial u^1} & \dfrac{\partial f^j}{\partial u^1} \\ \dfrac{\partial f^i}{\partial u^2} & \dfrac{\partial f^j}{\partial u^2} \end{vmatrix} \qquad (i, j = 1, 2, 3; i \neq j),$$

that is, A^{ij} is the jacobian of f^i and f^j with respect to u^1 and u^2. If A^{12} is identically equal to zero, then there is a functional relation between x^1 and x^2, that is, $\varphi_1(x^1, x^2) = 0$.* If also $A^{13} = 0$, there is a functional relation $\varphi_2(x^1, x^3) = 0$, and thus the locus is a curve and not a surface (see §1). It is readily shown that if $A^{12} = A^{13} = 0$ then $A^{23} = 0$. Hence we have

[10.1] *Three equations* (10.1) *are equations of a surface, if the jacobian matrix*

(10.4) $$\begin{Vmatrix} \dfrac{\partial f^1}{\partial u^1} & \dfrac{\partial f^2}{\partial u^1} & \dfrac{\partial f^3}{\partial u^1} \\ \dfrac{\partial f^1}{\partial u^2} & \dfrac{\partial f^2}{\partial u^2} & \dfrac{\partial f^3}{\partial u^2} \end{Vmatrix}$$

is of rank two, that is, not all of the determinants of order two are identically zero.

It is to be observed, as just remarked, that if two of these determinants are identically zero, the third also is, and the rank is less than two (see Ex. 6).

The definition of a surface by three equations in terms of two variables u^1 and u^2 as in (10.1) was introduced by Gauss.† Formerly a surface

* Fine, 1927, 1, p. 257.
† 1827, 1.

was defined by a single equation (10.2), until Monge used the particular form

(10.5) $x^3 = f(x^1, x^2)$.

The latter had advantages over the form (10.2) in the investigation of certain types of surfaces. However, the method of Gauss is in many respects superior to both the other methods. It is customary to refer to equations of the form (10.1) as *parametric equations of a surface*. In particular, the method of Monge is equivalent to the following equations:

(10.6) $x^1 = x^1$, $x^2 = x^2$, $x^3 = f(x^1, x^2)$,

where now x^1 and x^2 are the variables u^1 and u^2.

Another interpretation of theorem [10.1] is that u^1 and u^2 are independent variables, and thus that the locus is two dimensional. When in equations (10.1) u^1 and u^2 are given particular values, these equations give the coordinates of a point in the surface, as viewed from space in which the surface lies, that is, the *enveloping space*. However, one may consider the situation in the surface itself without reference to the enveloping space, and say that u^1 and u^2 are *coordinates in the surface of a point* in the surface. An example of this is the use of latitude and longitude as coordinates in the surface of the earth.

It should be remarked that, if one eliminates u^1 and u^2 from equations of the form (10.1) and obtains an equation of the form (10.2) it may be that equations (10.1) apply to only a portion of the surface with the resulting equation (10.2) (see Ex. 3). However, in consequence of theorem [10.1] it follows that two of the equations (10.1) can be solved for u^1 and u^2, namely two for which the corresponding A^{ij} is not identically zero, and when these values are substituted in the third one obtains an equation similar to (10.5). This equation defines the same surface or portion of a surface as equations (10.1), and thus for the coordinates x^i of a point in the surface u^1 and u^2 are uniquely defined.

When in equations (10.1) u^2 is given a constant value and u^1 varies, the locus is a curve, as viewed from the enveloping space, its equations being of the form (1.10) with u^1 as parameter; moreover, it is a curve in the surface. There is an infinity of such curves in the surface, one for each value of u^2; we call them the *coordinate curves* $u^2 = const.$, and also the u^1-*coordinate curves*. In like manner there is an infinity of *coordinate curves* $u^1 = const.$, and called also the u^2-*coordinate curves*. They are the analogue of the lines parallel to the coordinate axes in the plane referred to a cartesian system. When the plane is referred to polar coordinates, they are the analogue of the lines through the pole of the system, and of the circles with center at the pole (see Fig. 8).

The surface with equations of the form

(10.7) $x^1 = f^1(u^1),   x^2 = f^2(u^1),   x^3 = f^3(u^1, u^2)$

is a cylinder whose equation of the form $\varphi(x^1, x^2) = 0$ is obtained on the elimination of u^1 from the first two of equations (10.7); in this case the curves $u^1 = $ const. are the generators of the cylinder, and the character of the curves $u^2 = $ const. depends upon the form of the function f^3.

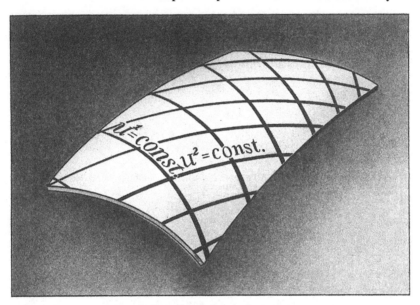

Fig. 8. Coordinate curves in a surface

In general the coordinate curves are not straight lines but curves. Accordingly some writers call u^1 and u^2 *curvilinear coordinates* in the surface, but we call them simply coordinates in the surface.

If in equations (10.1) we replace u^2 by

(10.8) $u^2 = \varphi(u^1),$

the resulting equations are equations of a curve, as viewed from the enveloping space. Since this curve is in the surface, an equation of the form (10.8) is an equation of a curve in the surface expressed in terms of the coordinates u^1 and u^2 in the surface, as is also an equation of the form

(10.9) $\varphi(u^1, u^2) = 0.$

For example, for a cylinder (10.7) the equation (see §3, Ex. 2)

$$f^3(u^1, u^2) - \cot \theta \int \sqrt{(f^{1'})^2 + (f^{2'})^2}\, du^1 = 0$$

is an equation of a curve in the cylinder which intersects the generators under the constant angle θ, that is, the curve is a cylindrical helix. Also each of the equations

$$f^i(u^1, u^2) = c^i$$

as i takes the values 1, 2, 3 and the c's are constants, is an equation of the family of curves in which the corresponding planes $x^i = c^i$ in space cut the surface with equations (10.1).

If in (10.1) we substitute for u^1 and u^2 the expressions

$$(10.10) \qquad\qquad u^\alpha = \varphi^\alpha(u'^1, u'^2) \qquad\qquad (\alpha = 1, 2),$$

where the φ's are independent functions of u'^1 and u'^2, that is,

$$(10.11) \qquad\qquad \frac{\partial(\varphi^1, \varphi^2)}{\partial(u'^1, u'^2)} \not\equiv 0,$$

we obtain another set of parametric equations of the surface, say

$$(10.12) \qquad\qquad x^i = f'^i(u'^1, u'^2) \qquad\qquad (i = 1, 2, 3).$$

Consequently there is great generality in the choice of two coordinates in terms of which a surface is defined by means of equations of the form (10.1).

We refer to (10.10) as a *transformation of coordinates in the surface*. If the jacobian (10.11) is identically zero, this means that there is a functional relation between the φ's, say $F(\varphi^1, \varphi^2) = 0$, and then from (10.10) it follows that the coordinates u'^α for $\alpha = 1$, 2 apply only to the curve $F(u^1, u^2) = 0$ and not to the whole surface. Another significance of the condition (10.11) is that equations (10.10) can be solved for u'^α,* say

$$(10.13) \qquad\qquad u'^\alpha = \varphi'^\alpha(u^1, u^2),$$

which equations give the coordinates u'^α of a point of coordinates u^α in the u-system. Equations (10.13) define the *inverse* of the transformation (10.10).

It should be remarked that even if (10.11) holds, there may be values of u'^α for which the jacobian is equal to zero. For such values there is

* Fine, 1927, 1, p. 334.

not a unique inverse (10.13), and special consideration has to be given to such cases. However, the discussion which follows presupposes that for the domain considered the condition (10.11) holds. Clearly for such a domain no two of the u^1- or u^2-coordinate curves can intersect.

From (10.10) it follows that the coordinate curves $u^\alpha = c^\alpha$ in the u-system have the equations

$$\varphi^\alpha(u'^1, u'^2) = c^\alpha$$

in the u'-system, and from (10.13) that the coordinate curves $u'^\alpha = c'^\alpha$ in the u'-system have the equations

$$\varphi'^\alpha(u^1, u^2) = c'^\alpha$$

in the u-system. However, for a transformation

(10.14) $$u^1 = \varphi^1(u'^1), \qquad u^2 = \varphi^2(u'^2)$$

or

(10.15) $$u^1 = \varphi^1(u'^2), \qquad u^2 = \varphi^2(u'^1)$$

the *net of coordinate curves*, that is, the two families of curves $u^1 =$ const. and $u^2 =$ const., is not changed but the coordinates are changed, as is readily verified.

EXERCISES

1. The equations
$$x^1 = a \sin u^1 \cos u^2, \qquad x^2 = a \sin u^1 \sin u^2, \qquad x^3 = a \cos u^1,$$
where a is a constant, are parametric equations of a sphere of radius a. What are the coordinate curves $u^1 =$ const. and $u^2 =$ const.?

2. A surface with the equations

(i) $$x^1 = u^1 \cos u^2, \qquad x^2 = u^1 \sin u^2, \qquad x^3 = \varphi(u^1)$$

is the surface generated when the plane curve with equations

$$x^3 = \varphi(x^1), \qquad x^2 = 0$$

is revolved about the x^3-axis; such a surface is called a *surface of revolution*. What are the coordinate curves $u^1 =$ const. and $u^2 =$ const.? The latter are called the *meridian curves*; by what change of coordinates can the equations of Ex. 1 be given the form (i)?

3. The equations

$$x^i = \pm \sqrt{\frac{a_i(a_i - u^1)(a_i - u^2)}{(a_i - a_j)(a_i - a_k)}}$$

in which the a's are constants, and i, j, k take the values 1, 2, 3, cyclically, are equations of a central quadric; it is

an ellipsoid when $a_1 > u^1 > a_2 > u^2 > a_3 > 0$,

an hyperboloid of one sheet when $a_1 > u^1 > a_2 > 0 > a_3 > u^2$,

an hyperboloid of two sheets when $a_1 > 0 > a_2 > u^1 > a_3 > u^2$.

4. A surface which is the locus of a line perpendicular to a fixed line, called the axis, and satisfies a further condition is called a *right conoid*; equations of a right conoid are

$$x^1 = u^1 \cos u^2, \qquad x^2 = u^1 \sin u^2, \qquad x^3 = \varphi(u^2);$$

when $\varphi(u^2) = a \cot u^2 + b$, where a and b are constants, the conoid is a hyperbolic paraboloid.

5. Find equations of a right conoid whose axis is the x^3-axis and which contains the ellipse

$$x^1 = a, \qquad \frac{(x^2)^2}{b^2} + \frac{(x^3)^2}{c^2} = 1.$$

6. When the rank of the jacobian matrix (10.4) is two and one of the determinants of the second order is identically zero, the surface is a cylinder.

11. TANGENT PLANE TO A SURFACE

The tangent at a point P to a curve upon a surface

$$(11.1) \qquad x^i = f^i(u^1, u^2) \qquad (i = 1, 2, 3)$$

is called a *tangent to the surface* at P. It is evident that there are an infinity of tangent lines to a surface at a point. We shall show that ordinarily these lines lie in a plane, called the *tangent plane* to the surface at the point.

If we define a curve through a point x^i by the equations

$$(11.2) \qquad u^\alpha = \varphi^\alpha(t) \qquad (\alpha = 1, 2),$$

we have from (11.1)

$$\frac{dx^i}{dt} = \frac{\partial f^i}{\partial u^1} \frac{du^1}{dt} + \frac{\partial f^i}{\partial u^2} \frac{du^2}{dt}.$$

Hence equations of the tangent at x^i are (see §3)

$$X^i - x^i = \left(\frac{\partial f^i}{\partial u^1} \frac{du^1}{dt} + \frac{\partial f^i}{\partial u^2} \frac{du^2}{dt} \right) l,$$

where l is a parameter. Eliminating $l \dfrac{du^1}{dt}$ and $l \dfrac{du^2}{dt}$ from these equations, we obtain

(11.3)
$$\begin{vmatrix} X^1 - x^1 & X^2 - x^2 & X^3 - x^3 \\ \dfrac{\partial f^1}{\partial u^1} & \dfrac{\partial f^2}{\partial u^1} & \dfrac{\partial f^3}{\partial u^1} \\ \dfrac{\partial f^1}{\partial u^2} & \dfrac{\partial f^2}{\partial u^2} & \dfrac{\partial f^3}{\partial u^2} \end{vmatrix} = 0.$$

For particular values of u^1 and u^2, that is, at a point on the surface, this is an equation of a plane. Since the equation is independent of the functions φ^1 and φ^2 in (11.2), it follows that this plane contains all the tangent lines to the surface at the point, and consequently is an equation of the tangent plane at the point. At points of the surface, if any, for which the cofactors of the elements of the first row in the determinant (11.3) are simultaneously equal to zero, the equation (11.3) is not defined. Such points are called *singular points* of the surface, and all other points are called *ordinary points*. Hence we have

[11.1] *The tangents at an ordinary point on a surface to the curves on the surface through the point lie in a plane; when the surface is defined by parametric equations* (11.1), *equation* (11.3) *is an equation of the tangent plane at the point* x^i.

For the tangent surface to a curve with the equations (9.1) the equation (11.3) reduces to

$$\begin{vmatrix} X^1 - x^1 & X^2 - x^2 & X^3 - x^3 \\ \alpha^1 & \alpha^2 & \alpha^3 \\ \beta^1 & \beta^2 & \beta^3 \end{vmatrix} = 0,$$

which by means of (5.7) is

$$\sum_i (X^i - x^i)\gamma^i = 0.$$

Since this equation does not involve u, the tangent plane at one point of the surface is tangent to the surface at each point of the generator through the given point. Comparing this equation with (3.13), we have in consequence of (4.7)

[11.2] *The tangent plane to the tangent surface of a curve is the same at all points of a generator; it is the osculating plane of the curve at the point where the generator is tangent to the curve.*

In order to obtain an equation of the tangent plane to the polar surface of a curve, we note that in consequence of equations (6.1) we have from (9.5)

$$\frac{\partial X^i}{\partial s} = \left(\frac{d\rho}{ds} + u\tau\right)\beta^i - \rho\tau\gamma^i, \qquad \frac{\partial X^i}{\partial u} = \gamma^i.$$

Hence in this case equation (11.3) is reducible to

$$\begin{vmatrix} X^1 - x^1 & X^2 - x^2 & X^3 - x^3 \\ \beta^1 & \beta^2 & \beta^3 \\ \gamma^1 & \gamma^2 & \gamma^3 \end{vmatrix} = 0,$$

where in this equation X^i are current coordinates. In consequence of (5.7) this equation reduces to

$$\sum_i (X^i - x^i)\alpha^i = 0.$$

From the form of the equation and the fact that it does not involve u' we have

[11.3]. *The tangent plane to the polar surface of a curve is the same at all points of a generator; it is the normal plane to the curve at the point whose polar line is this corresponding generator of the polar surface.*

In order to find an equation of the tangent plane to a surface defined by a single equation

(11.4) $f(x^1, x^2, x^3) = 0,$

we assume that x^i for the surface are expressed in the form (11.1) in terms of two coordinates u^1 and u^2. When these expressions are substituted in (11.4) the resulting equation is an identity in u^1 and u^2, and consequently the derivatives of this identity with respect to u^1 and u^2 are equal to zero. Hence we have

$$\sum_i \frac{\partial f}{\partial x^i}\frac{\partial f^i}{\partial u^1} = 0, \qquad \sum_i \frac{\partial f}{\partial x^i}\frac{\partial f^i}{\partial u^2} = 0.$$

From these equations it follows that the quantities $\frac{\partial f}{\partial x^i}$ are proportional to the cofactors of the elements of the first row in equation (11.3).* Consequently an equation of the tangent plane to the surface (11.4)

* C. G., p. 104.

at a point x^i is

(11.5) $$\sum_i \frac{\partial f}{\partial x^i}(X^i - x^i) = 0.$$

EXERCISES

1. Find an equation of the tangent plane to the sphere with equations of the form in §10, Ex. 1, and show therefrom that the tangent plane at a point is normal to the radius of the sphere at the point.

2. Find an equation of the tangent plane at a point of a cylinder with equations (10.7), and show that the tangent planes at all points of a generator are the same plane.

3. Find an equation of the tangent plane at a point of the cone $a_1x^{1^2} + a_2x^{2^2} + a_3x^{3^2} = 0$, where the a's are not all of the same sign, and show that the tangent planes at all points of a generator are the same plane.

4. For a surface with the equation (10.5) an equation of the tangent plane is

$$(X^1 - x^1)\frac{\partial f}{\partial x^1} + (X^2 - x^2)\frac{\partial f}{\partial x^2} - (X^3 - x^3) = 0;$$

this result follows also from (10.6) and (11.3).

5. The tangent plane to the right conoid (see §10, Ex. 4)

$$x^1 = u^1 \cos u^2, \qquad x^2 = u^1 \sin u^2, \qquad x^3 = a \sin u^2$$

at a point not on the x^3-axis meets the conoid in the generator through the point of tangency and in an ellipse.

6. The tangent planes at points of a generator of the right conoid

$$x^1 = u^1 \cos u^2, \qquad x^2 = u^1 \sin u^2, \qquad x^3 = a\sqrt{\tan u^2}$$

meet the plane $x^3 = 0$ in parallel lines.

7. The normal to the tangent plane to a surface at the point of tangency is called the *normal to the surface* at the point; the normals to any right conoid (see §10, Ex. 4) at points of a generator are one family of rulings of a hyperbolic paraboloid.

8. A surface with equations of the form

$$x^1 = u^1 \cos u^2, \qquad x^2 = u^1 \sin u^2, \qquad x^3 = \varphi(u^1) + au^2,$$

where a is a constant, is called a *helicoid*; the coordinate curves $u^1 = $ const are circular cylindrical helices; find an equation of the tangent plane.

9. The distance of a point Q on a surface from the tangent plane to the surface at a nearby point P is of the second order at least in comparison with the length of the arc PQ.

12. DEVELOPABLE SURFACES. ENVELOPE OF A ONE-PARAMETER FAMILY OF SURFACES

From the form of the equation (11.3) of a tangent plane to a surface it follows that ordinarily this equation involves both of the parameters u^1 and u^2, and consequently in general there is a double infinity of

tangent planes to a surface. This is the case, for example, with a
sphere, the tangent plane at a point being normal to the radius of the
sphere at the point, as one sees geometrically (see §11, Ex. 1). However, we have seen that the tangent planes along a generator of the
tangent surface of a curve coincide, this tangent plane for any generator
being the osculating plane of the curve at the point of tangency of the
given generator. Likewise the tangent planes to a cylinder, or a cone,
along a generator are the same as is evident geometrically. Hence the
tangent planes to a tangent surface of a curve, to a cone, or to a cylinder
involve only one parameter, and consequently these surfaces are the
envelopes of a one-parameter family of planes. They are called
developable surfaces, since any such surface can be *developed* upon a
plane, that is, rolled out without stretching or contracting any part of
it. It is evident that this can be done with a cylinder or a cone.

We desire to show that with the exception of cylinders and cones
every developable surface is the tangent surface of some curve. We
consider first the more general problem of finding the envelope of a
one-parameter family of surfaces.

An equation

$$(12.1) \qquad\qquad f(x^1, x^2, x^3; u^1) = 0$$

involving a parameter u^1 as well as the x's is an equation of a one-
parameter family of surfaces, each of which is defined by (12.1) when
u^1 is assigned a particular value. Consider now the curve of intersection of the surfaces (12.1) and $f(x^1, x^2, x^3; u^1 + \Delta u^1) = 0$ for particular
values of u^1 and Δu^1. This curve is the curve of intersection also of
(12.1) and the surface

$$\frac{f(x^1, x^2, x^3; u^1 + \Delta u^1) - f(x^1, x^2, x^3; u^1)}{\Delta u^1} = 0.$$

There is such a curve unless $f(x^1, x^2, x^3; u^1) = \varphi(x^1, x^2, x^3) + \psi(u^1)$, in
which case the preceding equation does not involve x^i. We assume,
therefore, that u^1 does not enter in this manner. As Δu^1 approaches
zero this curve approaches a limiting curve defined by (12.1) and

$$(12.2) \qquad\qquad \frac{\partial f}{\partial u^1} = 0.$$

The curves defined by (12.1) and (12.2) as u^1 takes all values are called
the *characteristics* of the family of surfaces (12.1). They form a surface
E, called the *envelope* of the surfaces (12.1). Each surface of the family
and the envelope have one of the characteristics in common, and, as we

shall show, are tangent to one another along this characteristic; that is, at each point of the characteristic the surface and envelope have the same tangent plane.

Let the coordinates x^i of a point on the envelope be expressed in terms of u^1 and some second parameter u^2, say $x^i = f^i(u^1, u^2)$. With this choice of coordinates the coordinate curves $u^1 = $ const. are the characteristics. When these expressions for x^i are substituted in (12.1), the resulting equations are identities in u^1 and u^2, and consequently the derivatives of this expression with respect to u^1 and u^2 are equal to zero. In consequence of (12.2) we have

$$\sum_i \frac{\partial f}{\partial x^i} \frac{\partial x^i}{\partial u^1} = 0, \qquad \sum_i \frac{\partial f}{\partial x^i} \frac{\partial x^i}{\partial u^2} = 0.$$

From these equations it follows that the cofactors of the elements of the first row in the equation (11.3) of the tangent plane to the envelope are proportional to $\dfrac{\partial f}{\partial x^i}$, and consequently an equation of the tangent plane to the envelope at the point x^i is

$$\sum_i (X^i - x^i) \frac{\partial f}{\partial x^i} = 0.$$

This is an equation of the tangent plane to a surface (12.1) for a given value of u^1, as follows from (11.5). Since u^2 does not enter in $\dfrac{\partial f}{\partial x^i}$, it follows that at each point of a characteristic the corresponding surface (12.1) and the envelope have the same tangent plane.

We consider now in addition to equations (12.1) and (12.2) the equation

(12.3) $$\frac{\partial^2 f}{\partial u^{1^2}} = 0.$$

We denote by

(12.4) $$x^i = f^i(u^1)$$

the common solution, if any, of equations (12.1), (12.2) and (12.3). Equations (12.4) are equations of a curve on the envelope of the surfaces (12.1) since they are solutions of (12.1) and (12.2), and for a value $u^1 = a^1$ the corresponding point of the curve (12.4) is a point of the characteristic $u^1 = a^1$. We wish to show that at this point the curve and the characteristic have a common tangent line.

If the expressions (12.4) are substituted in (12.1) and (12.2) the resulting equations are identities in u^1, and consequently their derivatives with respect to u^1 are equal to zero. In consequence of (12.2) and (12.3) the result of differentiating (12.1) and (12.2) with respect to u^1 is

$$(12.5) \qquad \sum_i \frac{\partial f}{\partial x^i} f^{i'} = 0, \qquad \sum_i \frac{\partial^2 f}{\partial u^1 \partial x^i} f^{i'} = 0,$$

where the prime indicates differentiation with respect to u^1. At a point of the curve (12.4) the quantities $f^{i'}$ are direction numbers of the tangent to the curve at the point.

A characteristic $u^1 = a^1$ is the intersection of the surfaces defined by (12.1) and (12.2) for this value of u^1. From (11.5) it follows that the tangent planes to these two surfaces at the point $x_1^i = f^i(a^1)$ are respectively

$$\sum_i (X^i - x_1^i) \frac{\partial f}{\partial x^i} = 0, \qquad \sum_i (X^i - x_1^i) \frac{\partial^2 f}{\partial u^1 \partial x^i} = 0.$$

Taken together these are equations of the line of intersection of these two planes, that is, the tangent at x_1^i to the characteristic $u^1 = a^1$. From these equations and (12.5) it follows that $f^{i''}$ are direction numbers of the tangent at x_1^i to the characteristic. Consequently the characteristics are tangent to the curve (12.4), and accordingly the latter is the envelope of the characteristics. It is called the *edge of regression* of the envelope E of the surfaces (12.1). Gathering together these results we have

[12.1] *The envelope of a one-parameter family of surfaces is a surface which is tangent to each surface of the family along a curve, the characteristic corresponding to the particular surface; the characteristics are tangent to a curve, the edge of regression of the envelope.*

When all the characteristics have one and only one point in common, that is, when equations (12.1), (12.2) and (12.3) admit a common solution only for a particular value of u^1, this point is a degenerate edge of regression.

We apply these results to a one-parameter family of planes with the equation

$$(12.6) \qquad a_i x^i + a = 0,$$

where a_i and a are functions of a parameter u^1 with the understanding that a_i are not proportional to a set of constants c_i, that is, that the planes of the family are not parallel.* In this case equations (12.2)

* C. G., p. 101.

and (12.3) are respectively

$$(12.7) \qquad a_i' x^i + a' = 0, \qquad a_i'' x^i + a'' = 0,$$

where the primes denote differentiation with respect to u^1. From (12.6) and the first of (12.7) it follows that the characteristics are straight lines. If the determinant A, defined by

$$(12.8) \qquad A = \begin{vmatrix} a_1 & a_2 & a_3 \\ a_1' & a_2' & a_3' \\ a_1'' & a_2'' & a_3'' \end{vmatrix},$$

is not identically zero, equations (12.6) and (12.7) admit a unique solution (12.4), and by theorem [12.1] the envelope is the tangent surface of the curve (12.4), unless the functions $f^i(u^1)$ in (12.4) are constants. If $f^i(u^1)$ are constants c^i, then from (12.6) we have $a = -a_i c^i$, in which case all of the planes of the family pass through the point c^i. When all the planes (12.6) have a point in common, the envelope is a cone, unless all the planes have a line in common. In the latter case the line of intersection of the planes (12.6) and the first of (12.7) must be independent of u^1; that is, that line and the one with equations (12.7) must coincide. The condition for this is that the matrix

$$(12.9) \qquad \begin{Vmatrix} a_1 & a_2 & a_3 & a \\ a_1' & a_2' & a_3' & a' \\ a_1'' & a_2'' & a_3'' & a'' \end{Vmatrix}$$

be of rank less than three.* Hence we have

[12.2] *The envelope of a one-parameter family of non-parallel planes (12.6) for which A, defined by (12.8), is not identically zero is the tangent surface of a curve, or a cone; in the latter case $a = -a_i c^i$, where the c's are constants, and c^i are the coordinates of the vertex of the cone.*

We consider next the case when $A = 0$. In this case there are quantities h^1, h^2, h^3 such that†

$$(12.10) \qquad h^i a_i = 0, \qquad h^i a_i' = 0, \qquad h^i a_i'' = 0.$$

Differentiating the first two of these equations, we obtain in consequence of the third

$$(12.11) \qquad h^{i'} a_i = 0, \qquad h^{i'} a_i' = 0.$$

* C. G., p. 125.
† C. G., p. 116.

From these equations and the first two of (12.10) it follows that a_i' are proportional to the corresponding a_i, and consequently a_i are proportional to constants c_i, the factor of proportionality being a function of u^1, in which case the planes (12.6) are parallel. Since this case has been excluded from this discussion, it follows that the h's are constants so that equations (12.11) do not exist. Since a_i are direction numbers of normals to the planes,* it follows from the first of (12.10) that all of the planes are parallel to a line of direction numbers h^i. Consequently the envelope is a cylinder, or all the planes have a line in common. The former case arises when the three equations (12.6) and (12.7) do not have a common solution, that is, when the matrix (12.9) is of rank three†. As remarked above, the planes pass through a line when the rank of the matrix is less than three. Hence we have

[12.3] *When for a family of non-parallel planes* $a_i x^i + a = 0$ *the determinant* (12.8) *is identically zero, the envelope of the planes is a cylinder or all the planes have a line in common according as the rank of the augmented matrix* (12.9) *is three or less than three.*

We seek now a necessary condition upon a function $f(x^1, x^2, x^3)$ in order that the surface

$$(12.12) \qquad\qquad f(x^1, x^2, x^3) = 0$$

shall be a developable surface. Assume that the surface is defined by the equations $x^i = \varphi^i(u^1, u^2)$, where $u^1 = $ const. are the generators and $u^2 = $ const. another set of coordinate curves. When these expressions for x^i are substituted in (12.12), the resulting equation is an identity in u^1 and u^2, and consequently

$$(12.13) \qquad\qquad f_i \frac{\partial x^i}{\partial u^1} = 0, \qquad f_i \frac{\partial x_i}{\partial u^2} = 0,$$

where $f_i \equiv \dfrac{\partial f}{\partial x^i}$, and the summation convention is applied. An equation of the tangent plane to the surface (12.12) at the point x^i is

$$(12.14) \qquad\qquad (X^i - x^i)f_i = 0.$$

Since equation (12.14) does not involve u^2 by hypothesis its derivative with respect to u^2 is equal to zero. Therefore, in consequence of the

* C. G., p. 92.
† C. G., p. 126.

second of (12.13) we have

$$(12.15) \qquad (X^i - x^i)f_{ij}\frac{\partial x^j}{\partial u^2} = 0,$$

where

$$f_{ij} \equiv \frac{\partial^2 f}{\partial x^i \partial x^j}.$$

Since $u^1 = $ const. are the generators of the surface by hypothesis, their equations are given in accordance with (12.1) and (12.2) by (12.14) and

$$\frac{\partial}{\partial u^1}[(X^i - x^i)f_i] = 0,$$

which in consequence of the first of (12.13) reduces to

$$(12.16) \qquad (X^i - x^i)f_{ij}\frac{\partial x^j}{\partial u^1} = 0.$$

On comparing equations (12.15) and (12.16) with (12.13), we see that

$$(12.17) \qquad (X^i - x^i)f_{ij} = tf_j,$$

where t is a factor of proportionality. In order that the three equations (12.17), and (12.14) be consistent, it is necessary* that the equation

$$(12.18) \qquad \begin{vmatrix} f_{11} & f_{12} & f_{13} & f_1 \\ f_{21} & f_{22} & f_{23} & f_2 \\ f_{31} & f_{32} & f_{33} & f_3 \\ f_1 & f_2 & f_3 & 0 \end{vmatrix} = 0$$

be satisfied in consequence of (12.12) or identically.

Conversely, if this condition is satisfied and $x^i = \varphi^i(u^1, u^2)$ are equations of the surface, that is, satisfy (12.12) identically, when these expressions for x^i are substituted in (12.18) the resulting equation is an identity in u^1 and u^2. Hence there exist functions a^i and a of u^1 and u^2 such that

$$(12.19) \qquad a^i f_{ij} + af_j = 0, \qquad a^i f_i = 0.$$

If any one of the quantities f_i is equal to zero identically, then (12.18) is satisfied identically and the surface is a cylinder whose generators are

---

* C. G., p. 139.

parallel to a coordinate axis. There remains to consider the case when each f_i is not identically zero, and we write the equation (12.14) of the tangent plane to the surface (12.12) in the form

(12.20) $$X^1 + X^2 \frac{f_2}{f_1} + X^3 \frac{f_3}{f_1} - \frac{x^i f_i}{f_1} = 0.$$

The quantities f_2/f_1, f_3/f_1, and $x^i f_i/f_1$ are functions of u^1 and u^2, their form depending upon the functions $\varphi^i(u^1, u^2)$. In order that the surface (12.12) be developable it is necessary and sufficient that there be co-ordinates u'^1 and u'^2 in the surface such that f_2/f_1, f_3/f_1, and $x^i f_i/f_1$ shall be functions of one of them, say u'^1. This means that these three quantities must be functions of the same function, say $\psi(u^1, u^2)$, in which case $u'^1 = \psi(u^1, u^2)$ and u'^2 is any other function of u^1 and u^2 such that the jacobian of this function and ψ is not identically zero. If this condition is satisfied, the surface is developable and the curves $\psi(u^1, u^2) = $ const. are its generators. Hence a sufficient condition that the surface be developable is that when (12.19) are satisfied the jacobian of each pair of the quantities f_2/f_1, f_3/f_1, and $x^i f_i/f_1$ with respect to u^1 and u^2 be identically zero. The jacobian of the first two of these quanti-ties is

$$\frac{1}{f_1^4} \begin{vmatrix} (f_1 f_{2i} - f_2 f_{1i}) \dfrac{\partial x^j}{\partial u^1} & (f_1 f_{3i} - f_3 f_{1i}) \dfrac{\partial x^j}{\partial u^1} \\[2ex] (f_1 f_{2i} - f_2 f_{1i}) \dfrac{\partial x^j}{\partial u^2} & (f_1 f_{3i} - f_3 f_{1i}) \dfrac{\partial x^j}{\partial u^2} \end{vmatrix},$$

which is equal to*

$$\frac{1}{f_1^3} \begin{vmatrix} f_1 & f_2 & f_3 \\[2ex] f_{1j} \dfrac{\partial x^j}{\partial u^1} & f_{2j} \dfrac{\partial x^j}{\partial u^1} & f_{3j} \dfrac{\partial x^j}{\partial u^1} \\[2ex] f_{1j} \dfrac{\partial x^j}{\partial u^2} & f_{2j} \dfrac{\partial x^j}{\partial u^2} & f_{3j} \dfrac{\partial x^j}{\partial u^2} \end{vmatrix}.$$

In consequence of (12.19) and (12.13) this determinant is equal to zero. Because of this result and (12.13) the jacobian of $x^i f_i/f_1$ and each of the quantities f_2/f_1 and f_3/f_1 can be shown to be equal to zero. Hence we have

[12.4] *A necessary and sufficient condition that* $f(x^1, x^2, x^3) = 0$ *be an equation of a developable surface is that equation* (12.18) *be satisfied in consequence of the equation* $f(x^1, x^2, x^3) = 0$ *or identically.*

* Fine, 1904, 1, p. 505.

Consider the equation

(12.21) $$c^i \frac{\partial f}{\partial x^i} = 0,$$

where the c's are constants. Evidently any solution of this equation satisfies (12.18) identically. Also (12.21) is the condition that the normals to such a developable surface be perpendicular to a line with direction numbers c^i. Hence we have

[12.5] *A necessary and sufficient condition that a surface* $f(x^1, x^2, x^3) = 0$ *be a cylinder is that f be a solution of an equation* (12.21) *in which the* c's *are constants.*

EXERCISES

1. The envelope of the planes normal to a curve is the polar surface (see §11); the polar surface is also called the *polar developable*.

2. The envelope of the plane normal to the principal normal to a curve at a point of a curve is called the *rectifying developable* of the curve; equations of its characteristics are

$$X^i = x^i + (\tau \alpha^i - \kappa \gamma^i)t,$$

where t is a parameter, and equations of its edge of regression are

$$X^i = x^i + \frac{(\tau \alpha^i - \kappa \gamma^i)\kappa}{\tau \dfrac{d\kappa}{ds} - \kappa \dfrac{d\tau}{ds}}.$$

3. If each of the generators of a developable surface, other than a cone or cylinder, is revolved through the same angle about the tangent to an orthogonal trajectory of the generators at the point of intersection, the locus of the resulting lines is a developable surface whose edge of regression is an evolute of the given trajectory.

4. The edge of regression of the family of planes

$$(1 - u^2)x^1 + i(1 + u^2)x^2 + 2ux^3 + f(u) = 0,$$

where $i = \sqrt{-1}$ and u is the parameter, is a minimal curve; the envelope is called an *isotropic* developable surface.

5. Find the edge of regression of the developable surface which envelopes the hyperbolic paraboloid $ax^3 = x^1x^2$ along the curve in which the paraboloid is cut by the parabolic cylinder $x^{1^2} = bx^2$.

6. The curvature and torsion of the edge of regression of the family of planes $a_i x^i + a = 0$, where the a's are functions of a parameter u, are given by

$$\kappa = \frac{\Delta^3}{D \sum_i a_i'^2}, \qquad \tau = \frac{\Delta^2}{D},$$

where

$$\Delta = \begin{vmatrix} a_1 & a_2 & a_3 \\ a_1' & a_2' & a_3' \\ a_1'' & a_2'' & a_3'' \end{vmatrix}, \qquad D = \begin{vmatrix} a_1 & a_2 & a_3 & a \\ a_1' & a_2' & a_3' & a' \\ a_1'' & a_2'' & a_3'' & a'' \\ a_1''' & a_2''' & a_3''' & a''' \end{vmatrix},$$

primes denoting differentiation with respect to u.

7. For a family of spheres

$$\sum_i (x^i - a^i)(x^i - a^i) - a^2 = 0,$$

in which the a's are functions of the arc of the curve C of centers, the edge of regression consists of two parts with corresponding points symmetric with respect to the corresponding osculating plane of C, unless

(i) $$a^2(1 - a'^2) - \rho^2[1 - (aa')']^2 = 0,$$

where ρ is the radius of curvature of C; when the condition (i) is satisfied, the edge is a single curve, C_1, its points lying in the osculating planes of C, and the spheres of the family are the osculating spheres of C_1.

CHAPTER II

Transformation of Coordinates. Tensor Calculus

13. TRANSFORMATION OF COORDINATES. CURVILINEAR COORDINATES

In the preceding chapter $x^i(i = 1, 2, 3)$ denote cartesian coordinates of space referred to rectangular axes. There are many such coordinate systems, and the relation between two such systems x^i and x'^i is given by equations of the form

$$(13.1) \qquad x^i = a^i_j x'^j + b^i.$$

The coordinates x'^i refer to a set of rectangular axes whose origin in the x-system is b^i, and a^1_j, a^2_j, a^3_j are the direction cosines of the x'^j-axis with respect to the x-system. These direction-cosines satisfy the conditions*

$$(13.2) \qquad \sum_i a^i_j a^i_k = \delta_{jk}, \qquad \sum_i a^j_i a^k_i = \delta^{jk},$$

where the quantities δ_{jk} and δ^{jk}, called *Kronecker deltas*, are defined by

$$(13.3) \qquad \delta_{jk} \text{ and } \delta^{jk} = 1 \text{ or } 0 \text{ according as } j = k \text{ or } j \neq k.$$

Moreover the determinant of the quantities a^i_j, that is,

$$(13.4) \qquad |a^i_j| \equiv \begin{vmatrix} a^1_1 & a^1_2 & a^1_3 \\ a^2_1 & a^2_2 & a^2_3 \\ a^3_1 & a^3_2 & a^3_3 \end{vmatrix}$$

is equal to $+1$.†

If we denote by a'^j_i the cofactor of a^i_j in the determinant (13.4) divided by the determinant, then

$$(13.5) \qquad a^i_k a'^k_j = \delta^i_j, \qquad a^l_j a'^i_l = \delta^i_j,$$

where δ^i_j, also called *Kronecker deltas*, are defined by

$$(13.6) \qquad \delta^i_j = 1 \text{ or } 0 \text{ according as } i = j \text{ or } i \neq j.$$

* C. G., pp. 161–164.
† C. G., p. 162.

The second set of equations (13.5) is equivalent to the statement that the sum of the products of the elements of the jth column and the co-factors of the corresponding elements of the ith column is equal to the determinant or zero according as $i = j$ or $i \neq j$.* If then we multiply equation (13.1) by $a_i'^k$ and sum with respect to i, we obtain

$$(13.7) \qquad a_i'^k(x^i - b^i) = a_i'^k a_j^i x'^j = \delta_j^k x'^j = x'^k.$$

Hence the loci $x'^k = $ const. are parallel planes for which the quantities $a_i'^k$ are direction numbers of the normals to these planes in the x-system.† However, if the first set of equations (13.2) for $j \neq k$ are not satisfied, the axes of the x'-system are not mutually perpendicular, as follows from the remark following equations (13.1). Hence in this case the x'-system is an *oblique* system of coordinates.

We consider next the case, when the determinant (13.4) is equal to zero. In this case equations (13.1) in the x''s are consistent only when the three determinants obtained on replacing the elements of a column in the determinant (13.4) by $x^1 - b^1$, $x^2 - b^2$, $x^3 - b^3$ respectively are equal to zero.‡ Consequently in this case the quantities x'^i are defined only at points of the locus with these equations, and thus are not co-ordinates for the space.

Equations (13.7) define the *inverse* of the transformation (13.1). Moreover, equations (13.1) define the inverse of the transformation (13.7) as follows from (13.5).

Polar coordinates constitute another type of coordinates frequently used in space, particularly in astronomy. With reference to a cartesian coordinate system x^i polar coordinates x'^i are defined by

$$(13.8) \qquad x^1 = x'^1 \sin x'^2 \cos x'^3, \quad x^2 = x'^1 \sin x'^2 \sin x'^3, \quad x^3 = x'^1 \cos x'^2.$$

The inverse of this transformation is

$$(13.9) \quad x'^1 = \sqrt{\sum_i x^i x^i}, \quad x'^2 = \cos^{-1} \frac{x^3}{\sqrt{\sum_i x^i x^i}}, \quad x'^3 = \tan^{-1} \frac{x^2}{x^1}.$$

From these equations it follows that x'^1 is the distance of the point $P(x^i)$ from the origin O of the x-system; x'^2 is the angle which the line OP makes with the x^3-axis; and x'^3 is the angle which the projection of OP on the x^1x^2-plane makes with the positive x^1-axis.

From (13.9) it follows that the loci $x'^1 = $ const. are spheres with O

---

* C. G., p. 110.
† C. G., pp. 92–93.
‡ C. G., pp. 124.

as center; the loci $x'^2 = $ const. are right circular cones each of which has its vertex at O and the x^3-axis for axis; and the loci $x'^3 = $ const. are planes through the x^3-axis. We call these surfaces the *coordinate surfaces* in the x'-system. In like manner the coordinate surfaces of the x'-system defined by (13.1) and (13.7) are planes, which as shown are mutually orthogonal only in case equations (13.5) are satisfied.

Equations (13.1) and (13.8) are particular cases of equations

$$(13.10) \qquad x^i = \varphi^i(x'^1, x'^2, x'^3),$$

where the φ's are one-valued functions of x'^1, x'^2, x'^3. For any such functions φ^i, these are equations of transformation to a general set of coordinates x'^1, x'^2, x'^3, provided that the functions φ^i are independent. If they were not independent there would be one, or two, relations of the form $F(\varphi^1, \varphi^2, \varphi^3) = 0$, and in that case (13.10) are not equations of a transformation of space; they have meaning only at points of the locus with this equation, or equations. A necessary and sufficient condition that the functions φ^i be independent is that their jacobian, namely

$$(13.11) \qquad \left| \frac{\partial \varphi}{\partial x'} \right| \equiv \begin{vmatrix} \dfrac{\partial \varphi^1}{\partial x'^1} & \dfrac{\partial \varphi^1}{\partial x'^2} & \dfrac{\partial \varphi^1}{\partial x'^3} \\[2mm] \dfrac{\partial \varphi^2}{\partial x'^1} & \dfrac{\partial \varphi^2}{\partial x'^2} & \dfrac{\partial \varphi^2}{\partial x'^3} \\[2mm] \dfrac{\partial \varphi^3}{\partial x'^1} & \dfrac{\partial \varphi^3}{\partial x'^2} & \dfrac{\partial \varphi^3}{\partial x'^3} \end{vmatrix} = e^{ijk} \frac{\partial \varphi^1}{\partial x'^i} \frac{\partial \varphi^2}{\partial x'^j} \frac{\partial \varphi^3}{\partial x'^k}$$

be not identically zero* (see (1.19)).

When this condition is satisfied, there may be particular values of x'^i for which the jacobian is zero, but in general this is not the case, that is, about a point x'^i for which $\left| \dfrac{\partial \varphi}{\partial x'} \right| \neq 0$ there is a domain for which this inequality holds. For such values of x'^i equations (13.10) can be solved for x'^i; we denote such a solution by

$$(13.12) \qquad x'^i = \varphi'^i(x^1, x^2, x^3).†$$

These equations define the *inverse* of the transformation (13.10).

For equations (13.1) the jacobian (13.11) is the determinant (13.4). When this determinant is not equal to zero, the $a_j'^i$ exist and equations

---

* Fine, 1927, 1, p. 253.
† Fine, 1927, 1, p. 253.

(13.7) are the inverse of equations (13.1) and give the coordinates x'^i for a point in terms of the x's of the point.

For equations (13.8) the jacobian (13.11) is found to be equal to $(x'^1)^2 \sin x'^2$. Since this quantity is not identically zero, polar coordinates apply to all points of space, but they are not uniquely defined when $x'^1 = 0$ or when $\sin x'^2 = 0$, that is, at O or on the x^3-axis. Thus at O, $x'^1 = 0$ and x'^2 and x'^3 can take any values, that is, all the coordinate surfaces x'^2-const. and x'^3-const. pass through O. For any point on the x^3-axis, x'^3 can take any value, that is, all the coordinate planes $x'^3 = $ const. meet in the x^3-axis. In these cases the coordinate surfaces are degenerate, that is, $x'^1 = 0$ is a point, and $x'^2 = 0$, π, a line. But for all other points equations (13.9) hold and are the equations of the inverse.

When the expressions (13.12) for x'^i are substituted in (13.10) we have

$$x^i - \varphi^i(\varphi'^1, \varphi'^2, \varphi'^3) = 0,$$

which are identities in the x's, as follows from the definition of (13.12). Since they are identities, the left-hand member does not vary with any of the x's, and consequently the derivative with respect to each x^i is equal to zero. Hence we have

$$\frac{\partial x^i}{\partial x^j} = \frac{\partial \varphi^i}{\partial \varphi'^k} \frac{\partial \varphi'^k}{\partial x^j},$$

where k is a dummy index. Since x^i and x^j for $i \neq j$ are independent, the left-hand member is $+1$ or 0 according as $i = j$ or $i \neq j$. Accordingly these equations may be written

(13.13)
$$\delta_j^i = \frac{\partial x^i}{\partial x'^k} \frac{\partial x'^k}{\partial x^j},$$

where δ_j^i are defined by (13.6). Since i and j take the values 1, 2, 3 there are nine equations in the set (13.13).

If we consider the expressions (13.10) for x^i substituted in (13.12), we obtain analogously to (13.13)

(13.14)
$$\frac{\partial x'^i}{\partial x^k} \frac{\partial x^k}{\partial x'^j} = \delta_j^i.$$

When the jacobian of the transformation (13.10), namely (13.11) denoted by $\left| \dfrac{\partial x}{\partial x'} \right|$, is multiplied by the jacobian of the inverse (13.12), one obtains in consequence of (13.13) a determinant in which each element of the main diagonal is $+1$ and every other element is zero. Hence we have

$$(13.15) \qquad \left| \frac{\partial x}{\partial x'} \right| \cdot \left| \frac{\partial x'}{\partial x} \right| = 1,$$

that is, either jacobian is the reciprocal of the other.   Thus the relation between equations (13.10) and (13.12) is such that either is the set of equations of the inverse transformation of the other.

If in equations (13.13) we give i a fixed value and let j take the values 1, 2, 3, we have three equations of the first degree in $\frac{\partial x^i}{\partial x'^1}$, $\frac{\partial x^i}{\partial x'^2}$ and $\frac{\partial x^i}{\partial x'^3}$. Solving these equations for these quantities, we obtain (see Ex. 1)

$$(13.16) \qquad \frac{\partial x^i}{\partial x'^j} = \frac{\text{cofactor of } \dfrac{\partial x''^j}{\partial x^i} \text{ in } \left| \dfrac{\partial x'}{\partial x} \right|}{\left| \dfrac{\partial x'}{\partial x} \right|}.$$

Although we started this section, interpreting x^i as cartesian coordinates in space, and took equations (13.10) as equations of a transformation from such coordinates to any other, in deriving the properties of the transformation (13.10) and the inverse (13.12), no use has been made of the fact that x^i were cartesian coordinates.   Therefore all the results of this section apply equally well when equations (13.10) give the relation between the coordinates of any two general systems whatever in space.   Thus, if x^i are any set of coordinates, equations (13.1) and (13.8) define a transformation of coordinates; but the geometric interpretation of the new coordinates in these two cases given above applies only to the case when x^i are cartesian coordinates.

We consider now in connection with a transformation (13.10) a second transformation

$$(13.17) \qquad x'^i = \psi^i(x''^1, x''^2, x''^3),$$

it being understood that the jacobian $\left| \dfrac{\partial x'}{\partial x''} \right|$ is not identically zero. When these expressions for x'^i are substituted in equations (13.10), the resulting equations denoted by

$$(13.18) \qquad x^i = x^i(x''^1, x''^2, x''^3)$$

define a transformation from x^i into x'''^i, called the *product* of the transformations (13.10) and (13.17).   Since

$$\frac{\partial x^i}{\partial x''^j} = \frac{\partial x^i}{\partial x'^k} \frac{\partial x'^k}{\partial x''^j},$$

it follows from the rule of multiplication of determinants that

$$\left| \frac{\partial x}{\partial x''} \right| = \left| \frac{\partial x}{\partial x'} \right| \cdot \left| \frac{\partial x'}{\partial x''} \right|,$$

and hence that the jacobian of the product of two transformations of coordinates is the product of the jacobians of these transformations. We say that transformations of coordinates have the *group property* by which we mean that the product of any two such transformations is a transformation of coordinates.

If x^i are cartesian coordinates, each of equations (13.12) for a particular value of an x' is evidently an equation of a surface in space, and for three particular values of x'^1, x'^2 and x'^3 these are equations of three surfaces in space intersecting in the point with these coordinates in the x'-system, as discussed above in the case of linear equations (13.1) and polar coordinates (13.8). In like manner for particular values of cartesian coordinates x^i, equations (13.10) are equations in the co-ordinates x''^i of the planes through the point x^i parallel to the coordinate planes $x^i = 0$.

If x^i and x''^i are any coordinates whatever, equations (13.12) for particular values of x'^1, x'^2, x'^3 are equations of three surfaces, defined in terms of the coordinates x^i, which pass through the point (x'^1, x'^2, x'^3); and similarly for equations (13.10).

Another way of stating the above remarks is that in any coordinate system each of the equations

(13.19) $x^1 = c^1,$ $x^2 = c^2,$ $x^3 = c^3$

for particular values of the constants c is an equation of a surface for which one of the coordinates is a constant for all points of the surface, the values of the other two coordinates determining a particular point on the surface. Thus equations (13.19) are equations of surfaces which are the analogue of planes parallel to the coordinate planes of a rectangular coordinate system. As the constant c^i in any one of these equations takes on a continuum of real values, the equation is an equation of a family of surfaces. Thus each of these equations is an equation of an endless number of *coordinate surfaces*. In the x'-system defined by (13.10) equations of these coordinate surfaces are $\varphi^i(x'^1, x'^2, x'^3) = c^i$. There passes one and only one surface of each family through each point in space for which the jacobian of the transformation from cartesian coordinates to the coordinates in question is not zero. At points where the jacobian vanishes the new coordinates are not uniquely defined, as remarked in the case of equations (13.8).

Any two of equations (13.19) for particular values of the c's are equations of the curve of intersection of the corresponding coordinate surfaces. Along such a curve the remaining coordinate is a parameter; thus x^1 is a parameter for (any) one of the curves $x^2 = c^2$ and $x^3 = c^3$. We call these curves *coordinate* curves, and, in particular, an x^i-*coordinate curve* one for which x^i alone varies, and thus will serve as a parameter. These curves are the analogues of lines parallel to the coordinate axes in cartesian coordinates. Since these curves are not in general straight lines, the corresponding coordinates are called *curvilinear*.

From the above discussion it follows that whatever be the coordinates x^i in general the locus defined by an equation $f(x^1, x^2, x^3) = 0$ is a surface, and the locus defined by two independent equations $f_1(x^1, x^2, x^3) = 0$, $f_2(x^1, x^2, x^3) = 0$ is a curve. This is equivalent to the definition of a surface as a two-dimensional locus, and of a curve as a one-dimensional locus (see §1).

EXERCISES

1. It follows from (13.5) that

$$| a_j^i | \cdot | a_j'^i | = 1,$$

and that a_j^i is the cofactor of $a_i'^i$ in the determinant $| a_j'^i |$ divided by the determinant.

2. From (13.3) and (13.6) it follows that

$$\delta^{ij}\delta_{jk} = \delta_k^i ,$$

and consequently δ^{ij} is the cofactor of δ_{ij} in the determinant $| \delta_{ij} |$ (see (13.16))

3. From (13.6) it follows that

$$\delta_i^i \equiv \sum_i \delta_i^i = 3.$$

4. From (13.4) and (1.19) it follows that

$$e_{ijk}a_l^i a_m^j a_n^k = | a_j^i | e_{lmn} ;$$
$$e^{ijk}a_i^l a_j^m a_k^n = | a_j^i | e^{lmn} .$$

5. By giving to i and j different values one verifies that the cofactor of a_j^i in the determinant (13.4) is given by

$$\tfrac{1}{2}e^{jkl} e_{imn} a_k^m a_l^n .$$

6. Determine the coordinate surfaces $x'^i = $ const. for each of the following transformations, in which x^i are cartesian, and find the points for which the jacobian is equal to zero:

(i)　　$x^1 = x'^1 x'^2 \cos x'^3$,　　$x^2 = x'^1 x'^2 \sin x'^3$,　　$2x^3 = (x'^1)^2 - (x'^2)^2;$

(ii)　　$$x^i = \left[\frac{(x'^1 - a^i)(x'^2 - a^i)(x'^3 - a^i)}{(a^j - a^i)(a^k - a^i)} \right]^{\frac{1}{2}},$$

where i, j, k take the values 1, 2, 3 in cyclic order, that is 1, 2, 3; 2, 3, 1; 3, 1, 2, and where $a^1 > a^2 > a^3 > 0$.

7. Discuss the coordinate curves for each of the systems in Ex. 6 and also for (13.8).

14. THE FUNDAMENTAL QUADRATIC FORM OF SPACE

We have seen in §2 that for a curve in space defined by equations of the form

$$(14.1) \qquad x^i = f^i(t),$$

the x's being cartesian, the differential of length of the curve is given by

$$ds^2 = \sum_i \left(\frac{df^i}{dt}\right)^2 dt^2.$$

Accordingly we say that the *element of length*, or *linear element*, ds of space is given by

$$(14.2) \qquad ds^2 = (dx^1)^2 + (dx^2)^2 + (dx^3)^2,$$

by which we mean that, when the differentials dx^i from (14.1) are substituted in (14.2), the expression for ds obtained therefrom is the differential of arc length of the curve (14.1). The right-hand member of (14.2) is called the *fundamental quadratic form* of space.

In terms of any other coordinate system x'^i we have from (13.10)

$$(14.3) \qquad dx^i = \frac{\partial x^i}{\partial x'^j} dx'^j,$$

and from (14.2)

$$(14.4) \qquad ds^2 = \sum_i \left(\frac{\partial x^i}{\partial x'^j} dx'^j\right)\left(\frac{\partial x^i}{\partial x'^k} dx'^k\right) = a'_{jk} dx'^j dx'^k,$$

where

$$(14.5) \qquad a'_{jk} = \sum_i \frac{\partial x^i}{\partial x'^j} \frac{\partial x^i}{\partial x'^k},$$

from which it follows that $a'_{jk} = a'_{kj}$.

Since the coordinate system x'^i is any whatever, it follows that in any coordinate system x^i the fundamental quadratic form is

$$(14.6) \qquad ds^2 = a_{ij} dx^i dx^j,$$

where a_{ij} is *symmetric in the indices i and j*, that is, $a_{ij} = a_{ji}$. When, in particular, the coordinates x^i are cartesian, we have (14.2), and consequently

$$(14.7) \qquad a_{ij} = \delta_{ij}, \qquad ds^2 = \delta_{ij} dx^i dx^j.$$

In the case of polar coordinates (13.8) we have from (14.5)

$$(14.8) \qquad a'_{11} = 1, \qquad a'_{22} = (x'^1)^2, \qquad a'_{33} = (x'^1)^2 \sin^2 x'^2,$$

$$a'_{ij} = 0 \text{ for } i \neq j.$$

Hence in polar coordinates the fundamental form is

$$(14.9) \qquad ds^2 = (dx'^1)^2 + (x'^1)^2 ((dx'^2)^2 + \sin^2 x'^2 (dx'^3)^2).$$

We desire now to find the relation between the coefficients a_{ij} and a'_{ij} of the fundamental form in any two coordinate systems x^i and x'^i. Since the element of length ds does not depend upon a coordinate system, the fundamental forms in any two coordinate systems are equal.   Hence we have

$$a_{ij} dx^i dx^j \;=\; a'_{kl} dx'^k dx'^l.$$

Substituting for dx^i and dx^j from equations of the form (14.3), we have

$$\left(a_{ij} \frac{\partial x^i}{\partial x'^k} \frac{\partial x^j}{\partial x'^l} - a'_{kl} \right) dx'^k dx'^l = 0.$$

Since this equation must hold for arbitrary values of the dx''s, and the expression in parentheses is symmetric in k and l, it follows that

$$(14.10) \qquad a'_{kl} = a_{ij} \frac{\partial x^i}{\partial x'^k} \frac{\partial x^j}{\partial x'^l}.$$

In order to obtain this result, one takes $dx'^1 \neq 0$, $dx'^2 = dx'^3 = 0$, and gets equation (14.10) for $k = l = 1$.   In like manner, as one takes every other two of the differentials equal to zero, one gets (14.10) for $k = l = 1, 2, 3$.   In order to obtain the remaining three equations (14.10) for $k \neq l$, one takes one of the differentials equal to zero at a time, and the others not equal to zero.   That there are only three of these equations follows from the fact that in any coordinate system $a_{kl} = a_{lk}$, as previously shown.

If we multiply (14.10) by $\dfrac{\partial x'^k}{\partial x^h} \dfrac{\partial x'^l}{\partial x^m}$ and sum with respect to k and l, we have on making use of (13.13)

$$a'_{kl} \frac{\partial x'^k}{\partial x^h} \frac{\partial x'^l}{\partial x^m} = a_{ij} \frac{\partial x^i}{\partial x'^k} \frac{\partial x'^k}{\partial x^h} \frac{\partial x^j}{\partial x'^l} \frac{\partial x'^l}{\partial x^m}$$

$$= a_{ij} \delta^i_h \delta^j_m = a_{hm}.$$

Hence, on changing indices, we have the following equations

$$(14.11) \qquad a_{ij} = a'_{kl} \frac{\partial x'^k}{\partial x^i} \frac{\partial x'^l}{\partial x^j},$$

connecting the a's and a'''s which are equivalent to (14.10), but in *inverse form*.

If we denote by a' and a the determinants of the quantities a'_{ij} and a_{ij} respectively, we have from the rule for the multiplication of determinants

$$(14.12) \qquad a' \equiv |a'_{kl}| = \left| a_{ij} \frac{\partial x^i}{\partial x'^k} \frac{\partial x^j}{\partial x'^l} \right| = \left| a_{ij} \frac{\partial x^i}{\partial x'^k} \right| \cdot \left| \frac{\partial x}{\partial x'} \right| = a \left| \frac{\partial x}{\partial x'} \right|^2.$$

For the values δ_{ij} of the a's, that is, when the coordinates are cartesian, the determinant is $+1$. From this result and (14.12) we have

[14.1] *The determinant of the coefficients of the fundamental form of euclidean space in any coordinate system is positive.*

Since $a \neq 0$, functions a^{ik} are defined uniquely by

$$(14.13) \qquad\qquad a^{ik} a_{kj} = \delta^i_j .$$

In fact, on solving these equations for a^{ik} in the manner which led to (13.16) from (13.13) we have

$$(14.14) \qquad\qquad a^{ik} = \frac{\text{cofactor of } a_{ki} \text{ in } a}{a}.$$

Since $a_{ij} = a_{ji}$, it follows from (14.14) that $a^{ik} = a^{ki}$, that is, the quantities a^{ik} are symmetric in the indices, in this case superscripts. In particular, when $a_{ij} = \delta_{ij}$, we have $a^{ij} = \delta^{ij}$ (see §13, Ex. 2).

If $\bar{a}$ denotes the determinant of the quantities a^{ik}, it follows from (14.13) that the product of the determinants a and $\bar{a}$ is equal to the determinant for which each element of the main diagonal is $+1$, and every other element is zero. Hence

$$(14.15) \qquad\qquad \bar{a} \equiv |a^{ij}| = \frac{1}{a}.$$

For any other coordinate system x'^i the coefficients a'_{ij} of the fundamental form are given by (14.10). By means of equations of the form (14.13) functions a'^{ik} are uniquely defined. We shall show that the following relations hold between the functions a'^{ij} and a^{kl}:

$$(14.16) \qquad\qquad a'^{ij} = a^{kl} \frac{\partial x'^i}{\partial x^k} \frac{\partial x'^j}{\partial x^l}.$$

In fact, if we take the equations

$$a'^{lk} a_{kh} = \delta^l_h ,$$

which are of the form (14.13), and substitute for a_{kh} expressions of the form (14.11), we obtain

$$a^{kl} a'_{im} \frac{\partial x'^i}{\partial x^k} \frac{\partial x'^m}{\partial x^h} = \delta_h^l.$$

If now we multiply by $\dfrac{\partial x'^j}{\partial x^l}$ and sum with respect to l, we have

$$a^{kl} \frac{\partial x'^i}{\partial x^k} \frac{\partial x'^j}{\partial x^l} a'_{im} \frac{\partial x'^m}{\partial x^h} = a^{kl} a_{kh} \frac{\partial x'^j}{\partial x^l} = \delta_h^l \frac{\partial x'^j}{\partial x^l} = \frac{\partial x'^j}{\partial x^h}.$$

Since $\dfrac{\partial x'^j}{\partial x^h}$ is equal to $\delta_m^j \dfrac{\partial x'^m}{\partial x^h}$, we can write the above equations in the form

$$\left(a^{kl} \frac{\partial x'^i}{\partial x^k} \frac{\partial x'^j}{\partial x^l} a'_{im} - \delta_m^j \right) \frac{\partial x'^m}{\partial x^h} = 0,$$

(we say equations, because j and h being free indices there are 9 of these equations). For a fixed value of j as h takes the values $1, 2, 3$ we have three equations, linear and homogeneous in the three expressions in parentheses as m takes the values $1, 2, 3$. Since $\left| \dfrac{\partial x'}{\partial x} \right| \neq 0$, we have on changing indices

$$a^{kl} \frac{\partial x'^h}{\partial x^k} \frac{\partial x'^j}{\partial x^l} a'_{hm} - \delta_m^j = 0.$$

Multiplying by a'^{mi} and summing with respect to m, we obtain finally (14.16).

EXERCISES

1. Show that $\sqrt{a_{11}}\, dx^1$ is the differential of length of an x^1-coordinate curve; apply this result to the case when the x's are polar coordinates.

2. Show that when in (14.13) $a_{ij} = 0$ for $i \neq j$, then

$$a^{ii} = \frac{1}{a_{ii}}, \qquad a^{ij} = 0 \qquad (i \neq j).$$

3. Show that in any coordinate system x^i the expression

$$a_{11}(dx^1)^2 + 2a_{12}\, dx^1\, dx^2 + a_{22}(dx^2)^2$$

is the square of the differential of length of a curve on any coordinate surface $x^3 = $ const., the curve being defined by $x^1 = f^1(t)$, $x^2 = f^2(t)$.

4. When the coordinates x^i are cartesian, the coefficients a'_{ij} of the fundamental form in the coordinates x'^i for a transformation (13.1) are given by

$$a'_{ij} = \sum_h a_i^h a_j^h,$$

and the determinant a' of these quantities is given by

$$a' = |\, a_j^i \,|^2;$$

consequently the coefficients $a'_{1,j}$ are constants.

5. Show that the fundamental quadratic forms of space in the coordinates x'^i of §13, Ex. 6 are respectively

(i) $[(x'^1)^2 + (x'^2)^2]\,[(dx'^1)^2 + (dx'^2)^2] + (x'^1 x'^2)^2 (dx'^3)^2;$

(ii) $\dfrac{1}{4} \sum\limits_i \dfrac{(x'^i - x'^i)\,(x'^k - x'^i)\,(dx'^i)^2}{(x'^i - a^1)\,(x'^i - a^2)\,(x'^i - a^3)},$

where j and k are the numbers 1, 2, 3 other than i.

15. CONTRAVARIANT VECTORS.   SCALARS

From the equations (13.12) of a transformation of coordinates, namely

(15.1) $x'^j = \varphi'^j(x^1, x^2, x^3),$

we have

(15.2) $dx'^j = \dfrac{\partial x'^j}{\partial x^h}\, dx^h.$

Thus whatever be the coordinates in two systems in transforming from one system to the other differentials undergo a linear homogeneous transformation, the coefficients being in general functions of the coordinates. In this sense a transformation of coordinates *induces* a linear homogeneous transformation of differentials of the coordinates. If equations (15.2) are multiplied by $\dfrac{\partial x^i}{\partial x'^j}$ and summed with respect to j, we have in consequence of (13.13)

(15.3) $\dfrac{\partial x^i}{\partial x'^j}\, dx'^j = \dfrac{\partial x^i}{\partial x'^j}\dfrac{\partial x'^j}{\partial x^h}\, dx^h = \delta_h^i\, dx^h = dx^i,$

as the inverse of (15.2).   Equations (15.3) are, in fact, equations (14.3), which were obtained from the inverse of (15.1).   If in (15.3) we replace dx'^j by $dx'^j = \dfrac{\partial x'^j}{\partial x''^k}\, dx''^k$ obtained from (13.17), we have

(15.4) $dx^i = \dfrac{\partial x^i}{\partial x'^j}\dfrac{\partial x'^j}{\partial x''^k}\, dx''^k = \dfrac{\partial x^i}{\partial x''^k}\, dx''^k.$

Hence the set of induced transformations has the group property (see §13).

If x^i in (15.1) are cartesian coordinates, the differentials dx^i are direction numbers of a direction in space, the direction being that of the line segment with end points x^i and $x^i + dx^i$.* The differentials dx'^j given by (15.2) determine the same direction in the x'-system, since a direction is independent of a coordinate system. However, whereas when the x's are cartesian coordinates the direction determined by fixed dx^i is the same everywhere in space, the corresponding values of dx'^i given by (15.2) depend upon the point at which the direction is considered, unless

$$\frac{\partial x'^j}{\partial x^h} = a_h^j,$$

the a's being constants. From these equations we obtain in this exceptional case by integration equations (13.1), in which case the coordinates x'^i are cartesian or oblique according as equations (13.2) hold or not. Consequently, although for any transformation of cartesian coordinates into other coordinates by equations not of the form (13.1) differentials dx'^i determine a direction at a point, they are not direction numbers of a line in space in the sense that the same values of dx'^i determine the direction of the line at every point of it, as is the case with differentials of cartesian coordinates.

In §3 we saw that for a curve with equations

(15.5) $x^i = f^i(t),$

where the coordinates x^i are cartesian, the quantities

$$\xi^i(t) = \frac{dx^i}{dt}$$

evaluated at a point of the curve are direction numbers of the tangent vector at the point. If for a coordinate system x'^i we put

$$\xi'^i(t) = \frac{dx'^i}{dt},$$

we have in consequence of (15.2)

(15.6) $\xi'^i(t) = \xi^j(t) \dfrac{\partial x'^i}{\partial x^j}.$

Thus the quantities $\xi'^i(t)$ determine at each point of the curve its tangent vector in the x'-system.

We consider now three functions of cartesian coordinates x^i which we denote by $\lambda^i(x)$. The values of these functions at each point in space

* C. G., p. 84.

may be taken as direction numbers of a vector at the point. Such a set of vectors is called a *vector-field*. If we put

$$(15.7) \qquad dx^i = \rho\lambda^i,$$

where ρ is any function of the x's, these values of the differentials at each point of space are also direction numbers of the vector determined by λ^i at the point. Consider for any other coordinate system x'^i the functions of the x''s denoted by $\lambda'^i(x')$ and defined by

$$(15.8) \qquad \lambda'^i(x') = \lambda^j(x)\,\frac{\partial x'^i}{\partial x^j},$$

when the x's in the right-hand members are replaced by their expressions in the x''s which define the transformation of coordinates. When we compare these equations and (15.7) with (15.2), we see that

$$(15.9) \qquad dx'^i = \rho\lambda'^i(x'),$$

that is, the quantities $\lambda'^i(x')$ define in the x''s the vector-field defined by $\lambda^i(x)$ in the x's. Just as equations (15.3) were obtained from (15.2), so equations (15.8) are equivalent to

$$(15.10) \qquad \lambda^i(x) = \lambda'^j(x')\,\frac{\partial x^i}{\partial x'^j},$$

and thus equations (15.8) and (15.10) are reciprocal in character.

For a third coordinate system x''^i we have analogously to (15.10)

$$\lambda^i(x) = \lambda''^k(x'')\,\frac{\partial x^i}{\partial x''^k}.$$

From these two sets of equations we have

$$\lambda'^j(x')\,\frac{\partial x^i}{\partial x'^j} = \lambda''^k(x'')\,\frac{\partial x^i}{\partial x''^k}.$$

If we multiply this equation by $\dfrac{\partial x'^h}{\partial x^i}$ and sum with respect to i, we have in consequence of (13.14)

$$\lambda'^h(x') = \lambda''^k(x'')\,\frac{\partial x'^h}{\partial x''^k}.$$

Hence the equations (15.10) have the group property, and consequently in any two coordinate systems x^i and x'^i the quantities $\lambda^i(x)$ and $\lambda'^i(x')$, which in the respective systems determine at each point the vector of a vector-field, are in the relation (15.10). The quantities λ^i and λ'^i are called the *components* in the x-system and x'-system respectively of a *contravariant vector*. As thus defined there is a contravariant vector at each point in space, that is, a vector-field. When the coordinates are cartesian, they are direction numbers of the vector; their geometric significance in any other coordinate system is shown in §16.

From the foregoing discussion it is seen that a contravariant vector is entirely defined by its components in any one coordinate system, and then its components in any other system are determined. Hence one may assign arbitrary functional expressions to λ^i in the x-system. For example, one may take λ^i as constants, and then in general the components in another coordinate system are not constants. If the coordinates x^i are cartesian and the λ^i are constants, all the vectors are parallel, since they have the same direction numbers. But in a general coordinate system constant components do not define parallel vectors.

At times in the consideration of a geometric problem one arrives at equations of the form (15.10) connecting certain quantities λ^i and λ'^i in any two coordinate systems. In this case we say that the geometric entity thus defined analytically is a contravariant vector whose components are λ^i and λ'^i in the respective coordinate systems.

If functions λ^i and λ'^i defined at points of a curve and not throughout space, satisfy equations of the form (15.8), λ^i are said to be the components of a contravariant vector at points of the curve; an example is afforded by equations (15.6).

Consider the differential equations

$$(15.11) \qquad \frac{dx^1}{\lambda^1} = \frac{dx^2}{\lambda^2} = \frac{dx^3}{\lambda^3},$$

where λ^i are the components of a contravariant vector. From the theory of such equations it follows that their integral is given by two equations of the form

$$(15.12) \qquad f_1(x^1, x^2, x^3) = c_1, \qquad f_2(x^1, x^2, x^3) = c_2,$$

where the c's are arbitrary constants,* and that f_1 and f_2 are two independent functions which are solutions of the equation

$$(15.13) \qquad \lambda^i \frac{\partial f}{\partial x^i} = 0.$$

For each pair of values of c_1 and c_2 equations (15.12) are equations of a curve. Through a point in space for which the functions f_1 and f_2

* Fine, 1927, 1, pp. 322, 325.

are single-valued there passes one and only one curve of the family. Such a two-parameter family of curves is called a *congruence*.

From the discussion of equations (15.7) and (15.9) it follows that in any other coordinate system x'^i the equations (15.11) are

$$\frac{dx'^1}{\lambda'^1} = \frac{dx'^2}{\lambda'^2} = \frac{dx'^3}{\lambda'^3},$$

and that their integral consists of the equations in the x''s obtained from (15.12), when the x's are replaced by the functions of the x''s which define the transformation of coordinates.

From (15.11) and (15.13) we have

$$\frac{\partial f_1}{\partial x^i}\, dx^i = 0, \qquad \frac{\partial f_2}{\partial x^i}\, dx^i = 0.$$

From these equations it follows that when the x's are cartesian dx^i, and consequently λ^i, are direction numbers of the tangent vector at a point to the curve of the congruence (15.12) through the point. Hence in any coordinate system λ^i are the components of such a tangent vector.

If f is any function of x^i and f' the function of x''^i obtained from f when x' are replaced by the functions of x''^i which define the transformation of coordinates, we have

(15.14) $f(x^1, x^2, x^3) = f'(x'^1, x'^2, x'^3).$

Either of these functions is called the *transform* of the other. From this equation we have

(15.15) $\dfrac{\partial f}{\partial x^i} = \dfrac{\partial f'}{\partial x'^j} \dfrac{\partial x'^j}{\partial x^i}.$

From this result and (15.10) we have

(15.16) $\lambda^i \dfrac{\partial f}{\partial x^i} = \lambda''\dfrac{\partial x^i}{\partial x'^j}\dfrac{\partial f'}{\partial x'^k}\dfrac{\partial x'^k}{\partial x^i} = \lambda''\dfrac{\partial f'}{\partial x'^k}\delta_j^k = \lambda'^j\dfrac{\partial f'}{\partial x'^j}.$

Hence the transforms of the solutions of an equation (15.13) are solutions of the equation

(15.17) $\lambda'^i\dfrac{\partial\theta}{\partial x'^i} = 0.$

Any function $f(x^1, x^2, x^3)$ and its transform in any other coordinate system define in their respective coordinate systems an entity called a *scalar*. Whenever in considering a problem one arrives in two coordinate systems at quantities which are transforms of one another in the sense of (15.14), one says that the entity so defined is a scalar. For example, since either of the first and last members in (15.16) is a transform of the other, we say that $\lambda^i \dfrac{\partial f}{\partial x^i}$ is a scalar, meaning that its transform in any other coordinate system is the analogous expression, in this case $\lambda'^j \dfrac{\partial f'}{\partial x'^j}$.

EXERCISES

1. For a linear transformation $x^i = a^i_j x'^j$, where the a's are constants, the coordinates are components of a contravariant vector in the two coordinate systems; in consequence of Euler's theorem the most general transformation for which this is true is when φ^i in (13.10) are homogeneous functions of the first degree in x'^i.

2. Show that in any coordinate systems $\lambda^1_{1|}$, 0, 0; 0, $\lambda^2_{2|}$, 0; 0, 0, $\lambda^3_{3|}$ are components of the tangents to the coordinate curves, the subscript of a λ indicating that it applies to the curve of parameter x^i, where i has the value of this subscript; also that these components may be written

(i) $$\lambda^i_{h|} = \delta^i_h \varphi_h,$$

where the subscript $h|$ for $h = 1, 2, 3$ denotes the vector, the δ's are defined by (13.6), and the φ's are non-zero functions of the x's.

3. Show that in any other coordinate system x'^i the components of the vectors (i) of Ex. 2 are given by

$$\lambda'^i_{h|} = \varphi_h \frac{\partial x'^i}{\partial x^h}.$$

4. What are the components in any coordinate system x'^i of the vector whose components in the x's are

$$\lambda^i = \frac{\partial f^i}{\partial x^1},$$

where f^i are any functions of x^1, x^2, x^3 which involve x^1 at least?

5. Three contravariant vectors of components $\lambda^i_{h|}$ in which $h|$ for $h = 1, 2, 3$ indicates the vector, are *independent*, that is, there are no functions a_i of the x's such that

(i) $$a_1 \lambda^i_{1|} + a_2 \lambda^i_{2|} + a_3 \lambda^i_{3|} = 0,$$

if and only if the determinant $\left| \lambda^i_{h|} \right|$ is not identically zero. Show that in this case any vector λ^i is expressible in the form

$$\lambda^i = b^1 \lambda^i_{1|} + b^2 \lambda^i_{2|} + b^3 \lambda^i_{3|} = b^h \lambda^i_{h|}$$

for suitable values of the b's as functions of the x's.

6. Show that when the determinant in Ex. 5 is equal to zero there exist functions a_1, a_2, a_3 such that equations (i) hold; discuss the geometric meaning of the cases when all the a's are different from zero and when one of them is equal to zero.

7. Given a contravariant vector λ^i in any coordinate system x^i; show that, if φ^1 and φ^2 are independent solutions of the equation $\lambda^i \dfrac{\partial \varphi}{\partial x^i} = 0$, and φ^3 is a solution of the equation $\lambda^i \dfrac{\partial \varphi}{\partial x^i} = 1$, the jacobian $\left| \dfrac{\partial \varphi}{\partial x} \right| \neq 0$; also that in the coordinate system x'^i defined by $x'^i = \varphi^i(x)$ the components of the given vector are 0, 0, 1.

16. LENGTH OF A CONTRAVARIANT VECTOR. ANGLE BETWEEN TWO VECTORS

Let λ^i be the components of a contravariant vector in any coordinate system x^i and consider the quantity $a_{ij}\lambda^i\lambda^j$, where a_{ij} are the coefficients of the fundamental form (14.6).   In consequence of (15.10) and (14.11) for any other coordinate system x'^i we have, using (13.14),

(16.1)
$$a_{ij}\lambda^i\lambda^j = a'_{kl}\lambda'^h\lambda'^m \frac{\partial x'^k}{\partial x^i}\frac{\partial x^i}{\partial x'^h}\frac{\partial x'^l}{\partial x^j}\frac{\partial x^j}{\partial x'^m}$$
$$= a'_{kl}\lambda'^h\lambda'^m \delta^k_h \delta^l_m = a'_{kl}\lambda'^k\lambda'^l.$$

Hence for any contravariant vector λ^i the expression $a_{ij}\lambda^i\lambda^j$ is a scalar (see §15).   In cartesian coordinates $a_{ij}\lambda^i\lambda^j$ assumes the form $\sum_i \lambda^i\lambda^i$, as follows from (14.7).   At each point it is the square of the length of the line-segment whose orthogonal projections upon the coordinate axes are λ^i; that is, λ^i are the *rectangular components* of the vector.   Hence we have

[16.1] *If λ^i are components in any coordinate system of a contravariant vector and a_{ij} are the coefficients of the fundamental form in this coordinate system, the quantity $a_{ij}\lambda^i\lambda^j$ is a scalar; it is the square of the length of the line-segment whose rectangular components are the corresponding λ's in a cartesian coordinate system.*

Hence in any coordinate system a set of components λ^i define at each point in space a vector whose length and direction are determined by λ^i.   Presently we shall see what the geometric significance of the components λ^i is in any coordinate system.

When, in particular, λ^i are such that

(16.2)
$$a_{ij}\lambda^i\lambda^j = 1,$$

we say that the vector is a *unit vector*.   In this case the components λ^i in a certesian system are direction cosines of the vector at each point.

For two vectors λ^i_1 and λ^i_2 we have

(16.3)
$$a_{ij}\lambda^i_1\lambda^j_2 = a'_{kl}\lambda'^k_1\lambda'^l_2,$$

as follows directly when we proceed as in the case of (16.1). Thus $a_{ij}\lambda_1^i\lambda_2^j$ is a scalar. If the coordinates are cartesian, in which case the components are direction numbers, this scalar is $\sum_i \lambda_1^i\lambda_2^i$. If this quantity is divided by $\sqrt{(\sum_i \lambda_1^i\lambda_1^i)(\sum_j \lambda_2^j\lambda_2^j)}$, the resulting expression is the cosine of the angle of the two vectors.* Hence we have:

[16.2] *If λ_1^i and λ_2^i are the components of two contravariant vectors in any coordinate system, the angle ($\leq 180°$) between the two vectors at a point is given by*

(16.4) $$\cos\theta = \frac{a_{ij}\lambda_1^i\lambda_2^j}{\sqrt{(a_{ij}\lambda_1^i\lambda_1^j)(a_{kl}\lambda_2^k\lambda_2^l)}}.$$

As a corollary we have:

[16.3] *A necessary and sufficient condition that at each point the contravariant vectors λ_1^i and λ_2^i be perpendicular is that the equation*

(16.5) $$a_{ij}\lambda_1^i\lambda_2^j = 0$$

be an identity in x^i.

When the vectors λ_1^i and λ_2^i are not perpendicular at every point in space, that is, when (16.5) is not an identity, they are perpendicular at each point of the surface whose equation in the x's is given by (16.5).

For an x^k-coordinate curve $dx^i = 0$ for $i \neq k$; consequently δ_k^i for k fixed and $i = 1, 2, 3$ are components of the contravariant vector tangent to the curve. From (16.4) we have that the angle at a point between the tangents to the x^k- and x^l-coordinate curves is given by

$$\frac{a_{ij}\delta_k^i\delta_l^j}{\sqrt{(a_{ij}\delta_k^i\delta_k^j)(a_{hm}\delta_l^h\delta_l^m)}} = \frac{a_{kl}}{\sqrt{a_{kk}a_{ll}}}.$$

Hence we have

[16.4] *In any coordinate system the cosine of the angle between an x^i-coordinate curve and an x^j-coordinate curve at any point is equal to the value of $a_{ij}/\sqrt{a_{ii}a_{jj}}$ at the point.*

If at a point $a_{ij} = 0$ for $i, j = 1, 2, 3$ and $i \neq j$, the three coordinate curves are mutually perpendicular, and consequently at the point the tangent planes to the three surfaces are mutually perpendicular. In this case we say that the coordinate surfaces through the point are *orthogonal* to one another at the point. If this situation exists at every

---

* C. G., p. 86.

point, we say that the coordinate surfaces form a *triply orthogonal family of surfaces*. Hence we have

[16.5] *The coordinate surfaces for a given coordinate system form a triply orthogonal system, if and only if the coefficients $a_{ij}(i \neq j)$ of the fundamental quadratic form in this coordinate system are equal to zero identically.*

Suppose now that we consider any vector λ^i and find the angle which it makes with each of the vectors

(16.6) $\lambda^1, 0, 0;$ $0, \lambda^2, 0;$ $0, 0, \lambda^3,$

which are tangent vectors to the x^1-, x^2-, and x^3-coordinate curves respectively. If we denote these respective angles by θ_i for $i = 1, 2, 3$, we have from (16.4)

(16.7) $$\sqrt{a_{ii}}\, \cos \theta_i = \frac{a_{ij}\lambda^j}{\sqrt{a_{ij}\lambda^i\lambda^j}}.$$

Since the lengths of the vectors (16.6) are

$$\sqrt{a_{11}}\,\lambda^1, \qquad \sqrt{a_{22}}\,\lambda^2, \qquad \sqrt{a_{33}}\,\lambda^3,$$

the sum of the projections of these lengths upon the line of the vector λ^i is in consequence of (16.7)

$$\frac{a_{ij}\lambda^i\lambda^j}{\sqrt{a_{ij}\lambda^i\lambda^j}},$$

that is, $\sqrt{a_{ij}\lambda^i\lambda^j}$ which is the length of the vector λ^i. This means that the vector λ^i at a point P is the diagonal from P of the parallelopiped whose sides are the lengths of the vectors (16.6). Hence we have

[16.6] *For a contravariant vector λ^i in any coordinate system the geometric significance of the components λ^i is that the length of the vector at any point P is the diagonal of the parallelopiped whose edges are line segments, with P as initial point, tangential to the coordinate curves at P and of the respective lengths $\sqrt{a_{ii}}\,\lambda^i$.*

When the coordinate curves through a point are mutually perpendicular, the parallelopiped is rectangular, but only when the coordinates are cartesian are λ^i the lengths of orthogonal projections of the length of the vector upon the tangents to the coordinate curves at the point.

EXERCISES

1. When space is referred to polar coordinates (13.8), the coordinate surfaces form a triply orthogonal system.

2. Determine the character of the coordinate surfaces $x'^i = $ const. for the transformation

$$x^1 = x'^1 \cos x'^2, \qquad x^2 = x'^1 \sin x'^2, \qquad x^3 = x'^3,$$

and show that they form a triply orthogonal system; the coordinates x'^i are called *cylindrical*.

3. If λ^i and μ^i are unit contravariant vectors perpendicular to one another,

$$(a_{hj} a_{ik} - a_{hk} a_{ij})\lambda^h \mu^i \lambda^j \mu^k = 1.$$

4. If μ^i and ν^i are contravariant vectors perpendicular to a contravariant vector λ^i, so also $u\mu^i + v\nu^i$ for any values of u and v are the components of a contravariant vector perpendicular to λ^i; for two sets of values u_1, v_1 and u_2, v_2 such that

$$u_1 u_2 a_{ij}\mu^i \mu^j + (u_1 v_2 + u_2 v_1)a_{ij}\mu^i \nu^j + v_1 v_2 a_{ij}\nu^i \nu^j = 0$$

the vectors $u_1\mu^i + v_1\nu^i$ and $u_2\mu^i + v_2\nu^i$ are perpendicular.

5. If $\lambda^i_{h|}$ are the components of three mutually perpendicular unit contravariant vectors, where h for $h = 1, 2, 3$ denotes the vector and i the components, then

$$\sum_h \lambda^i_{h|} \lambda^j_{h|} = a^{ij}.$$

17. COVARIANT VECTORS. CONTRAVARIANT AND COVARIANT COMPONENTS OF A VECTOR

Given any function $f(x^1, x^2, x^3)$ in any coordinate system, we have

$$(17.1) \qquad \frac{\partial f}{\partial x^i} = \frac{\partial f'}{\partial x'^j} \frac{\partial x'^j}{\partial x^i},$$

where f' is the transform of the given f for the transformation of the x's into any other coordinates x'^i. When we compare these equations with (15.10) we see that $\dfrac{\partial f}{\partial x^1}$ and $\dfrac{\partial f}{\partial x^2}$ are not components of a contravariant vector in their respective coordinate systems. However, they do belong to a new class of functions λ_i and λ'_i of the x's and x''s respectively (with indices as subscripts) related thus

$$(17.2) \qquad \lambda_i = \lambda'_j \frac{\partial x'^j}{\partial x^i}.$$

If one has two sets of functions so related and multiplies these equations by $\dfrac{\partial x^i}{\partial x'^k}$ and sums with respect to i, one obtains

$$(17.3) \qquad \lambda_i \frac{\partial x^i}{\partial x'^k} = \lambda'_j \frac{\partial x'^j}{\partial x^i} \frac{\partial x^i}{\partial x'^k} = \lambda'_j \delta^j_k = \lambda'_k,$$

which shows that the relation is reciprocal, as was shown to be the case with contravariant vectors (see Ex. 1).

Two sets of functions λ_i and λ_i' of the x's and x'''s related as in (17.2) are said to be the components in their respective coordinate systems of a *covariant vector*, there being a vector at each point of space. We observe that the indices of a covariant vector are written as subscripts, whereas for a contravariant vector they are written as superscripts, and that the partial derivatives enter in different manners in (15.10) and (17.2); but that in each case the dummy index applies to one coordinate system and the free index to the other. As remarked in the case of contravariant vectors, a covariant vector is completely determined by its components in one coordinate system, and its components in any other system are determined by the above equations. Also whenever one has, no matter how derived, two sets of functions satisfying equations (17.2) or (17.3), one concludes that the entity under consideration is a covariant vector. For example, from (17.1) it follows that for any function f of the x's the derivatives $\dfrac{\partial f}{\partial x^i}$ and the derivatives with respect to x''^i of the transform of f are components in their respective systems of a covariant vector; this covariant vector is called the *gradient* of f. Hence we have

[17.1] *The gradient of a scalar is a covariant vector.*

If λ^i and λ'^i are components of a contravariant vector in coordinate systems x^i and x'^i, we have from (14.10) and (15.8)

$$a_{kl}'\lambda'^k = a_{ij}\frac{\partial x^i}{\partial x'^k}\frac{\partial x^j}{\partial x'^l}\lambda^h\frac{\partial x'^k}{\partial x^h}$$

(17.4)

$$= a_{ij}\lambda^h\delta_h^i\frac{\partial x^j}{\partial x'^l} = a_{ij}\lambda^i\frac{\partial x^j}{\partial x'^l}.$$

We note that these equations are of the form (17.3), which means that the linear combinations $a_{ij}\lambda^i$ of the components of the given contravariant vector in the x-system, and the linear combination $a_{ij}'\lambda'^i$ in the x'-system are components in their respective systems of a covariant vector. In this sense we have

[17.2] *If λ^i are the components of a contravariant vector, $a_{ij}\lambda^i$ are the components of a covariant vector.*

From (14.16) and (17.3) we have

$$a''^{ih}\lambda_i' = a^{kl}\frac{\partial x'^i}{\partial x^k}\frac{\partial x'^h}{\partial x^l}\lambda_j\frac{\partial x^j}{\partial x'^i}$$

(17.5)

$$= a^{kl}\lambda_j\delta_k^j\frac{\partial x'^h}{\partial x^l} = a^{kl}\lambda_k\frac{\partial x'^h}{\partial x^l}.$$

Comparing this result with (15.8) we have

[17.3] *If λ_i and λ_i' are the components of a covariant vector in their respective coordinate systems, $a^{ij}\lambda_i$ and $a'^{ij}\lambda_i'$ are the components in these respective systems of a contravariant vector.*

When we apply this theorem to the covariant vector of components $a_{ji}\lambda^j$, we have in consequence of (14.13)

$$a^{ih}a_{ji}\lambda^j = \delta_j^h\lambda^j = \lambda^h,$$

that is, we obtain the vector λ^i from which the covariant vector was derived in accordance with theorem [17.2]. Similarly, if we start with a covariant vector λ_i, find the corresponding contravariant vector by theorem [17.3], and then from it the corresponding covariant vector by theorem [17.2], we obtain the original vector λ_i.

In view of the above results we say that λ^i and λ_i are the *contravariant and covariant components* respectively of the same vector, if

$$(17.6) \qquad\qquad \lambda_i = a_{ji}\lambda^j, \qquad \lambda^i = a^{ji}\lambda_j,$$

either of which set of equations, as we have seen, implies the other. When the coordinates are cartesian, in which case $a_{ij} = \delta_{ij}$, $a^{ij} = \delta^{ij}$, the corresponding contravariant and covariant components are equal (see Ex. 6) and are direction numbers of the vector, as shown in §15.

In consequence of the first of (17.6) and (14.13) we have

$$(17.7) \qquad a^{ij}\lambda_i\lambda_j = a^{ij}a_{hi}\lambda^h a_{kj}\lambda^k = \delta_h^j\lambda^h a_{kj}\lambda^k = a_{jk}\lambda^j\lambda^k.$$

Since the last of these quantities is a scalar, as shown in §16, the first quantity is a scalar. From this result and theorem [16.1] we have

[17.4] *The square of the length of a vector whose covariant components are λ_i is equal to the scalar $a^{ij}\lambda_i\lambda_j$.*

If $\lambda_{1|i}$ and $\lambda_{2|i}$ are the covariant components of two vectors, and $\lambda_{1|}^i$ and $\lambda_{2|}^i$ their respective contravariant components, by a procedure similar to that used in (17.7) we obtain

$$(17.8) \qquad\qquad a^{ij}\lambda_{1|i}\lambda_{2|j} = a_{ij}\lambda_{1|}^i\lambda_{2|}^j.$$

From this result, equation (17.7), and theorem [16.2] we have

[17.5] *The angle θ ($\leq 180°$) between the vectors at a point of two vectors whose covariant components are $\lambda_{1|i}$ and $\lambda_{2|i}$ is given by*

$$(17.9) \qquad\qquad \cos\theta = \frac{a^{ij}\lambda_{1|i}\lambda_{2|j}}{\sqrt{(a^{ij}\lambda_{1|i}\lambda_{1|j})(a^{kl}\lambda_{2|k}\lambda_{2|l})}}.$$

The geometric significance of the contravariant components of a vector is stated in theorem [16.6]. Now we derive the geometric significance of the covariant components. In consequence of the first of (17.6) we may write (16.7) in the form

$$(17.10) \qquad \sqrt{a_{jk}\lambda^j\lambda^k}\,\cos\theta_i = \frac{\lambda_i}{\sqrt{a_{ii}}},$$

where θ_i is the angle which the vector at a point makes with the x^i-coordinate curve through the point. The left-hand member of (17.10) is the length of the orthogonal projection upon the tangent to the x^i-coordinate curve of the length of the vector λ^i at the point. Hence we have

[17.6] *The geometric significance of the covariant components λ_i of a vector is that at each point P $\lambda_i/\sqrt{a_{ii}}$ is the length of the orthogonal projection of the vector upon the tangent at P to the x^i-coordinate curve through P.*

In §§15 and 16 we introduced the concept of a contravariant vector and derived properties of such vectors. At the beginning of the present section we defined the concept of a covariant vector. When one compares equations (15.10) and (17.2) giving the relations between the components of the two types of vectors in two coordinate systems, one observes that they are essentially different and might conclude that the two vectors are different entities. However, it has been shown with the aid of the coefficients a_{ij} of the fundamental form that the two entities are in fact identical, but that it is their determining components which have different geometric significance.

Were it not for the existence of the fundamental form $a_{ij}\,dx^i\,dx^j$, we should be compelled to treat contravariant and covariant vectors as different geometric entities. There are types of geometry in which no such form occurs as part of the theory, and in these geometries a distinction is made.

Consider in connection with the above remarks a covariant vector λ_i. From (17.2) and (15.3) it follows in consequence of (13.13) that

$$\lambda_i\,dx^i = \lambda'_j\,dx'^j,$$

that is, $\lambda_i\,dx^i$ is a scalar. Hence the equation

$$(17.11) \qquad \lambda_i\,dx^i = 0$$

is of the same form in any coordinate system. If d_1x^i and d_2x^i are two sets of differentials satisfying this equation which are not proportional, it follows from the equations

$$\lambda_i\,d_1x^i = 0, \qquad \lambda_i\,d_2x^i = 0,$$

and (17.11) that any solution of the latter is of the form*

$$dx^i = c_1 d_1 x^i + c_2 d_2 x^i.$$

Hence at each point in space every direction dx^i satisfying equation (17.11) is in the plane determined by the point and by the directions $d_1 x^i$ and $d_2 x^i$. Consequently we say that a covariant vector determines a plane at each point in space.

It is not true that every equation of the form (17.11) admits an integrating factor, that is, a function t of the x's such that (see Ex. 9)

(17.12) $$t \lambda_i dx^i = d\varphi,$$

where φ is some function of the x's. But when such an integrating factor exists, $\varphi = $ const. is an integral of the equation, and we have as equivalent to (17.11) the equation

$$\frac{\partial \varphi}{\partial x^i} dx^i = 0.$$

This is the condition that each set of differentials satisfying the original equation determines at a point a tangent to the surface $\varphi = $ const. through the point. Hence when the equation (17.11) admits an integrating factor, the planes determined by the vector λ_i are the tangent planes to a family of surfaces.

Suppose now that we invoke the metric properties of space based upon the fundamental form $a_{ij} dx^i dx^j$, and that we replace λ_i in (17.11) by $a_{ji}\lambda^j$, obtaining

$$a_{ji}\lambda^j dx^i = 0.$$

In consequence of theorem [16.3] we have that the directions dx^i satisfying equation (17.11) are perpendicular to the vector λ^i. Consequently at each point the given λ_i are the covariant components of the normal to the plane determined by λ_i. When furthermore the equation admits an integrating factor, λ_i are the covariant components of the normal to the surfaces $\varphi = $ const. Hence we have

[17.7] *When a covariant vector is the gradient of a function φ, or its components are proportional to the components of a gradient, the vector-field consists of vectors normal to the surfaces $\varphi = $ const.*

EXERCISES

1. Show that equations (17.2) possess the *group property*, as defined in §13 after equation (13.18).

---

* C. G., pp. 115–116.

2. If λ^i, μ_i and λ'^i, μ_i' are the components in the x-system and x'-system respectively of a contravariant and covariant vector, $\lambda^i \mu_i = \lambda'^i \mu_i'$, that is, $\lambda^i \mu_i$ is a scalar.

3. When x^i are polar coordinates, the covariant components of the vector with contravariant components λ^i are

$$\lambda^1, \qquad \frac{\lambda^2}{(x^1)^2}, \qquad \frac{\lambda^3}{(x^1 \sin x^2)^2}.$$

4. If for two sets of quantities $\mu_i(x)$ and $\mu_i'(x')$ we have

(i)
$$\mu_i' \lambda'^i_{h|} = \mu_l \lambda^l_{h|},$$

where the $\lambda^i_{h|}$ and $\lambda'^i_{h|}$ for $h = 1, 2, 3$ are the components of three independent contravariant vectors (see §15, Ex. 5), then the μ_i and μ_i' are the components in their respective coordinate systems of a covariant vector. Does this follow if equations (i) hold for fewer than three contravariant vectors?

5. If $\lambda^i_{h|}$ for $h = 1, 2, 3$ are three independent contravariant vectors (see §15, Ex. 5), and $\lambda^i_{h|}$ is the cofactor of $\lambda^i_{h|}$ in the determinant $|\lambda^i_{h|}|$ divided by the determinant, then $\lambda^{h|}_i$ are the components of a covariant vector for each value of h.

6. In order that corresponding contravariant and covariant components of a vector be equal in the coordinate system x^i, the contravariant components λ^i must satisfy the equations

(i)
$$(a_{ij} - \delta_{ij})\lambda^i = 0,$$

that is, the rank of the determinant $|a_{ij} - \delta_{ij}|$ must be less than 3. For equations (i) to hold for every vector the coordinate system must be cartesian.

7. From theorems [16.6] and [17.6] it follows that $\lambda_i / \sqrt{a_{ii}}$ is equal to the sum of the orthogonal projections upon the tangent to the x^i-coordinate curve of the lengths of the vectors (16.6).

8. Show that the equations

$$\frac{dx^1}{\lambda_1} = \frac{dx^2}{\lambda_2} = \frac{dx^3}{\lambda_3}$$

in one system of coordinates do not transform to equations of the same form in another system. Compare with equations (15.11).

9. From equations (17.12) one has $\dfrac{\partial \varphi}{\partial x^i} = t \lambda_i$ from which one obtains

$$\frac{\partial^2 \varphi}{\partial x^i \partial x^j} = t \frac{\partial \lambda_i}{\partial x^j} + \frac{\partial t}{\partial x^j} \lambda_i ;$$

noting that equations obtained from these by interchanging i and j must hold, and multiplying these equations by $e^{ijk}\lambda_k$ and summing with respect to i and j, one obtains

$$e^{ijk} \frac{\partial \lambda_i}{\partial x^j} \lambda_k = 0$$

as a necessary and sufficient condition that an equation (17.11) shall admit an integrating factor.

18. TENSORS. SYMMETRIC AND SKEW SYMMETRIC TENSORS

The equations (14.10) connecting the coefficients of the fundamental quadratic form in two coordinate systems are of the type

$$(18.1) \qquad b'_{kl}(x') = b_{ij}(x)\, \frac{\partial x^i}{\partial x'^k}\, \frac{\partial x^j}{\partial x'^l}.$$

There is a similarity between these equations and (17.3) in the sense that the indices are subscripts in both cases and that the derivatives occur in similar manner. We say that functions b'_{kl} and b_{ij} of the x''s and x's respectively related as in (18.1) are the components in their respective coordinate systems of a *covariant tensor* of the second order, and that a covariant vector is a covariant tensor of the first order, in each case the order being equal to the number of subscripts. Also we refer to the subscripts in each case as *covariant indices*.

Similarly equations (14.16) are of the type

$$(18.2) \qquad b'^{kl}(x') = b^{ij}(x)\, \frac{\partial x'^k}{\partial x^i}\, \frac{\partial x'^l}{\partial x^j}.$$

There is a similarity between these equations and (15.8) in the sense that the indices are superscripts in both cases and that the derivatives occur in similar manner. We say that functions b'^{kl} and b^{ij} related as in (18.2) are the components in their respective coordinate systems of a *contravariant tensor* of the second order, and that a contravariant vector is a contravariant tensor of the first order. Also we refer to the superscripts in each case as *contravariant indices*.

If λ^i and μ_i are the components of a contravariant and a covariant vector respectively, it follows from (15.8) and (17.3) that

$$\lambda'^k \mu'_l = \lambda^i \mu_j\, \frac{\partial x'^k}{\partial x^i}\, \frac{\partial x^j}{\partial x'^l}.$$

These equations are of the type

$$(18.3) \qquad b'^{k}{}_{l}(x') = b^{i}{}_{j}(x)\, \frac{\partial x'^k}{\partial x^i}\, \frac{\partial x^j}{\partial x'^l}.$$

We say that functions $b'^{k}{}_{l}$ and $b^{i}{}_{j}$ are the components in their respective coordinate systems of a *mixed tensor of the second order*. The components have one contravariant and one covariant index, and so we say that the tensor is contravariant of the first order and covariant of the first order.

From (18.1) we have

$$
b'_{kl} \frac{\partial x'^k}{\partial x^h} \frac{\partial x'^l}{\partial x^m} = b_{ij} \frac{\partial x^i}{\partial x'^k} \frac{\partial x^j}{\partial x'^l} \frac{\partial x'^k}{\partial x^h} \frac{\partial x'^l}{\partial x^m}
$$

(18.4)

$$
= b_{ij} \delta^i_h \delta^j_m = b_{hm}.
$$

In like manner, we have from (18.2) and (18.3)

(18.5)
$$
b'^{kl} \frac{\partial x^i}{\partial x'^k} \frac{\partial x^j}{\partial x'^l} = b^{ij},
$$

and

(18.6)
$$
b'^k{}_l \frac{\partial x^i}{\partial x'^k} \frac{\partial x'^l}{\partial x^j} = b^i{}_j.
$$

These results show the reciprocal character of the equations of tensors of the second order, that is, all coordinate systems are on a par.

Equations (17.3) and (18.1) are particular cases of the equations

(18.7)
$$
b'_{r_1 r_2 \cdots r_m} = b_{s_1 s_2 \cdots s_m} \frac{\partial x^{s_1}}{\partial x'^{r_1}} \frac{\partial x^{s_2}}{\partial x'^{r_2}} \cdots \frac{\partial x^{s_m}}{\partial x'^{r_m}},
$$

the b's and b'''s being functions of x's and x'''s respectively and having m indices (subscripts), where m is any positive integer. In this case the b's and b'''s satisfying these equations by means of the transformation equations are said to be the components in the x-system and x'-system respectively of a *covariant tensor of the mth order*.

Equations (15.8) and (18.2) are particular cases of the equations

(18.8)
$$
b'^{r_1 \cdots r_m} = b^{s_1 \cdots s_m} \frac{\partial x'^{r_1}}{\partial x^{s_1}} \frac{\partial x'^{r_2}}{\partial x^{s_2}} \cdots \frac{\partial x'^{r_m}}{\partial x^{s_m}},
$$

the b's and b'''s being functions of x's and x'''s respectively and having m indices (superscripts), where m is any positive integer. The b's and b'''s satisfying these equations are said to be the components in the x-system and x'-system respectively of a *contravariant tensor of the mth order*.

Equations (18.3) are a particular case of the equations

(18.9)
$$
b'^{r_1 \cdots r_m}_{p_1 \cdots p_n} = b^{s_1 \cdots s_m}_{q_1 \cdots q_n} \frac{\partial x'^{r_1}}{\partial x^{s_1}} \cdots \frac{\partial x'^{r_m}}{\partial x^{s_m}} \frac{\partial x^{q_1}}{\partial x'^{p_1}} \cdots \frac{\partial x^{q_n}}{\partial x'^{p_n}},
$$

the b's and b'''s being functions of x's and x'''s respectively and having m upper indices and n lower indices, where m and n are positive integers. The b's and b'''s satisfying these equations are said to be the components in the x-system and x'-system respectively of a *mixed tensor of order $m + n$, contravariant of order m and covariant of order n*.

We observe that in each set of equations (18.7), (18.8) and (18.9) the number of sets of partial derivatives entering in the equations is equal to the order of the tensor, but the way in which they enter depends upon whether the *indices are contravariant* (superscripts) or *covariant* (subscripts). In the case of a scalar, as defined in §15, there are no such sets of derivatives in the one equation (15.14) which expresses the equality of a function and its transform. Hence a scalar is called a *tensor of order zero*.

When one applies to a tensor of any order and type the processes used to obtain (18.4), (18.5), and (18.6) from (18.1), (18.2), and (18.3) respectively, one obtains equations which are the inverses of (18.7), (18.8), and (18.9) respectively. This shows that the equations giving the relations between the components of any tensor in two coordinate systems are reciprocal in character, and thus that no one coordinate system has a preferred position in the definition of a tensor.

From the above definition of tensors it follows that one may choose arbitrarily the components of a tensor in one coordinate system, and then the components in any other coordinate system are determined by (18.7), (18.8) or (18.9) (see Ex. 1). Frequently in the consideration of a geometric problem we deal with a geometric entity and find that its (analytical) components in two coordinate systems are related as in (18.7), (18.8) or (18.9). Then we say that we are dealing with a tensor. For example, at the beginning of this section we observed that equations (14.10) are of the type (18.1), and so we say that the coefficients of the fundamental quadratic form are components of a covariant tensor of the second order, or briefly that a_{ij} is a covariant tensor of the second order. We call it the *covariant metric tensor*, because, as we have seen, it enters into the determination of arc lengths, magnitudes of vectors and of angles. In like manner, we call a^{ij} the *contravariant metric tensor*.

Because of the linear homogeneous character of equations (18.7), (18.8) and (18.9) and their inverses we have

[18.1] *If all the components of a tensor are equal to zero in one coordinate system, they are equal to zero in every system.*

Such a tensor is called a *zero tensor*.

From the form of equations (18.7), (18.8) and (18.9) it is clear that the relative order (position) of the indices plays a role in these equations. It may be, however, that in the case of certain tensors when two contravariant (or covariant) indices are interchanged the new component is equal to the original one. In this case the relative order of these particular indices is immaterial. From the form of equations (18.7), (18.8) and (18.9) it follows that if this is true for certain indices in one coordinate system it is true for the corresponding indices in every system. For example, suppose that $b_{s_1 s_2 s_3 \cdots s_m} = b_{s_2 s_1 s_3 \cdots s_m}$, then from (18.7) we have

$$(18.10) \qquad b'_{r_1 r_2 r_3 \cdots r_m} = b_{s_2 s_1 s_3 \cdots s_m} \frac{\partial x^{s_1}}{\partial x'^{r_1}} \frac{\partial x^{s_2}}{\partial x'^{r_2}} \cdots \frac{\partial x^{s_m}}{\partial x'^{r_m}}$$

$$= b'_{r_2 r_1 r_3 \cdots r_m}.$$

When the relative order of two or more indices is immaterial, we say that the tensor is *symmetric with respect to these indices*. When the relative order of all the indices is immaterial, the tensor is said to be a *symmetric tensor*. Thus the metric tensors a_{ij} and a^{ij} are symmetric tensors. A general tensor of the second order, whether contravariant or covariant, has 9 different components, whereas a symmetric tensor has only 6 different components.

When for a tensor two components obtained from one another by the interchange of two particular indices, either contravariant or covariant, differ only in sign, the tensor is said to be *skew-symmetric with respect to these indices*. It can be shown that if a tensor is skew symmetric in any two indices in one coordinate system, it has this property in every system. For example, if the tensor $b_{s_1 s_2 \cdots s_m}$ is skew-symmetric in the first two indices we have (18.10) with a minus sign in the second and third members of these equations. When a tensor, whether contravariant or covariant, is skew-symmetric with respect to every pair of indices, it is called a *skew-symmetric tensor*.

From equation (14.12) and the definition of a scalar in §15 it follows that a is not a scalar. When one has two functions b' and b of the x''s and x's such that

$$b'(x') = b(\overset{.}{x}) \left| \frac{\partial x}{\partial x'} \right|^p,$$

we say that b is a *relative scalar of weight p*. Thus the determinant a of the components a_{ij} of the covariant metric tensor is a relative scalar of weight 2, and from (14.15) and (14.12) it follows that the determinant of the contravariant metric tensor is a relative scalar of weight -2. A relative scalar of weight 1 is called a *scalar density*.

In similar manner if the b''s and b's are such that instead of equations (18.7), (18.8) and (18.9), we have equations with the factor $\left| \frac{\partial x}{\partial x'} \right|^p$ in the right-hand members of these equations, we say that the b''s and b's are the components in their respective coordinate systems not of tensors but of *relative tensors of weight p* of the order and type determined by the character of their indices. In consequence of (13.15) and the above observation concerning the reciprocal character of equations (18.7), (18.8) and (18.9) it follows that the corresponding equations for relative tensors are reciprocal.

For a transformation of coordinates for which the jacobian $\left| \dfrac{\partial x}{\partial x'} \right|$ is positive, we have from (14.12)

$$(18.11) \qquad\qquad \sqrt{a'} = \sqrt{a} \left| \frac{\partial x}{\partial x'} \right|.$$

If the jacobian is negative, by a change of the sign in one of the equations of the transformation the resulting jacobian is positive. Hence there is no loss in generality in understanding that equation (18.11) holds. Thus $\sqrt{a}$ is a scalar density. From (18.11) and the definition of relative tensors we have:

[18.2] *If $b_{s_1 \cdots s_n}^{r_1 \cdots r_m}$ are the components of a relative tensor of weight p, then $b_{s_1 \cdots s_n}^{r_1 \cdots r_m}/a^{\frac{1}{2}p}$ are the components of a tensor.*

EXERCISES

1. If the Kronecker deltas δ_j^i are taken as the components of a mixed tensor in one coordinate system, the components in every other coordinate system are of the same kind; that is, δ_j^i is a mixed tensor of the second order.

2. If b_{ij} and c^{ij} are components of covariant and contravariant tensors respectively, the quantities $b_{ij}c^{ik}$ are components of a mixed tensor of the second order, and $b_{ij}c^{ij}$ is a scalar.

3. If b_{ij} and c_{ij} are two symmetric tensors such that

$$b_{ij}c_{kl} - b_{il}c_{jk} + b_{jk}c_{il} - b_{kl}c_{ij} = 0,$$

then $c_{ij} = \rho b_{ij}$, where ρ is a scalar.

4. How many functions are required to define a skew-symmetric covariant tensor of the second order?

5. For any skew-symmetric tensor all the components having two, or more, indices alike are equal to zero.

6. If b_{ij} is a skew-symmetric tensor and λ^i is a contravariant vector, then $b_{ij}\lambda^i\lambda^j = 0$; conversely, if $b_{ij}\lambda^i\lambda^j = 0$ for an arbitrary vector, b_{ij} is a skew-symmetric tensor.

7. Since $e_{hlm} \left| \dfrac{\partial x}{\partial x'} \right| = e_{ijk} \dfrac{\partial x^i}{\partial x'^h} \dfrac{\partial x^j}{\partial x'^l} \dfrac{\partial x^k}{\partial x'^m}$ (see §13, Ex. 4), the quantities e_{ijk} are the components of a relative tensor of weight -1; e^{ijk} are components of a relative tensor of weight $+1$.

8. The quantities

$$\epsilon_{ijk} = \sqrt{a}\, e_{ijk}, \qquad \epsilon^{ijk} = \frac{e^{ijk}}{\sqrt{a}}$$

are the components of covariant and contravariant tensors respectively of the third order, when the jacobian (13.11) is positive.

9. The rank of the determinant $|\, b_{ij}\,|$ of a tensor b_{ij} is called the *rank* of the tensor; show that the rank is invariant under any transformation of coordinates.

10. When the rank of a covariant tensor b_{ij} is three, the cofactors of b_{ij} in the determinant $|\, b_{ij}\,|$ are components of a relative contravariant tensor of weight two.

11. When the rank of a covariant tensor b_{ij} is two, there exist two relative contravariant vectors λ^i and μ^i each of weight one, such that the cofactor of b_{ij} in the determinant $|\, b_{ij}\,|$ is equal to $\lambda^i \mu^j$; when b_{ij} is symmetric, λ^i and μ^i are the same vectors.

19. ADDITION, SUBTRACTION AND MULTIPLICATION OF TENSORS. CONTRACTION

From the form of equations (18.7), (18.8) and (18.9) it follows that the sum or difference of two tensors of the same type and order is a tensor of the same type and order. The same is true of any linear homogeneous combination of tensors of the same type and order, the coefficients being constants or scalars.

If we take two tensors of any type and order, and form all possible products of a component of one tensor and a component of the other, we obtain a tensor whose order is the sum of the orders of the two given tensors. The number of contravariant, or covariant, indices is equal to the sum of the numbers of contravariant, or covariant, indices of the given tensors. For example, we have from (18.9)

$$(19.1) \qquad b'^{pq}{}_r c'^{s}{}_t = b^{ij}{}_k c^{l}{}_m \frac{\partial x'^p}{\partial x^i} \frac{\partial x'^q}{\partial x^j} \frac{\partial x'^s}{\partial x^l} \frac{\partial x^k}{\partial x'^r} \frac{\partial x^m}{\partial x'^t},$$

and thus $b^{ij}{}_k c^{l}{}_m$ are components of a mixed tensor of order 5, contravariant of order 3 and covariant of order 2. This process is general, so that by multiplying the components of any number of tensors we obtain a tensor, called the *product* or *outer product* of the given tensors, which is contravariant and covariant of orders which are the respective sums of the contravariant and covariant orders of the tensors multiplied.

Another process for obtaining a tensor from a given tensor, or a product of tensors, is called *contraction*. We have used this process in obtaining the covariant components of a contravariant vector in equations (17.4). Thus the product $a_{ij}\lambda^k$ of the metric tensor a_{ij} and the vector λ^i is a mixed tensor of the third order, contravariant of the first order and covariant of the second order. Each quantity $a_{ij}\lambda^i$ of this tensor for particular values of j is the sum of three components of this mixed tensor, and from (17.4) it follows that these quantities are components of a covariant vector; that is, by the summation of one contravariant and one covariant index we have obtained a tensor of order 2 less, the contravariant and covariant orders being each one less than for the original tensor. This process is called *contraction*. It applies also to any mixed tensor, and we have

[19.1] *By the contraction of any contravariant index with any covariant index there is obtained a tensor of order one less contravariant and one less covariant.*

For example,

(19.2)
$$b'^{pq}_{rps} = b^{ij}_{klm} \frac{\partial x'^p}{\partial x^i} \frac{\partial x'^q}{\partial x^j} \frac{\partial x^k}{\partial x'^r} \frac{\partial x^l}{\partial x'^p} \frac{\partial x^m}{\partial x'^s}$$
$$= b^{ij}_{klm} \delta^l_i \frac{\partial x'^q}{\partial x^j} \frac{\partial x^k}{\partial x'^r} \frac{\partial x^m}{\partial x'^s} = b^{ij}_{kim} \frac{\partial x'^q}{\partial x^j} \frac{\partial x^k}{\partial x'^r} \frac{\partial x^m}{\partial x'^s}.$$

When, in particular, contraction is applied to the product of two tensors, the resulting tensor is called an *inner product* of the two tensors.

The process of contraction may be applied in more than one way, and more than once. Thus from (19.2) we have

$$b'^{pq}_{qpr} = b^{ij}_{jim} \frac{\partial x^m}{\partial x'^r}.$$

When, in particular, we apply contraction twice to the product tensor $a_{ij}\lambda^k\lambda^l$, we obtain the scalar $a_{ij}\lambda^i\lambda^j$ which by theorem [16.1] is the square of the length of the vector λ^i. This is a particular case of the following theorem, which is a consequence of theorem [19.1]:

[19.2] *When as the result of contraction of one or more pairs of indices there remain no free indices, the resulting quantity is a scalar.*

An application of contraction, which is used frequently in tensor calculus, is what is called *lowering* a contravariant index by means of the covariant metric tensor a_{ij} and *raising* a covariant index by means of the contravariant metric tensor a^{ij}. This process was used in obtaining the covariant components of the vector λ^i, and the contravariant components of the vector λ_i. In carrying out this process it is important that the position of the index affected be not ambiguous.   For example, we have the following tensors derived from the tensor b_{ijk} ;

$$(19.3) \quad \begin{aligned} & b^l{}_{jk} = a^{il}b_{ijk} ; \quad b_i{}^l{}_k = a^{jl}b_{ijk} ; \quad b_{ij}{}^l = a^{kl}b_{ijk} ; \\ & b^{lm}{}_k = a^{il}a^{jm}b_{ijk} ; \; b^l{}_j{}^m = a^{il}a^{km}b_{ijk} ; \; b^{lmp} = a^{il}a^{jm}a^{kp}b_{ijk} . \end{aligned}$$

In similar manner we have

$$b_l{}^{jk} = a_{il}b^{ijk}; \qquad b_{lm}{}^k = a_{il}a_{jm}b^{ijk}; \qquad b_{lmp} = a_{il}a_{jm}a_{kp}b^{ijk}.$$

We remark that this process is reversible.   Thus from the first of (19.3) we have

$$a_{lm}b^l{}_{jk} = a_{lm}a^{il}b_{ijk} = \delta^i_m b_{ijk} = b_{mjk} ,$$

which is the tensor from which $b^l{}_{jk}$ was obtained.

At times in order to indicate the position from which an index has been raised or lowered a dot is placed in the original position of the index; thus the first of (19.3) would be $b^l_{.jk}$.   This notation emphasizes the position of an index.

Instead of referring to the quantities $b^l{}_{jk}$, $b_i{}^l{}_k$, $\cdots$, b^{lmp} in (19.3) as different tensors, we shall say that the quantities of each set are a set of components of the same tensor.   Any set of components determines the tensor, but in different manner according to the character of the indices, as was seen in §17 to be the case with the contravariant and covariant components of a vector.   When we apply this process to the tensor a_{ij} itself, we obtain

$$a^{jl}a^{ik}a_{ij} = \delta^k_j a^{jl} = a^{kl},$$

and hence we refer to a_{ij} and a^{ij} as the covariant and contravariant components respectively of the metric tensor.

Let now b^{ij}_{klm} and b'^{pq}_{rst} be functions of x^i and x'^i respectively, such that $b^{ij}_{klm}\lambda^l$ and $b'^{pq}_{rst}\lambda'^s$ are components in their respective coordinate systems of a tensor, and λ^i and λ'^i components of a contravariant vector.   From this hypothesis it follows that

$$b'^{pq}_{rst}\lambda'^s = b^{ij}_{klm}\lambda^l \frac{\partial x'^p}{\partial x^i} \frac{\partial x'^q}{\partial x^j} \frac{\partial x^k}{\partial x'^r} \frac{\partial x^m}{\partial x'^t} = b^{ij}_{klm}\lambda'^s \frac{\partial x^l}{\partial x'^s} \frac{\partial x'^p}{\partial x^i} \cdots \frac{\partial x^m}{\partial x'^t},$$

that is,

$$\left(b'^{pq}_{rst} - b^{ij}_{klm} \frac{\partial x'^p}{\partial x^i} \frac{\partial x'^q}{\partial x^j} \frac{\partial x^k}{\partial x'^r} \frac{\partial x^l}{\partial x'^s} \frac{\partial x^m}{\partial x'^t} \right) \lambda'^s = 0.$$

If these equations hold for every contravariant vector λ^i, and consequently for any three independent vectors, the quantity in parentheses is equal to zero, and hence b^{ij}_{klm} is a mixed tensor of the fifth order. By a similar argument we have

[19.3] *If a set of functions* $b^{r_1\cdots r_m}_{p_1\cdots p_n}$ *and* $b'^{s_1\cdots s_m}_{q_1\cdots q_n}$ *of* x^i *and* x'^i *respectively are such that* $b^{r_1\cdots r_m}_{p_1\cdots p_h\cdots p_n}\lambda^{p_h}$ *and* $b'^{s_1\cdots s_m}_{q_1\cdots q_h\cdots q_n}\lambda'^{q_h}$ *for any* p_h *and* q_h *are components of a tensor, where* λ^i *and* λ'^i *are components of an arbitrary vector in these respective coordinates, then the given functions are components of a tensor.*

A similar theorem holds if λ^i is replaced by any arbitrary tensor, and a covariant, or contravariant, index is contracted with a contravariant, or covariant, index of the given functions. Indeed, it suffices to take a tensor which is the product of distinct arbitrary vectors, in which case the result follows by repeated application of Theorem [19.3]. This is sometimes called the *quotient law of tensors.*

EXERCISES

1. Show that it follows from the identity

$$b_{ij} = \tfrac{1}{2}(b_{ij} + b_{ji}) + \tfrac{1}{2}(b_{ij} - b_{ji})$$

that any covariant tensor of the second order is the sum of a symmetric and a skew-symmetric tensor; is the same true of a contravariant tensor of the second order?

2. If $b_{ij} = b_{1|ij} + b_{2|ij}$, where $b_{1|ij}$ is symmetric in the indices and $b_{2|ij}$ is skew-symmetric, then

$$b_{ij}\, dx^i\, dx^j = b_{1|ij}\, dx^i\, dx^j.$$

3. If $b_{ij}\lambda^i\lambda^j$ is a scalar for λ^i an arbitrary contravariant vector, then $b_{ij} + b_{ji}$ are components of a tensor; if b_{ij} are symmetric in the indices, then b_{ij} are the components of a tensor.

4. If $b^h{}_{ij}\lambda^i\mu^j\nu_h$ is a scalar for arbitrary vectors λ^i, μ^i, and ν_i, $b^h{}_{ij}$ is a tensor.

5. If $b_{ijk}\, dx^i\, dx^j\, dx^k = 0$ for arbitrary values of the differentials, then

$$b_{123} + b_{231} + b_{312} + b_{132} + b_{321} + b_{213} = 0;$$

what is the condition when b_{ijk} is symmetric in i and j?

6. If b^{ij} are components of a tensor, and $ab^{ij} + cb^{ji} = 0$, where a and c are scalars, b^{ij} is either a symmetric or skew symmetric tensor.

7. When the coordinates are cartesian, the various components in (19.3) with the same values of the indices are equal.

8. Show that

$$\begin{vmatrix} a_{i_1j_1} & a_{i_1j_2} & a_{i_1j_3} \\ a_{i_2j_1} & a_{i_2j_2} & a_{i_2j_3} \\ a_{i_3j_1} & a_{i_3j_2} & a_{i_3j_3} \end{vmatrix} = e_{i_1i_2i_3}e_{j_1j_2j_3}\,|a|,$$

and consequently these determinants are the components of a covariant tensor of the sixth order which is the outer product of the tensor ϵ_{ijk} (§18, Ex. 8) with itself.

20. THE CHRISTOFFEL SYMBOLS.   THE RIEMANN TENSOR

At times equations involving the first derivatives of the components of the metric tensor are given simpler form by means of the following symbols:

(20.1) $$[ij, k] = \frac{1}{2}\left(\frac{\partial a_{ik}}{\partial x^j} + \frac{\partial a_{jk}}{\partial x^i} - \frac{\partial a_{ij}}{\partial x^k}\right),$$

(20.2) $$\left\{\begin{matrix} h \\ ij \end{matrix}\right\} = a^{hk}[ij, k].$$

Observe that from their definition $[ij, k]$ and $\left\{\begin{matrix} h \\ ij \end{matrix}\right\}$ are symmetric in i and j.   The symbols defined by (20.1) and (20.2) are called the *Christoffel symbols of the first and second kinds* respectively.*   We now derive equations involving these symbols which are of frequent use.

From (20.2) and (14.13), namely

(20.3) $$a^{ji}a_{ik} = \delta_k^j,$$

we have

(20.4) $$a_{lh}\left\{\begin{matrix} h \\ ij \end{matrix}\right\} = \delta_i^k[ij, k] = [ij, l].$$

Also from (20.1) we have

(20.5) $$\frac{\partial a_{ik}}{\partial x^j} = [ij, k] + [kj, i].$$

Differentiating (20.3) with respect to x^l, we have

$$a_{ik}\frac{\partial a^{ji}}{\partial x^l} + a^{ji}\frac{\partial a_{ik}}{\partial x^l} = 0.$$

Multiplying by a^{kh}, summing with respect to k, and substituting from (20.5), we have

$$\frac{\partial a^{jh}}{\partial x^l} = -a^{ji}a^{kh}\frac{\partial a_{ik}}{\partial x^l} = -a^{ji}a^{kh}([il, k] + [kl, i]),$$

from which we have, in consequence of (20.2),

(20.6) $$\frac{\partial a^{jh}}{\partial x^l} = -\left(a^{ji}\left\{\begin{matrix} h \\ il \end{matrix}\right\} + a^{kh}\left\{\begin{matrix} j \\ kl \end{matrix}\right\}\right).$$

* The forms of these symbols as defined by Christoffel, 1869, 1, p. 49, were $\left[\begin{matrix} ij \\ k \end{matrix}\right]$ and $\left\{\begin{matrix} ij \\ k \end{matrix}\right\}$, but we have adopted the above forms because they are in keeping with the summation convention.

If we denote by A^{ij} the cofactor of a_{ij} in the determinant $a \equiv |a_{ij}|$ and apply to this determinant the rule for the differentiation of a determinant, we have

$$\frac{\partial a}{\partial x^k} = \frac{\partial a_{ij}}{\partial x^k} A^{ij} = \frac{\partial a_{ij}}{\partial x^k} a^{ij} a,$$

the last expression being a consequence of (14.14). From this result and (20.5) we have

$$\frac{\partial a}{\partial x^k} = aa^{ij} \frac{\partial a_{ij}}{\partial x^k} = aa^{ij}([ik, j] + [jk, i]) = 2a \left\{ {i \atop ik} \right\},$$

and consequently

(20.7)
$$\frac{\partial \log \sqrt{a}}{\partial x^k} = \left\{ {i \atop ik} \right\},$$

where since i is repeated the summation convention applies, that is, the right-hand member is the sum of three symbols. Dummy indices often occur in expressions involving Christoffel symbols.

We now find the relation between the Christoffel symbols of the second kind in two coordinate systems. To this end we differentiate with respect to x'' the equations

(20.8)
$$a'_{pq} = a_{ij} \frac{\partial x^i}{\partial x'^p} \frac{\partial x^j}{\partial x'^q} \qquad (i, j, p, q = 1, 2, 3),$$

and obtain

(20.9)
$$\frac{\partial a'_{pq}}{\partial x'^r} = \frac{\partial a_{ij}}{\partial x^k} \frac{\partial x^i}{\partial x'^p} \frac{\partial x^j}{\partial x'^q} \frac{\partial x^k}{\partial x'^r} + a_{ij} \left(\frac{\partial x^i}{\partial x'^p} \frac{\partial^2 x^j}{\partial x'^q \partial x'^r} + \frac{\partial x^j}{\partial x'^q} \frac{\partial^2 x^i}{\partial x'^p \partial x'^r} \right).$$

By suitable changes of free and dummy indices in the above equation, we have the following equations

$$\frac{\partial a'_{rq}}{\partial x'^p} = \frac{\partial a_{kj}}{\partial x^i} \frac{\partial x^i}{\partial x'^p} \frac{\partial x^j}{\partial x'^q} \frac{\partial x^k}{\partial x'^r} + a_{ij} \left(\frac{\partial x^i}{\partial x'^r} \frac{\partial^2 x^j}{\partial x'^q \partial x'^p} + \frac{\partial x^j}{\partial x'^q} \frac{\partial^2 x^i}{\partial x'^p \partial x'^r} \right),$$

$$\frac{\partial a'_{pr}}{\partial x'^q} = \frac{\partial a_{ik}}{\partial x^j} \frac{\partial x^i}{\partial x'^p} \frac{\partial x^j}{\partial x'^q} \frac{\partial x^k}{\partial x'^r} + a_{ij} \left(\frac{\partial x^i}{\partial x'^p} \frac{\partial^2 x^j}{\partial x'^q \partial x'^r} + \frac{\partial x^j}{\partial x'^r} \frac{\partial^2 x^i}{\partial x'^p \partial x'^q} \right).$$

If from the sum of these equations we subtract (20.9) and divide the resulting equation by 2, we have in consequence of (20.1)

(20.10)
$$[pq, r]' = [ij, k] \frac{\partial x^i}{\partial x'^p} \frac{\partial x^j}{\partial x'^q} \frac{\partial x^k}{\partial x'^r} + a_{ij} \frac{\partial x^i}{\partial x'^r} \frac{\partial^2 x^j}{\partial x'^p dx'^q},$$

where $[pq, r]'$ is formed with respect to the tensor a'_{ij}.

From (14.16) we have

$$a''^{rs} \frac{\partial x^h}{\partial x'^s} = a^{hl} \frac{\partial x''^r}{\partial x^l}.$$

If we multiply the left- and right-hand members of equation (20.10) by $a''^{sr} \dfrac{\partial x^h}{\partial x'^s}$ and $a^{hl} \dfrac{\partial x''^r}{\partial x^l}$ respectively, and sum with respect to r, we have in consequence of (20.2)

$$\left\{ \begin{matrix} s \\ pq \end{matrix} \right\}' \frac{\partial x^h}{\partial x'^s} = [ij,\, k]\, a^{hl} \frac{\partial x^i}{\partial x'^p} \frac{\partial x^j}{\partial x'^q} \delta_l^k + a_{ij} a^{hl} \delta_l^i \frac{\partial^2 x^j}{\partial x'^p \partial x'^q},$$

which reduces to

(20.11) $$\frac{\partial^2 x^h}{\partial x'^p \partial x'^q} + \left\{ \begin{matrix} h \\ ij \end{matrix} \right\} \frac{\partial x^i}{\partial x'^p} \frac{\partial x^j}{\partial x'^q} = \left\{ \begin{matrix} s \\ pq \end{matrix} \right\}' \frac{\partial x^h}{\partial x'^s}.$$

These are the relations which we set out to obtain. On comparing (20.10) and (20.11) with (18.7), (18.8) and (18.9), we see that neither $[ij,\, k]$ nor $\left\{ \begin{matrix} h \\ ij \end{matrix} \right\}$ are components of a tensor (see Ex. 5).

If we differentiate equation (20.11) with respect to x'', and subtract the resulting equation from the one obtained therefrom on interchanging the indices q and r, we obtain

$$\left(\frac{\partial}{\partial x^i} \left\{ \begin{matrix} h \\ ik \end{matrix} \right\} - \frac{\partial}{\partial x^k} \left\{ \begin{matrix} h \\ ij \end{matrix} \right\} \right) \frac{\partial x^i}{\partial x'^p} \frac{\partial x^j}{\partial x'^q} \frac{\partial x^k}{\partial x'^r}$$

$$+ \left\{ \begin{matrix} h \\ ij \end{matrix} \right\} \left(\frac{\partial^2 x^i}{\partial x'^p \partial x'^q} \frac{\partial x^j}{\partial x'^r} - \frac{\partial^2 x^i}{\partial x'^p \partial x'^r} \frac{\partial x^j}{\partial x'^q} \right)$$

$$= \left(\frac{\partial}{\partial x'^q} \left\{ \begin{matrix} s \\ pr \end{matrix} \right\}' - \frac{\partial}{\partial x'^r} \left\{ \begin{matrix} s \\ pq \end{matrix} \right\}' \right) \frac{\partial x^h}{\partial x'^s}$$

$$+ \left\{ \begin{matrix} s \\ pr \end{matrix} \right\}' \frac{\partial^2 x^h}{\partial x'^s \partial x'^q} - \left\{ \begin{matrix} s \\ pq \end{matrix} \right\}' \frac{\partial^2 x^h}{\partial x'^s \partial x'^r}.$$

On substituting for the second derivatives in this equation their expressions from equations of the form (20.11), we obtain

(20.12) $$R^h{}_{ijk} \frac{\partial x^i}{\partial x'^p} \frac{\partial x^j}{\partial x'^q} \frac{\partial x^k}{\partial x'^r} = R'^s{}_{pqr} \frac{\partial x^h}{\partial x'^s},$$

where by definition

(20.13) $$R^h{}_{ijk} = \frac{\partial}{\partial x^i} \left\{ \begin{matrix} h \\ ik \end{matrix} \right\} - \frac{\partial}{\partial x^k} \left\{ \begin{matrix} h \\ ij \end{matrix} \right\} + \left\{ \begin{matrix} l \\ ik \end{matrix} \right\} \left\{ \begin{matrix} h \\ lj \end{matrix} \right\} - \left\{ \begin{matrix} l \\ ij \end{matrix} \right\} \left\{ \begin{matrix} h \\ lk \end{matrix} \right\},$$

(l being a dummy index indicates summation), and where $R'^{s}{}_{pqr}$ is the similar expression in the Christoffel symbols formed with respect to a'_{ij}.

If equations (20.12) be multiplied by $\dfrac{\partial x'^{t}}{\partial x^{h}}$ and summed with respect to h, we have

$$(20.14) \qquad R'^{t}{}_{pqr} = R^{h}{}_{ijk}\,\frac{\partial x^{i}}{\partial x'^{p}}\,\frac{\partial x^{j}}{\partial x'^{q}}\,\frac{\partial x^{k}}{\partial x'^{r}}\,\frac{\partial x'^{t}}{\partial x^{h}}\,.$$

From the form of these equations it follows that $R^{h}{}_{ijk}$, which are called *Riemann symbols of the second kind*, are the components of a tensor contravariant of the first order and covariant of the third order (see (18.9)). It is called the *Riemann tensor of the fourth order*. From (20.13) it follows that this tensor is skew-symmetric in the indices j and k.

The quantities R_{lijk} which are defined by

$$(20.15) \qquad R_{lijk} = a_{lh}R^{h}{}_{ijk}\,, \qquad R^{h}{}_{ijk} = a^{lh}R_{lijk}$$

are called *Riemann symbols of the first kind*. They are the components of the Riemann tensor with all four indices covariant.

When the coordinate system is cartesian, the coefficients of the fundamental form are constants, namely δ_{ij}. In this case the Christoffel symbols of either kind are zero, as follows from (20.1) and (20.2). From (20.13) it follows that in this case the components of the Riemann tensor are equal to zero, and by theorem [18.1] that the components are zero in every coordinate system. Hence we have

[20.1] *The Riemann tensor of euclidean 3-space is a zero tensor.*

This does not mean that the Christoffel symbols in any coordinate system are equal to zero, but that the functions (20.13) of these symbols are equal to zero.

If g_{ij} is any symmetric covariant tensor such that the determinant g of g_{ij} is not zero, that is,

$$(20.16) \qquad g \equiv |\,g_{ij}\,| \neq 0,$$

quantities g^{ij} are defined uniquely by

$$(20.17) \qquad g^{ij}g_{jk} = \delta^{i}_{k}\,.$$

As in the case of a^{ij} in §14 it can be shown that g^{ij} is a symmetric contravariant tensor.

Similarly to (20.1) and (20.2) we can define Christoffel symbols by means of g_{ij} and g^{ij} in which case it is advisable at times to use the

notation $\left\{ {i \atop jk} \right\}_g$ to indicate this fact.   Likewise a Riemann tensor in terms
of these symbols can be defined by (20.13).   Ordinarily the Riemann
tensor so defined is not a zero tensor (see §23).

EXERCISES

1. Show that

$$\frac{\partial a_{ij}}{\partial x^k} - \frac{\partial a_{ik}}{\partial x^i} = [jk, i] - [ij, k].$$

2. Show that if $a_{ij} = 0$ for $i \neq j$

$$\left\{ {i \atop ii} \right\} = \frac{1}{2} \frac{\partial \log a_{ii}}{\partial x^i}, \qquad \left\{ {i \atop ij} \right\} = \frac{1}{2} \frac{\partial \log a_{ii}}{\partial x^j},$$

$$\left\{ {i \atop jj} \right\} = -\frac{1}{2a_{ii}} \frac{\partial a_{jj}}{\partial x^i}, \qquad \left\{ {i \atop jk} \right\} = 0 \qquad\qquad (i, j, k \neq),$$

where a repeated index does not indicate summation and the notation $(i, j, k \neq)$
means that no two of i, j, k are equal.

3. When x^i are polar coordinates (see (14.8)) all the Christoffel symbols are
equal to zero except the following:

$$\left\{ {2 \atop 12} \right\} = \left\{ {3 \atop 13} \right\} = \frac{1}{x^1}, \qquad \left\{ {1 \atop 22} \right\} = -x^1, \qquad \left\{ {1 \atop 33} \right\} = -x^1 \sin^2 x^2,$$

$$\left\{ {3 \atop 23} \right\} = \cot x^2, \qquad \left\{ {2 \atop 33} \right\} = -\sin x^2 \cos x^2.$$

4. When x^i are cartesian, one has from (20.11)

$$\sum_h \frac{\partial^2 x^h}{\partial x'^p \partial x'^q} \frac{\partial x^h}{\partial x'^r} = [pq, r]'.$$

Show that from these equations one may derive equations of the form (20.11) in
general coordinate systems x'^i and x''^i.

5. For any linear transformation (13.1) the Christoffel symbols of the second
kind are related as components of a mixed tensor of the third order.

6. Show that for Christoffel symbols formed with respect to any tensor g_{ij}
satisfying (20.16)

$$g_{ih} \frac{\partial}{\partial x^j} \left\{ {h \atop ik} \right\} = \frac{\partial}{\partial x^j} [ik, l] - \left\{ {h \atop ik} \right\} ([hj, l] + [lj, h]).$$

7. Using (20.15), (20.13) and Ex. 6 show that

(i) $\qquad R_{lijk} = \frac{\partial}{\partial x^j} [ik, l] - \frac{\partial}{\partial x^k} [ij, l] + \left\{ {h \atop ij} \right\} [lk, h] - \left\{ {h \atop ik} \right\} [lj, h],$

where the symbols $[ij, k]$ are formed with respect to any g_{ij} satisfying (20.16); and that in terms of g_{ij} we have

(ii)
$$R_{lijk} = \frac{1}{2}\left(\frac{\partial^2 g_{lk}}{\partial x^i \partial x^j} + \frac{\partial^2 g_{ij}}{\partial x^l \partial x^k} - \frac{\partial^2 g_{lj}}{\partial x^i \partial x^k} - \frac{\partial^2 g_{ik}}{\partial x^l \partial x^j}\right)$$

$$+ g^{hm}([ij, m][lk, h] - [ik, m][lj, h]).$$

8. Show that it follows from Ex. 7 (ii) that

(i)
$$R_{lijk} = -R_{iljk} = -R_{likj} = R_{jkli} ,$$

and

(ii)
$$R_{lijk} + R_{ljki} + R_{lkij} = 0.$$

9. Show that

$$R^h{}_{hjk} = 0.$$

10. From (20.14) it follows that for the quantities R_{ij} defined by

$$R_{ij} = R^h{}_{ijh} = g^{hk}R_{hijk}$$

one has

$$R'_{pq} = R_{ij}\frac{\partial x^i}{\partial x'^p}\frac{\partial x^j}{\partial x'^q},$$

and consequently R_{ij} are the covariant components of a tensor of the second order, which is called the *Ricci tensor;*[*] from (20.13) and equations analogous to (20.7) one has

$$R_{ij} = \frac{\partial^2 \log \sqrt{g}}{\partial x^i \partial x^j} - \frac{\partial}{\partial x^h}\begin{Bmatrix} h \\ ij \end{Bmatrix} + \begin{Bmatrix} l \\ ih \end{Bmatrix}\begin{Bmatrix} h \\ lj \end{Bmatrix} - \begin{Bmatrix} l \\ ij \end{Bmatrix}\frac{\partial \log \sqrt{g}}{\partial x^l},$$

where g is defined by (20.16).

11. If $\begin{Bmatrix} i \\ jk \end{Bmatrix}_a$ and $\begin{Bmatrix} i \\ jk \end{Bmatrix}_g$ are Christoffel symbols formed with respect to symmetric tensors a_{ij} and g_{ij} such that $|a_{ij}| \neq 0$ and $|g_{ij}| \neq 0$, then the quantities $\begin{Bmatrix} i \\ jk \end{Bmatrix}_a - \begin{Bmatrix} i \\ jk \end{Bmatrix}_g$ are components of a tensor of the third order.

21. THE FRENET FORMULAS IN GENERAL COORDINATES

If the coordinates x'^i are cartesian and x^i are any general coordinates, equations (20.11) become

(21.1)
$$\frac{\partial^2 x^i}{\partial x'^j \partial x'^k} + \begin{Bmatrix} i \\ hl \end{Bmatrix}\frac{\partial x^h}{\partial x'^j}\frac{\partial x^l}{\partial x'^k} = 0,$$

[*] See 1904, 2, p. 1234.

since the a'''s are constants in the this case and consequently the Christoffel symbols $\begin{Bmatrix} i \\ jk \end{Bmatrix}'$ are all zero.

We consider now a curve with the equations $x'^i = f^i(t)$. When these expressions are substituted in the equations $x^i = \varphi^i(x'^1, x'^2, x'^3)$ connecting the two sets of coordinates x^i and x'^i, we obtain equations of the curve in terms of x^i and t. The derivatives $\dfrac{dx^i}{dt}$ are given by

$$(21.2) \qquad \frac{dx^i}{dt} = \frac{\partial x^i}{\partial x'^j} \frac{dx'^j}{dt},$$

from which it is seen that $\dfrac{dx^i}{dt}$ are the components of the contravariant vector whose components $\dfrac{dx'^i}{dt}$ in cartesian coordinates x'^i are direction numbers of the tangent to the curve. Differentiating equations (21.2) with respect to t we have

$$\frac{d^2 x^i}{dt^2} = \frac{\partial^2 x^i}{\partial x'^j \partial x'^k} \frac{dx'^j}{dt} \frac{dx'^k}{dt} + \frac{\partial x^i}{\partial x'^j} \frac{d^2 x'^j}{dt^2},$$

which in consequence of (21.1) may be written in the form

$$(21.3) \qquad \frac{d^2 x^i}{dt^2} + \begin{Bmatrix} i \\ hl \end{Bmatrix} \frac{dx^h}{dt} \frac{dx^l}{dt} = \frac{d^2 x'^j}{dt^2} \frac{\partial x^i}{\partial x'^j}.$$

Since for cartesian coordinates x'^i the symbols $\begin{Bmatrix} i \\ jk \end{Bmatrix}'$ are zero, $\dfrac{d^2 x'^j}{dt^2}$ may be given the same form as the left-hand member of (21.3), namely $\dfrac{d^2 x'^j}{dt^2} + \begin{Bmatrix} j \\ kl \end{Bmatrix}' \dfrac{dx'^k}{dt} \dfrac{dx'^l}{dt}$. Consequently on comparing this result with (15.6) and (15.10), we have

[21.1] *For a curve defined in terms of any coordinates x^i the quantities $\dfrac{d^2 x^i}{dt^2} + \begin{Bmatrix} i \\ jk \end{Bmatrix} \dfrac{dx^j}{dt} \dfrac{dx^k}{dt}$ are components of a contravariant vector, meaning that in any coordinate system the components have this form.*

If the arc s is the parameter we have from (21.2) and (3.2)

$$(21.4) \qquad \frac{dx^i}{ds} = \frac{\partial x^i}{\partial x'^j} \alpha'^j,$$

where α'^i are components in the x''s of the direction cosines of the tangent to the given curve, and consequently of the unit vector tangent to the curve. From equations (21.4) and (15.10) it follows that α^i defined by

$$(21.5) \qquad \alpha^i = \frac{dx^i}{ds}$$

are the contravariant components in the x's of the unit vector tangent to the curve. This result is in keeping with the observation in §15 that when a coordinate system is cartesian the contravariant components of a unit vector are direction cosines of the vector. Accordingly if β'^i and γ'^i are direction cosines of the principal normal and binormal of a curve defined in cartesian coordinates x'^i, that is, their contravariant components in these coordinates, the contravariant components of these respective unit vectors in any coordinate system x^i are given by

$$(21.6) \qquad \beta^i = \beta'^j \frac{\partial x^i}{\partial x'^j}, \qquad \gamma^i = \gamma'^j \frac{\partial x^i}{\partial x'^j}.$$

In cartesian coordinates x'^i we have from (4.7)

$$(21.7) \qquad \frac{d^2 x'^j}{ds^2} = \kappa \beta'^j,$$

where κ is the curvature of the curve. In any coordinates x^i in consequence of (21.6) and (21.7) we have from (21.3)

$$\frac{d^2 x^i}{ds^2} + \left\{ \begin{matrix} i \\ jk \end{matrix} \right\} \frac{dx^j}{ds} \frac{dx^k}{ds} = \kappa \beta^i,$$

which because of (21.5) may be written

$$(21.8) \qquad \frac{d\alpha^i}{ds} + \left\{ \begin{matrix} i \\ jk \end{matrix} \right\} \alpha^j \frac{dx^k}{ds} = \kappa \beta^i.$$

If we differentiate equations (21.6) with respect to s, and make use of the Frenet formulas (6.1) and equations (21.1) and (21.6), we obtain

$$\frac{d\beta^i}{ds} = - (\kappa \alpha'^j + \tau \gamma'^j) \frac{\partial x^i}{\partial x'^j} + \beta'^j \frac{\partial^2 x^i}{\partial x'^j \partial x'^k} \frac{dx'^k}{ds}$$

$$= - (\kappa \alpha'^j + \tau \gamma'^j) \frac{\partial x^i}{\partial x'^j} - \beta'^j \left\{ \begin{matrix} i \\ hl \end{matrix} \right\} \frac{\partial x^h}{\partial x'^j} \frac{\partial x^l}{\partial x'^k} \frac{dx'^k}{ds}$$

$$= - (\kappa \alpha^i + \tau \gamma^i) - \left\{ \begin{matrix} i \\ hl \end{matrix} \right\} \beta^h \frac{dx^l}{ds},$$

$$\frac{d\gamma^i}{ds} = \tau \beta'^j \frac{\partial x^i}{\partial x'^j} - \left\{ \begin{matrix} i \\ hl \end{matrix} \right\} \gamma'^j \frac{\partial x^h}{\partial x'^j} \frac{\partial x^l}{\partial x'^k} \frac{dx'^k}{ds} = \tau \beta^i - \left\{ \begin{matrix} i \\ hl \end{matrix} \right\} \gamma^h \frac{dx^l}{ds}.$$

From these results and (21.8) we have

[21.2] *When a curve is defined in terms of general coordinates* x^i, *the Frenet formulas are*

$$\frac{d\alpha^i}{ds} + \left\{\begin{matrix} i \\ jk \end{matrix}\right\} \alpha^j \frac{dx^k}{ds} = \kappa\beta^i,$$

(21.9)
$$\frac{d\beta^i}{ds} + \left\{\begin{matrix} i \\ jk \end{matrix}\right\} \beta^j \frac{dx^k}{ds} = -(\kappa\alpha^i + \tau\gamma^i),$$

$$\frac{d\gamma^i}{ds} + \left\{\begin{matrix} i \\ jk \end{matrix}\right\} \gamma^j \frac{dx^k}{ds} = \tau\beta^i,$$

where α^i, β^i, *and* γ^i *are the contravariant components of unit vectors having the directions of the tangent, principal normal, and binormal respectively of the curve, s being the arc of the curve, and* κ *and* τ *the curvature and torsion respectively.*

Consider now the equations of motion in cartesian coordinates x'^i

(21.10)
$$m \frac{d^2 x'^i}{dt^2} = -\frac{\partial V}{\partial x'^i}$$

of a particle of mass m in a field of potential V. From Theorem [21.1] it follows that in general coordinates the left-hand members of these equations are the expressions given in Theorem [21.1] multiplied by m, and that they are contravariant components of a vector. Consequently the right-hand members in general coordinates must appear as contravariant components of a vector. In §17 it was shown that in any coordinate system $\frac{\partial V}{\partial x^i}$ are covariant components of a vector and that $a^{ij}\frac{\partial V}{\partial x^j}$ are its contravariant components. Hence in general coordinates x^i equations (21.10) are

(21.11)
$$m\left(\frac{d^2 x^i}{dt^2} + \left\{\begin{matrix} i \\ jk \end{matrix}\right\} \frac{dx^j}{dt} \frac{dx^k}{dt}\right) = -a^{ij}\frac{\partial V}{\partial x^j}.$$

Another way of arriving at this result is to note that

$$m\left(\frac{d^2 x^i}{dt^2} + \left\{\begin{matrix} i \\ jk \end{matrix}\right\} \frac{dx^j}{dt} \frac{dx^k}{dt}\right) + a^{ij}\frac{\partial V}{\partial x^j}$$

are the components of a contravariant vector. From (21.10) we have that the components of this vector in cartesian coordinates are equal to zero, and consequently they are equal to zero in any coordinate system, and one obtains (21.11) in any coordinate system.

EXERCISES

1. In any coordinate system the equations

$$\frac{d^2 x^i}{ds^2} + \left\{ \begin{matrix} i \\ jk \end{matrix} \right\} \frac{dx^j}{ds} \frac{dx^k}{ds} = 0$$

are the differential equations of the straight lines in space.

2. When x^i are polar coordinates equations (21.11) are (see §20, Ex. 3)

$$m \left[\frac{d^2 x^1}{dt^2} - x^1 \left(\frac{dx^2}{dt} \right)^2 - x^1 \sin^2 x^2 \left(\frac{dx^3}{dt} \right)^2 \right] = -\frac{\partial V}{\partial x^1},$$

$$m \left[\frac{d^2 x^2}{dt^2} + \frac{2}{x^1} \frac{dx^1}{dt} \frac{dx^2}{dt} - \sin x^2 \cos x^2 \left(\frac{dx^3}{dt} \right)^2 \right] = -\frac{1}{(x^1)^2} \frac{\partial V}{\partial x^2},$$

$$m \left[\frac{d^2 x^3}{dt^2} + \frac{2}{x^1} \frac{dx^1}{dt} \frac{dx^3}{dt} + 2 \cot x^2 \frac{dx^2}{dt} \frac{dx^3}{dt} \right] = -\frac{1}{(x^1 \sin x^2)^2} \frac{\partial V}{\partial x^3}.$$

3. If $\lambda^i(s)$ are components of a contravariant vector at points of a curve $x^i(s)$, the quantities

(i) $$\frac{d\lambda^i}{ds} + \left\{ \begin{matrix} i \\ jk \end{matrix} \right\} \lambda^j \frac{dx^k}{ds}$$

are components of a contravariant vector at point of the curve.

4. If $\lambda^i(s)$ in Ex. 3 are such that the quantities (i) are identically zero, the vectors λ^i are parallel, that is, the components of the vectors in any cartesian system are constants.

22. COVARIANT DIFFERENTIATION

By theorem [17.1] we have that the partial derivatives of a scalar, that is, a tensor of order zero, are the components of a covariant vector, that is, a tensor of the first order. In this section it will be shown that this is the only case in which the derivatives of a tensor are components of a tensor, but at the same time we shall find expressions involving derivatives of a tensor which are components of a tensor.

We consider first a contravariant vector λ^i, and differentiate with respect to x^j the equations

(22.1) $$\lambda^i = \lambda'^p \frac{\partial x^i}{\partial x'^p},$$

with the result

(22.2) $$\frac{\partial \lambda^i}{\partial x^j} = \frac{\partial \lambda'^p}{\partial x'^q} \frac{\partial x'^q}{\partial x^j} \frac{\partial x^i}{\partial x'^p} + \lambda'^p \frac{\partial^2 x^i}{\partial x'^p \partial x'^q} \frac{\partial x'^q}{\partial x^j}.$$

In consequence of (20.11) the right-hand member of this equation is equal to

$$\frac{\partial x'^q}{\partial x^j}\left[\frac{\partial \lambda'^p}{\partial x'^q}\frac{\partial x^i}{\partial x'^p} + \lambda'^p\left(\left\{\begin{matrix} l \\ pq \end{matrix}\right\}'\frac{\partial x^i}{\partial x'^l} - \left\{\begin{matrix} i \\ hk \end{matrix}\right\}\frac{\partial x^h}{\partial x'^p}\frac{\partial x^k}{\partial x'^q}\right)\right]$$

$$= \frac{\partial x'^q}{\partial x^j}\left(\frac{\partial \lambda'^p}{\partial x'^q} + \lambda'^r\left\{\begin{matrix} p \\ rq \end{matrix}\right\}'\right)\frac{\partial x^i}{\partial x'^p} - \left\{\begin{matrix} i \\ hk \end{matrix}\right\}\lambda^h\frac{\partial x'^q}{\partial x^j}\frac{\partial x^k}{\partial x'^q}$$

$$= \left(\frac{\partial \lambda'^p}{\partial x'^q} + \lambda'^r\left\{\begin{matrix} p \\ rq \end{matrix}\right\}'\right)\frac{\partial x^i}{\partial x'^p}\frac{\partial x'^q}{\partial x^j} - \left\{\begin{matrix} i \\ hk \end{matrix}\right\}\lambda^h\delta^k_j.$$

Hence equation (22.2) may be written

(22.3) $$\frac{\partial \lambda^i}{\partial x^j} + \lambda^h\left\{\begin{matrix} i \\ hj \end{matrix}\right\} = \left(\frac{\partial \lambda'^p}{\partial x'^q} + \lambda'^r\left\{\begin{matrix} p \\ rq \end{matrix}\right\}'\right)\frac{\partial x^i}{\partial x'^p}\frac{\partial x'^q}{\partial x^j}.$$

Consequently, if we define $\lambda^i_{,j}$ by

(22.4) $$\lambda^i_{,j} \equiv \frac{\partial \lambda^i}{\partial x^j} + \lambda^h\left\{\begin{matrix} i \\ hj \end{matrix}\right\},$$

and similarly $\lambda'^p_{,q}$, equation (22.3) is

$$\lambda^i_{,j} = \lambda'^p_{,q}\frac{\partial x^i}{\partial x'^p}\frac{\partial x'^q}{\partial x^j}.$$

Hence by (18.6) $\lambda^i_{,j}$ and $\lambda'^p_{,q}$ are the components in their respective coordinate systems of a mixed tensor of the second order. We say that the component $\lambda^i_{,j}$ is obtained from the contravariant vector λ^i by *covariant differentiation* with respect to x^j.

If we differentiate the equations

$$\lambda'_p = \lambda_i\frac{\partial x^i}{\partial x'^p}$$

with respect to x'^q and make use of (20.11), we obtain

(22.5) $$\lambda'_{p,q} = \lambda_{i,j}\frac{\partial x^i}{\partial x'^p}\frac{\partial x^j}{\partial x'^q},$$

where

(22.6) $$\lambda_{i,j} \equiv \frac{\partial \lambda_i}{\partial x^j} - \lambda_h\left\{\begin{matrix} h \\ ij \end{matrix}\right\},$$

and similarly for $\lambda'_{p,q}$. Consequently $\lambda_{i,j}$ are components of a covariant tensor of the second order, which is said to be obtained from the covariant vector λ_i by covariant differentiation.

Consider next the covariant tensor b_{ij}, and differentiate with respect to x''^r the equations

$$(22.7) \qquad b'_{pq} = b_{ij} \frac{\partial x^i}{\partial x'^p} \frac{\partial x^j}{\partial x'^q}.$$

Making use of equations of the form (20.11), we obtain

$$\frac{\partial b'_{pq}}{\partial x'^r} = \frac{\partial b_{ij}}{\partial x^k} \frac{\partial x^i}{\partial x'^p} \frac{\partial x^j}{\partial x'^q} \frac{\partial x^k}{\partial x'^r} + b_{ij} \left[\frac{\partial x^j}{\partial x'^q} \left(\left\{ \begin{matrix} l \\ pr \end{matrix} \right\}' \frac{\partial x^i}{\partial x'^l} - \left\{ \begin{matrix} i \\ hk \end{matrix} \right\} \frac{\partial x^h}{\partial x'^p} \frac{\partial x^k}{\partial x'^r} \right) \right.$$

$$\left. + \frac{\partial x^i}{\partial x'^p} \left(\left\{ \begin{matrix} l \\ qr \end{matrix} \right\}' \frac{\partial x^j}{\partial x'^l} - \left\{ \begin{matrix} j \\ hk \end{matrix} \right\} \frac{\partial x^h}{\partial x'^q} \frac{\partial x^k}{\partial x'^r} \right) \right],$$

from which by means of (22.7) and suitable choice of dummy indices we have

$$\frac{\partial b'_{pq}}{\partial x'^r} - b'_{lq} \left\{ \begin{matrix} l \\ pr \end{matrix} \right\}' - b'_{pl} \left\{ \begin{matrix} l \\ qr \end{matrix} \right\}'$$

$$= \left(\frac{\partial b_{ij}}{\partial x^k} - b_{hj} \left\{ \begin{matrix} h \\ ik \end{matrix} \right\} - b_{ih} \left\{ \begin{matrix} h \\ jk \end{matrix} \right\} \right) \frac{\partial x^i}{\partial x'^p} \frac{\partial x^j}{\partial x'^q} \frac{\partial x^k}{\partial x'^r}.$$

Hence, if we put

$$(22.8) \qquad b_{ij,k} \equiv \frac{\partial b_{ij}}{\partial x^k} - b_{hj} \left\{ \begin{matrix} h \\ ik \end{matrix} \right\} - b_{ih} \left\{ \begin{matrix} h \\ jk \end{matrix} \right\}$$

and $b'_{pq,r}$ for the left-hand member of the above equation, we have that $b_{ij,k}$ are components of a covariant tensor of the third order, which we say is obtained from b_{ij} by covariant differentiation.

By proceeding in like manner one may show that

$$(22.9) \qquad b^{ij}{}_{,k} \equiv \frac{\partial b^{ij}}{\partial x^k} + b^{hj} \left\{ \begin{matrix} i \\ hk \end{matrix} \right\} + b^{ih} \left\{ \begin{matrix} j \\ hk \end{matrix} \right\},$$

and

$$(22.10) \qquad b^i{}_{j,k} \equiv \frac{\partial b^i{}_j}{\partial x^k} + b^h{}_j \left\{ \begin{matrix} i \\ hk \end{matrix} \right\} - b^i{}_h \left\{ \begin{matrix} h \\ jk \end{matrix} \right\}$$

are mixed tensors of the third order, which we say are obtained by covariant differentiation.

By referring to (22.4), (22.6), (22.8), (22.9) and (22.10), one observes that in every case covariant differentiation is indicated by a covariant index preceded by a comma; that each expression contains the derivative of the original tensor with respect to x^k, where k is this index; and that corresponding to each contravariant index there is added a term involving a Christoffel symbol, and corresponding to each covariant index there is subtracted such a term. All of these expressions are particular cases of the following general rule for covariant differentiation:

$$b^{r_1\cdots r_m}_{s_1\cdots s_p,\, i} = \frac{\partial b^{r_1\cdots r_m}_{s_1\cdots s_p}}{\partial x^i} + \sum_{\alpha}^{1,\cdots,m} b^{r_1\cdots r_{\alpha-1}j\,r_{\alpha+1}\cdots r_m}_{s_1\cdots\cdots\cdots\cdots s_p} \left\{ \begin{matrix} r_\alpha \\ ji \end{matrix} \right\}$$

(22.11)

$$ - \sum_{\beta}^{1,\cdots,p} b^{r_1\cdots\cdots\cdots\cdots\cdots r_m}_{s_1\cdots s_{\beta-1}j\,s_{\beta+1}\cdots s_p} \left\{ \begin{matrix} j \\ s_\beta i \end{matrix} \right\}.$$

When equations (20.5) are written in the form

(22.12) $$\frac{\partial a_{ij}}{\partial x^k} = a_{hj} \left\{ \begin{matrix} h \\ ik \end{matrix} \right\} + a_{ih} \left\{ \begin{matrix} h \\ jk \end{matrix} \right\},$$

we see from (22.8) that

(22.13) $$a_{ij,k} = 0.$$

Also when equations (20.6) are compared with (22.9), we see that

(22.14) $$a^{ij},_k = 0.$$

By §18 Ex. 1, δ^i_j are the components of a mixed tensor. From (22.10) it follows that

(22.15) $$\delta^i_{j,k} = 0.$$

Hence we have

[22.1] *The tensors a_{ij}, a^{ij}, and δ^i_j behave as constants in covariant differentiation.*

When the Christoffel symbols are formed with respect to any tensor g_{ij} such that the determinant $g \neq 0$, the above results concerning covariant differentiation hold equally well. However, it is advisable in such a case to use the term *covariant differentiation based upon g_{ij}*. Thus we should say that the results of the first part of this section involve differentiation based upon the metric tensor a_{ij} of space. Since we shall have occasion later to use covariant differentiation not based upon the metric tensor of space, it is understood in what follows that we are dealing with properties of covariant differentiation based upon a general tensor g_{ij}.

From the form of equations (22.11) it follows that the covariant derivative of the sum, or difference, of two tensors of the same order and type is equal to the sum, or difference, of the covariant derivatives of the two tensors. We consider next the covariant derivative of the product of two tensors, and in particular the following:

$$
(22.16) \quad (b_{ij}c^{kl})_{,m} = \frac{\partial}{\partial x^m}(b_{ij}c^{kl}) - c^{kl}\left(b_{hj}\begin{Bmatrix}h\\im\end{Bmatrix} + b_{ih}\begin{Bmatrix}h\\jm\end{Bmatrix}\right)
$$
$$
+ b_{ij}\left(c^{hl}\begin{Bmatrix}k\\hm\end{Bmatrix} + c^{kh}\begin{Bmatrix}l\\hm\end{Bmatrix}\right) = c^{kl}b_{ij,m} + b_{ij}c^{kl}{}_{,m},
$$

which is the same as the rule for ordinary differentiation of a product.

Since a tensor formed by multiplication and contraction is a sum of products, we have in particular

$$
(22.17) \quad (b_{ij}c^{jl})_{,m} = c^{jl}b_{ij,m} + b_{ij}c^{jl}{}_{,m}.
$$

The results for these particular cases illustrate the following general theorem:

[22.2] *Covariant differentiation of the sum, difference, outer and inner product of tensors obeys the same rules as ordinary differentiation.*

If we differentiate covariantly the tensor $\lambda_{i,j}$ defined by (22.6), we have

$$
\lambda_{i,jk} = \frac{\partial}{\partial x^k}\left(\frac{\partial \lambda_i}{\partial x^j} - \lambda_h\begin{Bmatrix}h\\ij\end{Bmatrix}\right) - \left(\frac{\partial \lambda_l}{\partial x^j} - \lambda_h\begin{Bmatrix}h\\lj\end{Bmatrix}\right)\begin{Bmatrix}l\\ik\end{Bmatrix} - \left(\frac{\partial \lambda_i}{\partial x^l} - \lambda_h\begin{Bmatrix}h\\il\end{Bmatrix}\right)\begin{Bmatrix}l\\jk\end{Bmatrix}
$$
$$
= \frac{\partial}{\partial x^k}\left(\frac{\partial \lambda_i}{\partial x^j}\right) - \frac{\partial \lambda_h}{\partial x^k}\begin{Bmatrix}h\\ij\end{Bmatrix} - \frac{\partial \lambda_l}{\partial x^j}\begin{Bmatrix}l\\ik\end{Bmatrix} - \frac{\partial \lambda_i}{\partial x^l}\begin{Bmatrix}l\\jk\end{Bmatrix}
$$
$$
- \lambda_h\left(\frac{\partial}{\partial x^k}\begin{Bmatrix}h\\ij\end{Bmatrix} - \begin{Bmatrix}h\\lj\end{Bmatrix}\begin{Bmatrix}l\\ik\end{Bmatrix} - \begin{Bmatrix}h\\il\end{Bmatrix}\begin{Bmatrix}l\\jk\end{Bmatrix}\right).
$$

Since

$$
(22.18) \quad \frac{\partial}{\partial x^k}\left(\frac{\partial \lambda_i}{\partial x^j}\right) = \frac{\partial}{\partial x^j}\left(\frac{\partial \lambda_i}{\partial x^k}\right),
$$

we have in consequence of (20.13)

$$
(22.19) \quad \lambda_{i,jk} - \lambda_{i,kj} = \lambda_h R^h{}_{ijk}.
$$

In like manner for a tensor b_{ij} we have

$$
(22.20) \quad b_{ij,kl} - b_{ij,lk} = b_{hj}R^h{}_{ikl} + b_{ih}R^h{}_{jkl},
$$

and in general

$$
(22.21) \quad b_{r_1\cdots r_m,kl} - b_{r_1\cdots r_m,lk} = \sum_{\alpha}^{1,\cdots,m} b_{r_1\cdots r_{\alpha-1}hr_{\alpha+1}\cdots r_m}R^h{}_{r_\alpha kl}.
$$

Thus far we have been dealing with covariant tensors. We shall now find the corresponding results for contravariant and mixed tensors. Instead of proceeding directly to do so we make use of the fact that $g^{il}_{,j} = 0$ by an argument similar to that which led to Theorem [22.1].

Thus, considering λ^i as the contravariant components of the covariant vector λ_i, we have

$$\lambda^i_{,jk} = (g^{il}\lambda_l)_{,jk} = (g^{il}\lambda_{l,j})_{,k} = g^{il}\lambda_{l,jk}.$$

From this result, (22.19), and (20.15) we have

$$\lambda^i_{,jk} - \lambda^i_{,kj} = g^{il}(\lambda_{l,jk} - \lambda_{l,kj}) = g^{il}\lambda_h g^{hm}R_{mljk} = g^{il}\lambda^m R_{mljk}.$$

From equations (ii) of §20 Ex. 7 and the second set of (20.15), we have

$$g^{il}R_{mljk} = -g^{il}R_{lmjk} = -R^i_{mjk}.$$

Consequently

(22.22) $$\lambda^i_{,jk} - \lambda^i_{,kj} = -\lambda^m R^i_{mjk}.$$

Similarly it can be shown that

$$b^{hi}_{,jk} - b^{hi}_{,kj} = -b^{mi}R^h_{mjk} - b^{hm}R^i_{mjk}.$$

From results of this type and (22.21) we have the following general formula:

(22.23)
$$b^{r_1\cdots r_m}_{s_1\cdots s_p,jk} - b^{r_1\cdots r_m}_{s_1\cdots s_p,kj} = \sum_\alpha^{1,\cdots,p} b^{r_1\cdots\cdots\cdots\cdots\cdots r_m}_{s_1\cdots s_{\alpha-1}ls_{\alpha+1}\cdots s_p}R^l_{s_\alpha jk}$$
$$- \sum_\beta^{1,\cdots,m} b^{r_1\cdots r_{\beta-1}lr_{\beta+1}\cdots r_m}_{s_1\cdots\cdots\cdots\cdots\cdots s_p}R^{r_\beta}_{ljk}.$$

The equations (22.19) to (22.23) are known as *Ricci identities* after Ricci to whom they are due.[*]

From the manner in which equations (22.19) were derived it follows that, when they are satisfied, equations (22.18) follow. Since similar results follow from equations (22.20)–(20.23), we have that when covariant differentiation is used, Ricci identities take the place of the ordinary conditions of integrability, that is, that equations such as (22.18) are satisfied (see §23).

Although the tensor calculus as developed thus far has been in terms of three coordinates the results apply in general to any number of coordinates. In the remaining two chapters the tensor calculus is applied in the case of two coordinates. In General Relativity it is

_____

[*] Cf. Ricci and Levi-Civita, 1901, 1, p. 143.

applied to spaces of four dimensions. Its use is fundamental in Riemannian Geometry of n dimensions.*

EXERCISES

1. Show that $\lambda_{i,j} = \lambda_{j,i}$, if and only if λ_i is a gradient.

2. Show that for covariant differentiation based upon the metric tensor a_{ij}

$$\lambda^i_{,i} = \frac{1}{\sqrt{a}} \frac{\partial}{\partial x^i} (\sqrt{a}\, \lambda^i).$$

The scalar $\lambda^i_{,i}$ is called the *divergence* of λ^i.

3. Show that a necessary and sufficient condition that the second covariant derivative of an arbitrary tensor be symmetric in the indices of covariant differentiation is that the differentiation be based upon the metric tensor a_{ij} of euclidean space.

4. Given any contravariant unit vector λ^i, we have along a curve

$$\lambda^i_{,j} \frac{dx^j}{ds} = \mu^i,$$

where μ^i is a contravariant vector; at each point of the integral curves of equations (15.11) μ^i is the principal normal vector to the curve (see (21.9)).

5. Even if the curve in Ex. 4 is not an integral curve of equations (15.11), the vector μ^i is perpendicular to the vector λ^i at each point of the curve.

6. For any scalar f the quantity $a^{ij}f_{,ij}$ is a scalar, and it is equal to the expression

$$\frac{1}{\sqrt{a}} \frac{\partial}{\partial x^i} \left(\sqrt{a}\, a^{ij} \frac{\partial f}{\partial x^j} \right).$$

7. In cartesian coordinates the equation

(i)
$$a^{ij}f_{,ij} = 0$$

is

$$\sum_i \frac{\partial^2 f}{\partial x^{i^2}} = 0,$$

and thus (i) is the equation of Laplace in general coordinates; in polar coordinates it is

$$\frac{\partial^2 f}{\partial x^{1^2}} + \frac{1}{(x^1)^2} \frac{\partial^2 f}{\partial x^{2^2}} + \frac{1}{(x^1 \sin x^2)^2} \frac{\partial^2 f}{\partial x^{3^2}} + \frac{2}{x^1} \frac{\partial f}{\partial x^1} + \frac{1}{(x^1)^2} \cot x^2 \frac{\partial f}{\partial x^2} = 0.$$

8. When λ^i satisfy the equations

$$\lambda^i_{,j} = 0,$$

they are the contravariant components of a field of parallel vectors.

---

* See, 1926, 1.

23. SYSTEMS OF PARTIAL DIFFERENTIAL EQUATIONS OF THE FIRST ORDER.   MIXED SYSTEMS

In this section we discuss the existence of solutions of certain systems of partial differential equations of the first order of the kind which arise in various geometric problems; that is, whether there exist functions of the independent variables which satisfy the equations identically.

Consider the system of equations

$$(23.1) \qquad \frac{\partial \theta^\alpha}{\partial x^i} = \psi_i^\alpha(\theta^1, \cdots, \theta^m; x^1, \cdots, x^n) \qquad \begin{pmatrix} \alpha = 1, \cdots, m; \\ i = 1, \cdots, n \end{pmatrix},$$

where the ψ's are functions of the θ's and the x's.   It is understood that the following treatment applies to a domain in which the functions ψ_i^α are continuous and have continuous derivatives up to the order entering in the treatment.   Equations (23.1) are equivalent to the system of *total* differential equations

$$(23.2) \qquad\qquad d\theta^\alpha = \psi_i^\alpha \, dx^i,$$

as is seen when equations (23.2) are written

$$\left(\frac{\partial \theta^\alpha}{\partial x^i} - \psi_i^\alpha\right) dx^i = 0,$$

and it is required that these equations hold for arbitrary values of the differentials.

Differentiating equations (23.1) with respect to x^j and in the result replacing the derivatives of the θ's by their expressions from (23.1), we obtain

$$\frac{\partial^2 \theta^\alpha}{\partial x^i \partial x^j} = \frac{\partial \psi_i^\alpha}{\partial x^j} + \frac{\partial \psi_i^\alpha}{\partial \theta^\beta} \psi_j^\beta.$$

Since the left-hand member of these equations is symmetric in i and j, it follows that the right-hand member must be equal to the expression obtained from it on interchanging i and j, that is, it is necessary that the functions ψ_i^α be such that by

$$(23.3) \qquad \frac{\partial \psi_i^\alpha}{\partial x^j} + \frac{\partial \psi_i^\alpha}{\partial \theta^\beta} \psi_j^\beta = \frac{\partial \psi_j^\alpha}{\partial x^i} + \frac{\partial \psi_j^\alpha}{\partial \theta^\beta} \psi_i^\beta \qquad \begin{pmatrix} \alpha, \beta = 1, \cdots, m; \\ i, j = 1, \cdots, n \end{pmatrix}.$$

The conditions imposed upon the functions ψ_i^α by equations (23.3) are called *conditions of integrability* of equations (23.1), meaning that if equations (23.1) are to admit solutions, either equations (23.3) are identities in θ^α and x^i, or there are relations between these quantities which must be satisfied by θ^α to be solutions of the system (23.1).

When equations (23.3) are identities in the θ^α and x' the system is said to be *completely integrable*, or to be a *complete system*. With regard to such systems Darboux* has established the theorem

[23.1] *A complete system of equations* (23.1) *admits one and only one set of solutions* θ^α *such that for arbitrary initial values* x_0^i *of the x's the functions* θ^α *reduce to arbitrary constants* c^α.

We consider next the case when equations (23.3) are not identities in the θ's and the x's, and refer to them as the set of equations E_1. If m of these equations, say $\varphi^1 = 0, \cdots, \varphi^m = 0$, are independent, that is, if the jacobian of the φ's with respect to the θ's is not identically zero, these equations can be solved for the θ's as functions of the x's, and the solution of these equations is unique.† If then any other of the set E_1 is independent of the above equations, and the expression for the θ's are substituted in this equation, there results a relation between the x's, which is contrary to the hypothesis that the x's are independent. Consequently, if there are more than m independent equations in the set E_1, equations (23.1) do not admit a solution. If there are exactly m independent equations $\varphi^1 = 0, \cdots, \varphi^m = 0$, we differentiate each of these equations with respect to $x^1, \cdots, x^n$, thus obtaining mn equations, and in them substitute for the derivatives of the θ's the expressions ψ_i^α from (23.1). If these equations are satisfied identically when the solutions θ^α of the equations $\varphi^1 = 0, \cdots, \varphi^m = 0$ are substituted, then θ^α constitute a solution of equations (23.1), and the only solution. If these equations are not satisfied identically, equations (23.1) do not have a solution.

If the number of independent equations in the set E_1 is less than m, we differentiate the independent ones with respect to the x's, substitute from (23.1) for the derivatives of the θ's and denote the resulting set of necessary conditions by E_2. If the number of independent equations in the sets E_1 and E_2 is greater than m, there is no solution of equations (23.1), as shown by the argument used above. If the number is m, by the process described above we determine whether there is a solution (which is unique) or not. If this number is less than m, we differentiate the equations of the set E_2, substitute from (23.1) and denote the resulting set of equations by E_3.

Proceeding in this manner we get a sequence of sets of equations.

* 1910, 2, pp. 326–335.
† Fine, 1927, 1, p. 253.

If all of the equations of one of these sets are not equivalent to equations of the preceding sets, the set introduces at least one additional condition upon the θ's. Consequently, if equations (23.1) are to admit a solution, there must be a positive integer N such that the equations of the $(N + 1)^{\text{th}}$ set are satisfied because of the equations of the preceding N sets; otherwise more than m independent equations would be obtained, and thus there would be a relation between the x's. Moreover, from the above argument it follows that $N \leq m$.

Suppose then that there is a number $N(\leq m)$ such that the equations of the sets

$$(23.4) \qquad\qquad E_1, \cdots, E_N$$

are consistent but that each of the sets introduces one or more conditions upon the θ's independent of the conditions imposed by the equations of the preceding sets, and that all of the equations of the set

$$(23.5) \qquad\qquad E_{N+1}$$

are satisfied in consequence of the equations of the sets (23.4). Suppose that there are $p(\leq m)$ independent conditions imposed by (23.4), say $G_j(\theta; x) = 0$ for $j = 1, \cdots, p$. This means that the jacobian matrix of the G's with respect to the θ's is of rank p, that is, if $p < m$ at least one of the determinants of order p of the matrix is not identically zero, and consequently the equations $G_j = 0$ can be solved for at least one set of p of the θ's as functions of the remaining θ's and the x's. If $p = m$, the equations $G_j = 0$ can be solved for the θ's in terms of the x's, and these θ's are a solution of equations (23.1), since they satisfy the equations of the sets (23.4) and (23.5).

We consider now the case when $p < m$. By a suitable renumbering of the θ's, if necessary, the equations $G_j = 0$ can be solved for $\theta^1, \cdots, \theta^p$, which result we write as follows:

$$(23.6) \qquad\qquad \theta^\nu - \varphi^\nu(\theta^{p+1}, \cdots, \theta^m; x) = 0 \qquad (\nu = 1, \cdots, p).$$

From these equations we have by differentiation

$$\frac{\partial \theta^\nu}{\partial x^i} - \frac{\partial \varphi^\nu}{\partial \theta^\sigma}\frac{\partial \theta^\sigma}{\partial x^i} - \frac{\partial \varphi^\nu}{\partial x^i} = 0 \qquad\qquad (\sigma = p+1, \cdots, m).$$

Replacing the derivatives of the θ's by means of (23.1), we have

$$(23.7) \qquad\qquad \psi_i^\nu - \frac{\partial \varphi^\nu}{\partial \theta^\sigma}\psi_i^\sigma - \frac{\partial \varphi^\nu}{\partial x^i} = 0,$$

which equations are satisfied in consequence of (23.4) and (23.5), as follows from the method by which the sets (23.4) and (23.5) were obtained. From (23.7) and the preceding set of equations we have by subtraction

$$(23.8) \qquad \frac{\partial \theta^\nu}{\partial x^i} - \psi_i^\nu - \frac{\partial \varphi^\nu}{\partial \theta^\sigma}\left(\frac{\partial \theta^\sigma}{\partial x^i} - \psi_i^\sigma\right) = 0 \qquad \left(\begin{matrix} \nu = 1, \cdots, p; \\ \sigma = p+1, \cdots, m \end{matrix}\right).$$

Consider now the set of equations

$$(23.9) \qquad \frac{\partial \theta^\sigma}{\partial x^i} - \bar{\psi}_i^\sigma = 0,$$

where $\bar{\psi}_i^\sigma$ are the functions of $\theta^{p+1}, \cdots, \theta^m$, and the x's obtained from ψ_i^σ when θ^ν, for $\nu = 1, \cdots, p$, are replaced by their expressions from (23.6). For any solution of equations (23.9) we have from (23.8)

$$(23.10) \qquad \frac{\partial \theta^\nu}{\partial x^i} - \bar{\psi}_i^\nu = 0,$$

where $\bar{\psi}_i^\nu$ are the functions obtained from ψ_i^ν when $\theta^1, \cdots, \theta^p$ are replaced by the functions in terms of $\theta^{p+1}, \cdots, \theta^m$, and x^i as given by (23.6). Consequently any given solution of (23.9) and θ^ν from (23.6) constitute a solution of (23.1).

We take up now the question of the solution of equations (23.9). The conditions of integrability of these equations result from

$$\frac{\partial \psi_i^\sigma}{\partial x^j} + \frac{\partial \psi_i^\sigma}{\partial \theta^\nu}\frac{\partial \theta^\nu}{\partial x^j} + \frac{\partial \psi_i^\sigma}{\partial \theta^\tau}\frac{\partial \theta^\tau}{\partial x^j} = \frac{\partial \psi_j^\sigma}{\partial x^i} + \frac{\partial \psi_j^\sigma}{\partial \theta^\nu}\frac{\partial \theta^\nu}{\partial x^i} + \frac{\partial \psi_j^\sigma}{\partial \theta^\tau}\frac{\partial \theta^\tau}{\partial x^i} \quad \left(\begin{matrix} \nu = 1, \cdots, p; \\ \sigma, \tau = p+1, \cdots, m \end{matrix}\right),$$

when θ^ν are replaced by their expressions from (23.6). From (23.3) for $\alpha = p+1, \cdots, m$ we have

$$\frac{\partial \psi_i^\sigma}{\partial x^j} + \frac{\partial \psi_i^\sigma}{\partial \theta^\nu}\psi_j^\nu + \frac{\partial \psi_i^\sigma}{\partial \theta^\tau}\psi_j^\tau = \frac{\partial \psi_j^\sigma}{\partial x^i} + \frac{\partial \psi_j^\sigma}{\partial \theta^\nu}\psi_i^\nu + \frac{\partial \psi_j^\sigma}{\partial \theta^\tau}\psi_i^\tau \quad (\sigma, \tau = p+1, \cdots, m),$$

and these are identities when the expressions (23.6) for θ^ν are substituted, since (23.6) satisfies all of the sets (23.4) and (23.5). Substituting the expressions (23.6) for θ^ν in both sets of the above equations and subtracting, we have

$$\frac{\partial \psi_i^\sigma}{\partial \theta^\nu}\left(\frac{\partial \theta^\nu}{\partial x^j} - \bar{\psi}_j^\nu\right) + \frac{\partial \psi_i^\sigma}{\partial \theta^\tau}\left(\frac{\partial \theta^\tau}{\partial x^j} - \bar{\psi}_j^\tau\right) = \frac{\partial \psi_j^\sigma}{\partial \theta^\nu}\left(\frac{\partial \theta^\nu}{\partial x^i} - \bar{\psi}_i^\nu\right) + \frac{\partial \psi_j^\sigma}{\partial \theta^\tau}\left(\frac{\partial \theta^\tau}{\partial x^i} - \bar{\psi}_i^\tau\right),$$

which equations are satisfied in consequence of (23.9) and (23.10). Hence the system (23.9) is a complete system, and by theorem [23.1] its solution involves $m - p$ arbitrary constants. Consequently we have

[23.2] *In order that a system of equations* (23.1) *which is not complete admit a solution, it is necessary and sufficient that there exist a positive integer* $N(\leq m)$ *such that the equations of the sets* $E_1, \cdots, E_N$ *are consistent for all values of* x^i *in the domain under consideration, and that the equations of the sets* E_{N+1} *are satisfied because of the equations of the preceding sets; if* $p(< m)$ *is the number of independent equations in the first* N *sets, the solution involves* $m - p$ *arbitrary constants; if* $p = m$, *the solution is completely determined.*

It is evident from the above discussion that when a positive integer N exists such that the conditions of the theorem are satisfied, they are satisfied also for any integer larger than N, since if the set E_{N+1} are satisfied so also are all subsequent ones. However, it is understood in the theorem and in any of its applications that N is the least integer for which the conditions are satisfied.

The above theorem applies also to the case when there are q functional relations between the θ's and x's which must be satisfied in addition to the differential equations (23.1), say

$$(23.11) \qquad\qquad f^\gamma(\theta; x) = 0 \qquad\qquad (\gamma = 1, \cdots, q).$$

Equations (23.1) and (23.11) are said to constitute a *mixed system*. In the case of a mixed system we denote the set of equations (23.11) by E_0 and include in the set E_1 of the theorem also such conditions as arise from (23.11) by differentiation and substitution from (23.1). Then the theorem holds with the difference in this case that we have the sets $E_0, E_1, \cdots, E_N$ instead of the sets $E_1, \cdots, E_N$, and that the equations of the set E_{N+1} are satisfied because of the equations of the preceding sets.

As an application of the preceding results, if we define quantities p_p^h by

$$(23.12) \qquad\qquad \frac{\partial x^h}{\partial x'^p} = p_p^h,$$

equations (20.11) may be written

$$(23.13) \qquad
\begin{aligned}
\frac{\partial p_p^h}{\partial x'^q} &= -\begin{Bmatrix} h \\ ij \end{Bmatrix} p_p^i p_q^j + \begin{Bmatrix} r \\ pq \end{Bmatrix}' p_r^h, \\[2ex]
\frac{\partial p_q^h}{\partial x'^p} &= -\begin{Bmatrix} h \\ ij \end{Bmatrix} p_p^i p_q^j + \begin{Bmatrix} r \\ pq \end{Bmatrix}' p_r^h
\end{aligned}$$

and (20.8) as

$$(23.14) \qquad\qquad a'_{pq} = a_{ij} p_p^i p_q^j.$$

Equations (23.12), (23.13) and (23.14) constitute a mixed system for which x^i and p_p^h are the dependent variables, that is, they are the θ's of equations (23.1), and x'^p are the independent variables. Moreover, equations (23.14) are the set E_0 referred to above.

From (23.12) it follows that we must have

$$\text{(23.15)} \qquad \frac{\partial p_p^h}{\partial x'^q} = \frac{\partial p_q^h}{\partial x'^p},$$

which are satisfied identically by (23.13), and the conditions of integrability of (23.13) are

$$\text{(23.16)} \qquad R^h{}_{ijk} p_p^i p_q^j p_r^k = R'^s{}_{pqr} p_s^h,$$

as follows from (20.12). Since equations (20.11), and consequently (23.13) were obtained by differentiating (23.14) it follows that no conditions are imposed by differentiating the equations of the set E_0 in this case, and consequently equations (23.16) are the set E_1.

Although the above results were developed from the fundamental tensor a_{ij} of euclidean 3-space, they apply equally well to any covariant tensor g_{ij}, provided only that the determinant g of the g_{ij}'s is not zero, as discussed at the close of §20.

From the definition (20.13) of the Riemannian tensor it is seen that only if the Riemann tensor is a zero tensor is it possible to have all the Christoffel symbols $\left\{ {r \atop pq} \right\}'$ equal to zero for some coordinate system x''. If there are coordinates x'^i for which $\left\{ {r \atop pq} \right\}'$ are zero, equations (23.13) become

$$\text{(23.17)} \qquad \frac{\partial p_p^h}{\partial x'^q} = \frac{\partial p_q^h}{\partial x'^p} = - \left\{ {h \atop ij} \right\} p_p^i p_q^j$$

and equations (23.16) reduce to

$$R^h{}_{ijk} p_p^i p_q^j p_r^k = 0.$$

Since the tensor $R^h{}_{ijk}$ is a zero tensor, these equations are satisfied identically, and consequently the system (23.17) and (23.12) is a complete system. This means that initial values of the x^i and $p_p^i \left(= \dfrac{\partial x^i}{\partial x'^p} \right)$ can be chosen arbitrarily, as follows from theorem [23.1], and then the

x's as functions of the x'''s are completely determined as solutions of the system of equations (23.17), which are in fact

$$\frac{\partial^2 x^h}{\partial x'^p \partial x'^q} + \begin{Bmatrix} h \\ ij \end{Bmatrix} \frac{\partial x^i}{\partial x'^p} \frac{\partial x^j}{\partial x'^q} = 0.$$

Furthermore it follows from the manner in which equations (20.11) were derived from (20.8) that the solutions are such that

$$g_{ij} \frac{\partial x^i}{\partial x'^p} \frac{\partial x^j}{\partial x'^q} = g'_{pq}$$

where the g'''s are constants. That they are constants follows from equations of the form (20.5) when the corresponding Christoffel symbols $\begin{Bmatrix} r \\ pq \end{Bmatrix}'$ are zero.

If $g > 0$ it follows from an equation of the form (14.12) that $g' > 0$. If the quantities $g'_{ij} > 0$ are not equal to δ_{ij}, it follows from the theory of algebraic quadratic forms that there exist linear transformations* with constant coefficients by means of which the quadratic form is transformed into (14.2). Hence we have

[23.3] *A symmetric covariant tensor g_{ij} for which the determinant is positive is the metric tensor of euclidean 3-space, if and only if the Riemann tensor formed with respect to g_{ij} is a zero tensor.*

Although this result has been derived for the case when $n = 3$, there is nothing in the proof which limits the result to this value for n, that is, it applies to any case for $n > 1$.

Although euclidean space of any dimensions n is characterized by the property that there exist coordinate systems for which the Christoffel symbols are equal to zero, we shall show that for any space there exist coordinate systems for which at a given point the Christoffel symbols are all equal to zero. In fact, if the space is referred to a general coordinate system x^i and $\begin{Bmatrix} i \\ jk \end{Bmatrix}_0$ denote the values of the Christoffel symbols for this coordinate system at the point x_0^i, and coordinates x'^i are defined by

$$x^i = x_0^i + x'^i - \tfrac{1}{2} \begin{Bmatrix} i \\ jk \end{Bmatrix}_0 x'^j x'^k,$$

* See Bôcher, 1907, 1, Chapter 10.

then the point $x'^i = 0$ is the point x_0^i, and at this point

$$\frac{\partial x^i}{\partial x'^p} = \delta_p^i, \qquad \frac{\partial^2 x^i}{\partial x'^p \partial x'^q} = -\left\{\begin{matrix} i \\ pq \end{matrix}\right\}_0.$$

Substituting these values in (20.11), we find that $\left\{\begin{matrix} r \\ pq \end{matrix}\right\}' = 0$ at the point, as was to be proved. Hence we have

[23.4] *Even when the Riemann tensor based upon a tensor g_{ij} is not a zero tensor, a coordinate system can be found such that at a given point all of the Christoffel symbols in this coordinate system are equal to zero.*

In this coordinate system at the origin, that is, $x'^i = 0$, the covariant derivative of a tensor is the ordinary derivative. In particular, we have

$$R'^l_{pqr,s} = \frac{\partial R'^l_{pqr}}{\partial x'^s} = \frac{\partial^2}{\partial x'^q \partial x'^s}\left\{\begin{matrix} l \\ pr \end{matrix}\right\}' - \frac{\partial^2}{\partial x'^r \partial x'^s}\left\{\begin{matrix} l \\ pq \end{matrix}\right\}',$$

as follows from (20.13). From these equations we have

$$R'^l_{pqr,s} + R'^l_{prs,q} + R'^l_{psq,r} = 0.$$

The left-hand members of these equations are components of a tensor. Since they are equal to zero at the point in one coordinate system, they are equal to zero at this point in any coordinate system. Moreover, this result can be obtained at each point of the space, and consequently we have

[23.5] *For any space and in any coordinate system*

(23.18) $R^i_{jkl,m} + R^i_{jlm,k} + R^i_{jmk,l} = 0.$

Since the covariant derivatives of g^{ik} are equal to zero, we have

$$R^i_{jkl,m} = (g^{ih}R_{hjkl})_{,m} = g^{ih}R_{hjkl,m}.$$

From this result and (23.18) it follows, since the determinant of g^{ih} is not equal to zero, that (on changing indices)

(23.19) $R_{ijkl,m} + R_{ijlm,k} + R_{ijmk,l} = 0.$

Equations (23.18) and (23.19) are called the *Bianchi identities*. They hold in any space of any dimensionality and in any coordinate system.*

* See 1902, 1, p. 351.

EXERCISES

1. Find the mixed system of differential equations of the type (23.1) for which

$$u = ax^1 + bx^2, \qquad v = ax^1x^2, \qquad w = b(x^1)^2 + a(x^2)^2$$

is the general solution, a and b being arbitrary constants; what is the system E_0 in this case?

2. When the functions ψ_i^α in (23.1) are linear and homogeneous in the θ's, the equations of the sets E_1, $\cdots$, E_N have this property; and in order that the equations (23.1) have a solution, it is necessary and sufficient that there exist a positive integer N ($\leqq m$) such that the rank of the matrix of the sets of equations E_1, $\cdots$, E_N is $m - q$ ($q \geqq 1$), and that this be the rank also of the equations of the sets E_1, $\cdots$, E_{N+1}.

3. Consider the Riemann tensor R_{ijkl} based upon a tensor g_{ij}, where $i, j = 1, \cdots, n$; because of equations (i) of §20, Ex. 8 there are $n_2 = n(n-1)/2$ ways in which the first pair of indices are like the second pair, and $n_2(n_2 - 1)/2$ ways in which the first and second pairs are different, and consequently there are $n_2(n_2 + 1)/2$ distinct components as regards equations (i); however, there are $n(n-1)(n-2)(n-3)/4!$ equations of the form (ii) of §20, Ex. 8; hence there are $n^2(n^2 - 1)/12$ distinct components of the tensor.

4. For $n = 3$ there are 6 distinct components of the tensor R_{ijkl}, and there are 6 equations

$$R_{jk} = g^{il} R_{ijkl} \; ;$$

show that the solutions for R_{ijkl} of these equations are given by

$$R_{ijkl} = g_{il} R_{jk} + g_{jk} R_{il} - g_{ik} R_{jl} - g_{jl} R_{ik} + \frac{R}{2} (g_{ik} g_{jl} - g_{il} g_{jk}),$$

where

$$R = g^{ij} R_{ij} \; .$$

5. Since

$$R_{jk} = R^i_{jki} \; ,$$

one has from (23.18) on contracting for i and l

$$R_{jk,m} - R_{jm,k} + R^i_{jmk,i} = 0;$$

if this equation is multiplied by g^{jk} and summed with respect to j and k, one obtains

$$R^i_{m,i} = \frac{1}{2} \frac{\partial R}{\partial x^m} \; .$$

CHAPTER III

Intrinsic Geometry of a Surface

24. LINEAR ELEMENT OF A SURFACE. FIRST FUNDAMENTAL QUADRATIC FORM OF A SURFACE. VECTORS IN A SURFACE

Consider upon a surface with the equations

$$(24.1) \qquad x^i = f^i(u^1, u^2) \qquad (i = 1, 2, 3),$$

x^i being cartesian coordinates in the space in which the surface is *imbedded*, a curve defined by expressing u^1 and u^2 as functions of a parameter t, thus

$$(24.2) \qquad u^\alpha = \varphi^\alpha(t) \qquad (\alpha = 1, 2).*$$

Looked upon as a curve in space, its element of length ds is given by (see §2).

$$(24.3) \qquad ds^2 = \sum_i dx^i \, dx^i \, .$$

From (24.1) we have

$$(24.4) \qquad dx^i = \frac{\partial f^i}{\partial u^\alpha} \, du^\alpha \equiv \frac{\partial x^i}{\partial u^\alpha} \, du^\alpha,$$

where as follows from (24.2)

$$(24.5) \qquad du^\alpha = \frac{d\varphi^\alpha}{dt} \, dt.$$

Substituting from (24.4) in (24.3) we have

$$(24.6) \qquad ds^2 = g_{\alpha\beta} \, du^\alpha \, du^\beta,$$

where the quantities $g_{\alpha\beta}$ are defined by

$$(24.7) \qquad g_{\alpha\beta} = \sum_i \frac{\partial x^i}{\partial u^\alpha} \frac{\partial x^i}{\partial u^\beta}.$$

---

* In general in what follows Latin indices take the values 1, 2, 3 and Greek indices the values 1, 2.

The expression for ds given by equation (24.6) is called the *linear element* of the surface in terms of coordinates u^1 and u^2. The right-hand member of (24.6) is called the *first fundamental quadratic form* of the surface. Although the linear element was derived in seeking the element of length of a curve on the surface looked upon as a curve in the euclidean 3-space in which the surface is imbedded, the significance of the linear element of the surface is that once it has been obtained for a surface the arc of any curve (24.2) is given by

$$ s = \int_{t_0}^{t} \sqrt{g_{\alpha\beta} \frac{du^\alpha}{dt} \frac{du^\beta}{dt}}\, dt = \int_{t_0}^{t} \sqrt{g_{\alpha\beta} \varphi^{\alpha'} \varphi^{\beta'}}\, dt. $$

In particular, for the coordinate curves $u^2 = $ const. and $u^1 = $ const. the respective elements of length are given by

(24.8) $$ ds_1^2 = g_{11}(du^1)^2, \qquad ds_2^2 = g_{22}(du^2)^2. $$

Consequently $g_{11} > 0$ and $g_{22} > 0$ unless the respective coordinate curves are minimal curves (see §2). Hence, if we are dealing with real surfaces and real coordinates, we have that g_{11} and g_{22} are positive functions.

A change of coordinates to a new set u'^1, u'^2 is defined by means of equations

(24.9) $$ u^\alpha = \psi^\alpha(u'^1, u'^2) \qquad (\alpha = 1, 2), $$

provided the jacobian $\dfrac{\partial(\psi^1, \psi^2)}{\partial(u'^1, u'^2)}$ is not identically zero. From (24.9) we have

(24.10) $$ du^\alpha = \frac{\partial u^\alpha}{\partial u'^\gamma}\, du'^\gamma \qquad (\alpha, \gamma = 1, 2). $$

When these expressions are substituted in (24.6) we obtain

(24.11) $$ ds^2 = g'_{\gamma\delta}\, du'^\gamma\, du'^\delta, $$

where

(24.12) $$ g'_{\gamma\delta} = g_{\alpha\beta} \frac{\partial u^\alpha}{\partial u'^\gamma} \frac{\partial u^\beta}{\partial u'^\delta}. $$

From (24.7) it follows that $g_{\alpha\beta} = g_{\beta\alpha}$, that is, the g's are symmetric in the indices. Since equations (24.12) are of the form (18.1) we say that $g_{\alpha\beta}$ are the components of, or that $g_{\alpha\beta}$ is, the *covariant metric tensor* of the surface; it is symmetric and of the second order.

If we denote by g the determinant of the g's, that is

$$(24.13) \qquad g = \begin{vmatrix} g_{11} & g_{12} \\ g_{12} & g_{22} \end{vmatrix} \equiv |g_{\alpha\beta}|,$$

we have from (24.7), analogously to (5.1),

$$(24.14) \qquad g \equiv |g_{\alpha\beta}| = (A^{12})^2 + (A^{23})^2 + (A^{31})^2,$$

where A^{ij} are defined by (10.3).

On the understanding that we are dealing only with real quantities it follows from (24.14) that g is never a negative quantity. It is equal to zero only in case

$$(24.15) \qquad A^{12} = A^{23} = A^{31} = 0.$$

These equations cannot be identities in the u's, as follows from theorem [10.1]. However, equations (24.15) may be satisfied by certain values of u^1 and u^2. The points having such values for coordinates are called *singular points*, as remarked in §11; they may be isolated points, or constitute one or more *singular curves* on the surface. Only *ordinary* points, that is, non-singular points, on a surface are considered in this book, unless the contrary is stated.

Since $g \neq 0$ functions $g^{\alpha\beta}$ are uniquely defined by

$$(24.16) \qquad g^{\alpha\beta} g_{\beta\gamma} = \delta^\alpha_\gamma,$$

where

$$(24.17) \qquad \delta^\alpha_\gamma = 1 \text{ or } 0 \text{ according as } \alpha = \gamma \text{ or } \alpha \neq \gamma.$$

In fact, equations (24.16) are equivalent to

$$(24.18) \qquad g^{11} = \frac{g_{22}}{g}, \qquad g^{12} = g^{21} = -\frac{g_{12}}{g}, \qquad g^{22} = \frac{g_{11}}{g}.$$

As in §14 it can be shown that it follows from (24.12) that

$$(24.19) \qquad g^{\alpha\beta} = g'^{\gamma\delta} \frac{\partial u^\alpha}{\partial u'^\gamma} \frac{\partial u^\beta}{\partial u'^\delta},$$

where $g'^{\alpha\beta}$ bear to $g'_{\alpha\beta}$ the relation (24.18). Thus $g^{\alpha\beta}$ is a contravariant tensor; it is called the *contravariant metric tensor* of the surface and is of the second order.

We observe that equations (24.12) and (24.19) are of the same form as (14.10) and (14.16) respectively, which refer to the fundamental quadratic form of euclidean 3-space in two general systems of coordinates x^i and x'^i.

As stated in §10, the use of two coordinates u^α in the definition of a surface, as in equations (24.1), was introduced by Gauss. He used p and q as coordinates and wrote (24.6) in the form

$$ds^2 = E\, dp^2 + 2F\, dp\, dq + G\, dq^2.$$

This notation, at times with u and v used in place of p and q, has been followed generally, but in what follows we use (24.6), since it enables one to write equations in simpler form with the use of the summation convention.

From (24.4) it follows that at each point of the surface differentials du^1 and du^2 determine in the enveloping space a direction tangential to the surface (see §11) for which dx^i are direction numbers. Accordingly we say that at each point of a surface differentials du^α determine *a vector in the surface*. Moreover, since equations (24.10) for a transformation of coordinates in the surface are of the form (15.10), we say that du^α are the contravariant components of the vector at each point. One would expect from geometric considerations that for a curved surface vectors with the same components du^α at different points would have different directions as viewed from the enveloping space. This expectation is verified analytically by (24.4), since ordinarily the quantities $\dfrac{\partial x^i}{\partial u^\alpha}$ are functions of the u's, and consequently vary from point to point in the surface.

Consider two sets of functions $\lambda^\alpha(u^1, u^2)$ and $\lambda'^\alpha(u'^1, u'^2)$ in two coordinate systems u^α and u'^α in the surface related as follows

$$(24.20) \qquad \lambda^\alpha = \lambda'^\beta \frac{\partial u^\alpha}{\partial u'^\beta},$$

which are equivalent to

$$(24.21) \qquad \lambda'^\alpha = \lambda^\beta \frac{\partial u'^\alpha}{\partial u^\beta}$$

in view of the identities

$$(24.22) \qquad \frac{\partial u^\alpha}{\partial u'^\beta}\frac{\partial u'^\beta}{\partial u^\gamma} = \delta^\alpha_\gamma, \qquad \frac{\partial u'^\alpha}{\partial u^\beta}\frac{\partial u^\beta}{\partial u'^\gamma} = \delta^\alpha_\gamma,$$

these being analogues of (13.13) and (13.14). Since (24.20) and (24.21) are of the form (15.10) and (15.8) respectively, we say that λ^α and λ'^α are the components in their respective coordinate systems of a *contravariant vector* in the surface, there being one such vector at each point of the surface.

If we define quantities ξ^i by

(24.23)
$$\xi^i = \lambda^\alpha \frac{\partial x^i}{\partial u^\alpha},$$

it follows from (24.20) that

(24.24)
$$\xi^i = \lambda'^\beta \frac{\partial x^i}{\partial u'^\beta}.$$

The quantities $\dfrac{\partial x^i}{\partial u^1}$ and $\dfrac{\partial x^i}{\partial u^2}$ are direction numbers in the enveloping space of the tangents to the coordinate curves $u^2 = $ const. and $u^1 = $ const. respectively in the surface, that is, the u^1-*coordinate curves* and the u^2-*coordinate curves*. Since ξ^i are linear homogeneous combinations of these direction numbers, they are direction numbers of a line tangent to the surface, as follows from the results of §11, and consequently are the components in the enveloping space of the contravariant vector whose components in the surface are λ^α.

Two sets of functions $\lambda_\alpha(u^1, u^2)$ and $\lambda'_\alpha(u'^1, u'^2)$ which are related as follows

(24.25)
$$\lambda_\alpha = \lambda'_\beta \frac{\partial u'^\beta}{\partial u^\alpha}, \qquad \lambda'_\alpha = \lambda_\beta \frac{\partial u^\beta}{\partial u'^\alpha}$$

(either set of equations being equivalent to the other in consequence of (24.22)) are the components in their respective coordinate systems of a *covariant vector* in the surface (see §17). From the results of §19, which apply to any number of coordinates, it follows that if λ^α are components of a contravariant vector in the surface, then

(24.26)
$$\lambda_\alpha = g_{\alpha\beta}\lambda^\beta$$

are the components of a covariant vector, since $g_{\alpha\beta}$ is a covariant tensor. Also if λ_α are the components of a covariant vector, then

(24.27)
$$\lambda^\alpha = g^{\alpha\beta}\lambda_\beta$$

are the components of a contravariant vector, since $g^{\alpha\beta}$ is a contravariant tensor. As in §17, we say that two sets of functions λ^α and λ_α of u^α related as in (24.26) and (24.27) are the *contravariant and covariant components* respectively of the same vector in the surface. The geometric significance of these components is shown in §25.

From (24.23), (24.7), and (24.27) we have

(24.28)
$$\sum_i \xi^i \xi^i = \lambda^\alpha \lambda^\beta g_{\alpha\beta} = g^{\alpha\gamma}\lambda_\gamma g^{\beta\delta}\lambda_\delta g_{\alpha\beta} = g^{\alpha\gamma}\lambda_\gamma \lambda_\delta \delta^\delta_\alpha = g^{\alpha\gamma}\lambda_\gamma \lambda_\alpha.$$

The first member of these equations is the square of the length of the vector as viewed from the enveloping space (see §16 following (16.1)). From §19 it follows that $g_{\alpha\beta}\lambda^\alpha\lambda^\beta$ and $g^{\alpha\gamma}\lambda_\alpha\lambda_\gamma$ are scalars. Hence we have

[24.1] *For a vector whose contravariant and covariant components are* λ^α *and* λ_α *respectively each of the scalars* $g_{\alpha\beta}\lambda^\alpha\lambda^\beta$ *and* $g^{\alpha\beta}\lambda_\alpha\lambda_\beta$ *is equal to the square of the length of the vector.*

As a corollary we have

[24.2] *A necessary and sufficient condition that* λ^α *and* λ_α *are the contravariant and covariant components of a unit vector in the surface is that*

$$(24.29) \qquad g_{\alpha\beta}\lambda^\alpha\lambda^\beta = 1, \qquad g^{\alpha\beta}\lambda_\alpha\lambda_\beta = 1.$$

EXERCISES

1. For a sphere with equations of the form in §10, Ex. 1 the coefficients $g_{\alpha\beta}$ of the linear element are given by

$$g_{11} = a^2, \qquad g_{12} = 0, \qquad g_{22} = a^2 \sin^2 u^1.$$

2. When equations of the tangent surface of a curve with equations $x^i = f^i(u^1)$, where u^1 is the arc, are written in the form (see (8.5))

$$X^i = x^i + (u^2 - u^1)\frac{dx^i}{du^1},$$

the curves $u^2 = $ const. are the orthogonal trajectories of the generators, and in consequence of (2.8), (2.9), and (4.3) one has

$$g_{11} = \kappa^2(u^2 - u^1)^2, \qquad g_{12} = 0, \qquad g_{22} = 1.$$

3. For any surface of revolution, when the equations are given in the form in §10, Ex. 2,

$$g_{11} = 1 + \varphi'^2, \qquad g_{12} = 0, \qquad g_{22} = u^{1^2},$$

where the prime indicates the derivative.

4. The equations

$$x^1 = u^1 \cos u^2, \qquad x^2 = u^1 \sin u^2, \qquad x^3 = a \cosh^{-1}\frac{u^1}{a}$$

are equations of the surface of revolution of the catenary $x^1 = a \cosh \dfrac{x^3}{a}$ about the x^3-axis (see §10, Ex. 2); this surface is called the *catenoid*. In this case $\sqrt{1 + \varphi'^2} = \dfrac{u^1}{\sqrt{u^{1^2} - a^2}}$ so that for the coordinates u'^1, u^2 where $u'^1 = \sqrt{u^{1^2} - a^2}$

$$g_{11} = 1, \qquad g_{12} = 0, \qquad g_{22} = a^2 + (u'^1)^2.$$

Does this change of coordinates change the coordinate curves?

5. For a right conoid with equations of the form in §10, Ex. 4,

$$g_{11} = 1, \qquad g_{12} = 0, \qquad g_{22} = u^{1^2} + \varphi'^2.$$

6. On a surface with the equations

$$x^1 = u^1 \cos u^2, \qquad x^2 = u^1 \sin u^2, \qquad x^3 = \varphi(u^1) + au^2,$$

where a is a constant, the curves $u^1 =$ const. are circular helices (see §3, Ex. 1), and

$$g_{11} = 1 + \varphi'^2, \qquad g_{12} = a\varphi', \qquad g_{22} = u^{1^2} + a^2;$$

such a surface is called a *helicoid*; all the curves $u^2 =$ const. are congruent plane curves; a is called the *helicoidal parameter*.

7. When in Ex. 6 $\varphi(u^1) = 0$, the helicoid is a right-conoid as follows from §10, Ex. 4, and

$$g_{11} = 1, \qquad g_{12} = 0, \qquad g_{22} = a^2 + u^{1^2};$$

this surface for any non-zero value of a is called a *skew helicoid*.

8. Along any curve on a surface the quantities $\dfrac{du^\alpha}{ds}$ are the contravariant components of the unit tangent vector.

25. ANGLE OF TWO INTERSECTING CURVES IN A SURFACE. ELEMENT OF AREA

We have seen in §24 that values of du^α determine at each point of a surface a vector tangent to the surface whose components dx^i in the enveloping space are given by (24.4). From the standpoint of the enveloping space the direction cosines α^i of this tangent vector are $\dfrac{dx^i}{ds}$ (see §3). In consequence of (24.4) and (24.6) we have

$$(25.1) \qquad\qquad \alpha^i = \frac{dx^i}{ds} = \frac{\dfrac{\partial x^i}{\partial u^\alpha} du^\alpha}{\sqrt{g_{\alpha\beta} du^\alpha du^\beta}}.$$

Since dx^i are direction numbers of such a tangent to the surface, and the quantities $\dfrac{\partial x^i}{\partial u^\alpha}$ are functions of the u's, it follows from (24.4) that the direction of a vector of components du^α depends not only upon these components but also upon the point of tangency. However, values of the differentials du^α determine a direction at each point of the surface.

For a curve on the surface defined by equations of the form (24.2) we can find from (25.1) the expressions for α^i in terms of t, giving the direction cosines in the enveloping space of the tangent at any point of the curve. Equations (24.5) give values in the surface of the components du^α of this vector.

Suppose now that at a point on a surface we have two vectors in the surface of components d_1u^α and d_2u^α, and we denote by α_1^i and α_2^i the direction cosines of the tangent vectors so determined in the enveloping space. An angle θ between these tangents is given by

$$\cos \theta = \sum_i \alpha_1^i \alpha_2^i = \frac{\sum_i \dfrac{\partial x^i}{\partial u^\alpha} d_1 u^\alpha \dfrac{\partial x^i}{\partial u^\beta} d_2 u^\beta}{\sqrt{(g_{\alpha\beta} d_1 u^\alpha d_1 u^\beta)(g_{\gamma\delta} d_2 u^\gamma d_2 u^\delta)}},$$

which because of (24.7) becomes

$$(25.2) \qquad \cos \theta = \frac{g_{\alpha\beta} d_1 u^\alpha d_2 u^\beta}{\sqrt{(g_{\alpha\beta} d_1 u^\alpha d_1 u^\beta)(g_{\gamma\delta} d_2 u^\gamma d_2 u^\delta)}}.$$

In terms of α_1^i and α_2^i we have

$$\sin^2 \theta = 1 - \cos^2 \theta = \begin{vmatrix} \sum_i \alpha_1^i \alpha_1^i & \sum_i \alpha_1^i \alpha_2^i \\ \sum_i \alpha_1^i \alpha_2^i & \sum_i \alpha_2^i \alpha_2^i \end{vmatrix}$$

$$= \begin{vmatrix} \alpha_1^1 & \alpha_1^2 \\ \alpha_2^1 & \alpha_2^2 \end{vmatrix}^2 + \begin{vmatrix} \alpha_1^2 & \alpha_1^3 \\ \alpha_2^2 & \alpha_2^3 \end{vmatrix}^2 + \begin{vmatrix} \alpha_1^3 & \alpha_1^1 \\ \alpha_2^3 & \alpha_2^1 \end{vmatrix}^2. \;\;*$$

From (25.1) we have

$$\begin{vmatrix} \alpha_1^i & \alpha_1^j \\ \alpha_2^i & \alpha_2^j \end{vmatrix} = \begin{vmatrix} \dfrac{\partial x^i}{\partial u^\alpha} \dfrac{d_1 u^\alpha}{d_1 s} & \dfrac{\partial x^j}{\partial u^\alpha} \dfrac{d_1 u^\alpha}{d_1 s} \\ \dfrac{\partial x^i}{\partial u^\alpha} \dfrac{d_2 u^\alpha}{d_2 s} & \dfrac{\partial x^j}{\partial u^\alpha} \dfrac{d_2 u^\alpha}{d_2 s} \end{vmatrix} = \begin{vmatrix} \dfrac{\partial x^i}{\partial u^1} & \dfrac{\partial x^j}{\partial u^1} \\ \dfrac{\partial x^i}{\partial u^2} & \dfrac{\partial x^j}{\partial u^2} \end{vmatrix} \cdot \begin{vmatrix} \dfrac{d_1 u^1}{d_1 s} & \dfrac{d_1 u^2}{d_1 s} \\ \dfrac{d_2 u^1}{d_2 s} & \dfrac{d_2 u^2}{d_2 s} \end{vmatrix}.$$

From this result and (24.14) we have

$$\sin^2 \theta = g \begin{vmatrix} \dfrac{d_1 u^1}{d_1 s} & \dfrac{d_1 u^2}{d_1 s} \\ \dfrac{d_2 u^1}{d_2 s} & \dfrac{d_2 u^2}{d_2 s} \end{vmatrix}^2.$$

* See remark following (5.1); also C. G., p. 80.

Thus far θ is one of two angles whose sum is 360°, and the formula (25.2) does not distinguish between these two angles. This ambiguity is removed, if we take

$$(25.3) \qquad \sin \theta = \sqrt{g} \; \frac{d_1 u^1 \, d_2 u^2 - d_1 u^2 \, d_2 u^1}{\sqrt{(g_{\alpha\beta} \, d_1 u^\alpha \, d_1 u^\beta)(g_{\gamma\delta} \, d_2 u^\gamma \, d_2 u^\delta)}}.$$

The significance of this choice is shown later.

From (25.2), and also from (25.3), it follows that the angle between tangents of components $d_1 u^\alpha$ and $d_2 u^\alpha$ depends not only upon these components but also upon the point of tangency unless $g_{\alpha\beta}$ are constants. We show later that $g_{\alpha\beta}$ are constants only in case of developable surfaces, and then only for particular sets of coordinates.

If two curves defined by equations

$$u^\alpha = \varphi_1^\alpha(t_1), \qquad u^\alpha = \varphi_2^\alpha(t_2)$$

have a point in common, for values of t_1 and t_2 at the point we can obtain from (25.2) and (25.3) the angle θ between the tangents to the curves at the point of intersection, that is, the *angle of the two intersecting curves*.

In order to find at a point the angle ω of the coordinate curves $u^2 = $ const. and $u^1 = $ const., for the directions in which u^1 and u^2 respectively are increasing, we take

$$(25.4) \qquad d_1 u^1 > 0, \qquad d_1 u^2 = 0, \qquad d_2 u^1 = 0, \qquad d_2 u^2 > 0,$$

and obtain from (25.2) and (25.3)

$$(25.5) \qquad \cos \omega = \frac{g_{12}}{\sqrt{g_{11} g_{22}}}, \qquad \sin \omega = \frac{\sqrt{g}}{\sqrt{g_{11} g_{22}}}.$$

Thus ω is the angle formed by the vectors at the point in the directions in which u^1 and u^2 respectively are increasing and such that $0 < \omega < 180°$.

From the first of these equations we have

[25.1] *A necessary and sufficient condition that at each point of a surface the coordinate curves meet at right angles, that is, are orthogonal to one another, is that $g_{12} = 0$.*

In this case the coordinate curves are said to form an *orthogonal net* (see §24, Exs. 1–5, 7).

We next apply equations (25.2) and (25.3) to find the angle θ_0 in which a curve upon the surface meets a coordinate curve $u^2 = $ const. If at the point of intersection the components of the tangent to the curve are denoted by du^α and we use the first two of (25.4), replacing $d_2 u^\alpha$ in (25.2) and (25.3) by du^α, we obtain

$$(25.6) \qquad \cos \theta_0 = \frac{g_{1\beta}\, du^\beta}{\sqrt{g_{11} g_{\gamma\delta}\, du^\gamma\, du^\delta}}, \qquad \sin \theta_0 = \frac{\sqrt{g}\, du^2}{\sqrt{g_{11} g_{\gamma\delta}\, du^\gamma\, du^\delta}},$$

and consequently

$$(25.7) \qquad \tan \theta_0 = \frac{\sqrt{g}\, du^2}{g_{1\beta}\, du^\beta}$$

From the second of (25.6) it follows that the angle between the directions for which u^1 is increasing on a curve $u^2 = $ const., and u^2 is increasing on the given curve lies between $0°$ and $180°$. Thus θ_0 is the angle $(0 < \theta_0 < 180°)$ through which the vector $du^1 > 0$, $du^2 = 0$ must be rotated to be brought into coincidence with the tangent vector to the curve in the direction for which $du^2 > 0$. This is the sense in which the first of these vectors is rotated through the angle ω to be brought into coincidence with the vector $du^1 = 0$, $du^2 > 0$, and may be called the *positive sense* of rotation.

We shall show that the angle θ through which the vector $d_1 u^\alpha$ must be rotated in the positive sense to be brought into coincidence with the vector $d_2 u^\alpha$ is given by (25.3). We say that this is *the angle which the vector $d_2 u^\alpha$ makes with the vector $d_1 u^\alpha$*. In fact, if θ_{1_0} and θ_{2_0} denote the angles made at a point by the vectors $d_1 u^\alpha$ and $d_2 u^\alpha$ with the vector $du^1 > 0$, $du^2 = 0$, it follows from (25.6) that

$$\sin (\theta_{2_0} - \theta_{1_0}) = \frac{\sqrt{g}}{g_{11}}\, \frac{d_2 u^2 g_{1\beta} d_1 u^\beta - d_1 u^2 g_{1\gamma} d_2 u^\gamma}{\sqrt{(g_{\alpha\beta} d_1 u^\alpha d_1 u^\beta)(g_{\gamma\delta} d_2 u^\gamma d_2 u^\delta)}}.$$

The right-hand member of this equation reduces to the right-hand member of (25.3). If now $\theta_{1_0} < \theta_{2_0}$, $\theta = \theta_{2_0} - \theta_{1_0}$, and if $\theta_{1_0} > \theta_{2_0}$, $\theta = 2\pi - \theta_{1_0} + \theta_{2_0}$; in either case the left-hand member of the above equation is $\sin \theta$.

For a transformation of coordinates (24.9) we have from (24.12) that

$$(25.8) \qquad g' = g \left[\frac{\partial(u^1, u^2)}{\partial(u'^1, u'^2)} \right]^2,$$

where g' and g are the determinants of $g'_{\alpha\beta}$ and $g_{\alpha\beta}$ respectively. Also for this transformation of coordinates equation (25.3) becomes

$$\sin \theta = \sqrt{g}\, \begin{vmatrix} \dfrac{d_1 u'^1}{d_1 s} & \dfrac{d_1 u'^2}{d_1 s} \\[2mm] \dfrac{d_2 u'^1}{d_2 s} & \dfrac{d_2 u'^2}{d_2 s} \end{vmatrix} \cdot \frac{\partial(u^1, u^2)}{\partial(u'^1, u'^2)} = \pm\sqrt{g'}\, \begin{vmatrix} \dfrac{d_1 u'^1}{d_1 s} & \dfrac{d_1 u'^2}{d_1 s} \\[2mm] \dfrac{d_2 u'^1}{d_2 s} & \dfrac{d_2 u'^2}{d_2 s} \end{vmatrix},$$

the plus or minus sign to be used according as the jacobian is positive or negative. If we denote by ω' the angle between the coordinate curves u'^α = const., and if θ denotes this angle in the u-system, we have from (25.5)

$$\sin \theta = \pm \sin \omega'.$$

This means that positive sense with respect to the u-system and u'-system is the same or opposite according as the jacobian is positive or negative. Hence we say that a *transformation of coordinates is positive or negative* according as the jacobian is positive or negative. However, if a transformation is negative, and if in the equations of the transformation one replaces u'^1 or u'^2 by $-u'^1$ or $-u'^2$, the sign of the jacobian is changed, and the new transformation is positive. This change does not change the coordinate curves u'^α = const., but merely the positive sense on one family of coordinate curves. Hence there is no loss of generality, if we understand in what follows that a transformation is positive unless there is a statement to the contrary.

Consider now two vectors of contravariant components $\lambda_{1|}^\alpha$ and $\lambda_{2|}^\alpha$. If at a point on the surface we take differentials $d_1 u^\alpha$ and $d_2 u^\alpha$ given by

$$d_1 u^\alpha = \lambda_{1|}^\alpha, \qquad d_2 u = \lambda_{2|}^\alpha,$$

we have from (25.2) and (25.3)

(25.9)
$$\cos \theta = \frac{g_{\alpha\beta} \lambda_{1|}^\alpha \lambda_{2|}^\beta}{\sqrt{(g_{\alpha\beta} \lambda_{1|}^\alpha \lambda_{1|}^\beta)(g_{\gamma\delta} \lambda_{2|}^\gamma \lambda_{2|}^\delta)}},$$

$$\sin \theta = \sqrt{g} \, \frac{\lambda_{1|}^1 \lambda_{2|}^2 - \lambda_{1|}^2 \lambda_{2|}^1}{\sqrt{(g_{\alpha\beta} \lambda_{1|}^\alpha \lambda_{1|}^\beta)(g_{\gamma\delta} \lambda_{2|}^\gamma \lambda_{2|}^\delta)}},$$

θ being the angle which the vector $\lambda_{2|}^\alpha$ makes with the vector $\lambda_{1|}^\alpha$. We have, on changing dummy indices,

$$g_{\alpha\beta} \lambda_{1|}^\alpha \lambda_{2|}^\beta = g_{\alpha\beta} g^{\alpha\gamma} \lambda_{1|\gamma} g^{\beta\delta} \lambda_{2|\delta} = g^{\alpha\gamma} \lambda_{1|\gamma} \lambda_{2|\delta} \delta_\alpha^\delta = g^{\alpha\beta} \lambda_{1|\alpha} \lambda_{2|\beta},$$

and

$$\lambda_{1|}^1 \lambda_{2|}^2 - \lambda_{1|}^2 \lambda_{2|}^1 = g^{1\alpha} \lambda_{1|\alpha} g^{2\beta} \lambda_{2|\beta} - g^{2\alpha} \lambda_{1|\alpha} g^{1\beta} \lambda_{2|\beta}$$

$$= (g^{11} g^{22} - (g^{12})^2)(\lambda_{1|1} \lambda_{2|2} - \lambda_{1|2} \lambda_{2|1}).$$

From these results, (25.9), (24.18) and theorem [24.1] we have

(25.10)
$$\cos \theta = \frac{g^{\alpha\beta} \lambda_{1|\alpha} \lambda_{2|\beta}}{\sqrt{(g^{\alpha\beta} \lambda_{1|\alpha} \lambda_{1|\beta})(g^{\gamma\delta} \lambda_{2|\gamma} \lambda_{2|\delta})}},$$

$$\sin \theta = \frac{1}{\sqrt{g}} \frac{\lambda_{1|1} \lambda_{2|2} - \lambda_{1|2} \lambda_{2|1}}{\sqrt{(g^{\alpha\beta} \lambda_{1|\alpha} \lambda_{1|\beta})(g^{\gamma\delta} \lambda_{2|\gamma} \lambda_{2|\delta})}}.$$

From the first of equations (25.9) and (25.10) we have

[25.2] *A necessary and sufficient condition that at a point the vectors $\lambda_{1|}^\alpha$ and $\lambda_{2|}^\alpha$ be perpendicular is*

$$g_{\alpha\beta}\lambda_{1|}^\alpha\lambda_{2|}^\beta \;=\; \lambda_{1\beta}\lambda_{2|}^\beta \;=\; \lambda_{1|}^\alpha\lambda_{2|\alpha} \;=\; g^{\alpha\beta}\lambda_{1|\alpha}\lambda_{2|\beta} \;=\; 0,$$

$\lambda_{1|\alpha}$ *and* $\lambda_{2|\alpha}$ *being the covariant components of the respective vectors.*

When, in particular, we denote by θ_1 and θ_2 the angles which the vector λ^α makes with the vectors λ^1, 0 and 0, λ^2 respectively, that is, the angles which the vector λ^α makes with tangents to the curves u^2-const. and $u^1 = $ const. respectively, we have from (25.9)

$$(25.11) \qquad \cos\theta_1 = \frac{g_{\alpha 1}\lambda^\alpha}{\sqrt{g_{11}g_{\alpha\beta}\lambda^\alpha\lambda^\beta}}, \qquad \cos\theta_2 = \frac{g_{\alpha 2}\lambda^\alpha}{\sqrt{g_{22}g_{\alpha\beta}\lambda^\alpha\lambda^\beta}}.$$

If then in the tangent plane at a point P, we lay off the vectors λ^α; $\lambda^1, 0; 0, \lambda^2$, and note that the lengths of the last two vectors are $\sqrt{g_{11}}\,\lambda^1$ and $\sqrt{g_{22}}\,\lambda^2$ respectively, we have from (25.11) that the projection of these lengths upon the line of the vector λ^α are respectively

$$\frac{g_{\alpha 1}\lambda^\alpha\lambda^1}{\sqrt{g_{\alpha\beta}\lambda^\alpha\lambda^\beta}}, \qquad \frac{g_{\alpha 2}\lambda^\alpha\lambda^2}{\sqrt{g_{\alpha\beta}\lambda^\alpha\lambda^\beta}}.$$

The sum of these two vectors is $\sqrt{g_{\alpha\beta}\lambda^\alpha\lambda^\beta}$, that is, the length of the vector λ^α. Again from (25.11) we have that the projections of the vector λ^α upon the tangents to the curves $u^2 = $ const. and $u^1 = $ const. at P are respectively

$$\frac{g_{\alpha 1}\lambda^\alpha}{\sqrt{g_{11}}}, \qquad \frac{g_{\alpha 2}\lambda^\alpha}{\sqrt{g_{22}}}.$$

Noting that the numerators of these expressions are the covariant components of the vector λ^α, we have (see theorems [16.6] and [17.6])

[25.3] *At any point P of a surface the vector λ^α is the diagonal from P of the parallelogram whose sides are segments of the tangents at P to the curves $u^2 = $ const. and $u^1 = $ const. of the respective lengths $\sqrt{g_{11}}\,\lambda^1$ and $\sqrt{g_{22}}\,\lambda^2$; and the projections of the vector upon these tangents are the covariant components of the vector divided by $\sqrt{g_{11}}$ and $\sqrt{g_{22}}$ respectively.*

We define two sets of numbers $e_{\alpha\beta}$ and $e^{\alpha\beta}$ as follows:

$$(25.12) \quad e_{11} = e_{22} = e^{11} = e^{22} = 0, \qquad e_{12} = e^{12} = 1, \qquad e_{21} = e^{21} = -1.$$

For a transformation of coordinates in the surface we have

$$e_{\alpha\beta} \frac{\partial u^\alpha}{\partial u'^1} \frac{\partial u^\beta}{\partial u'^1} = e_{\alpha\beta} \frac{\partial u^\alpha}{\partial u'^2} \frac{\partial u^\beta}{\partial u'^2} = 0,$$

$$e_{\alpha\beta} \frac{\partial u^\alpha}{\partial u'^1} \frac{\partial u^\beta}{\partial u'^2} = \frac{\partial(u^1, u^2)}{\partial(u'^1, u'^2)} = - e_{\alpha\beta} \frac{\partial u^\alpha}{\partial u'^2} \frac{\partial u^\beta}{\partial u'^1}.$$

If the transformation is positive, we have

(25.13)
$$\sqrt{g'} = \sqrt{g}\, \frac{\partial(u^1, u^2)}{\partial(u'^1, u'^2)},$$

consequently

$$e_{\gamma\delta} \sqrt{g'} = e_{\alpha\beta} \sqrt{g}\, \frac{\partial u^\alpha}{\partial u'^\gamma} \frac{\partial u^\beta}{\partial u'^\delta}.$$

In like manner it can be shown that

$$\frac{e^{\gamma\delta}}{\sqrt{g'}} = \frac{e^{\alpha\beta}}{\sqrt{g}}\, \frac{\partial u'^\gamma}{\partial u^\alpha} \frac{\partial u'^\delta}{\partial u^\beta}.$$

Hence we have

[25.4] *The quantities*

(25.14)
$$\epsilon_{\alpha\beta} = e_{\alpha\beta}\sqrt{g}, \qquad \epsilon^{\alpha\beta} = \frac{e^{\alpha\beta}}{\sqrt{g}}$$

are the components under a positive transformation of a skew-symmetric covariant and contravariant tensor of the second order respectively.

From (25.14) and (25.12) we have

(25.15)
$$\epsilon^{\alpha\gamma}\epsilon_{\beta\gamma} = \epsilon^{\gamma\alpha}\epsilon_{\gamma\beta} = \delta_\beta^\alpha.$$

Also the second of equations (25.9) may be written

(25.16)
$$\sin\theta = \frac{\epsilon_{\alpha\beta}\lambda_{1|}^\alpha \lambda_{2|}^\beta}{\sqrt{(g_{\alpha\beta}\lambda_{1|}^\alpha \lambda_{1|}^\beta)(g_{\gamma\delta}\lambda_{2|}^\gamma \lambda_{2|}^\delta)}}.$$

Hence we have

[25.5] *A necessary and sufficient condition that a unit vector* $\lambda_{2|}^\alpha$ *makes a right angle with a unit vector* $\lambda_{1|}^\alpha$ *is that*

(25.17)
$$\epsilon_{\alpha\beta}\lambda_{1|}^\alpha\lambda_{2|}^\beta = 1.$$

Consider a unit vector λ^α and the vector whose covariant components μ_α are given by

$$(25.18) \qquad \mu_\alpha = \epsilon_{\beta\alpha}\lambda^\beta.$$

Since $\mu_\alpha\lambda^\alpha = 0$ it follows from theorem [25.2] that μ_α is perpendicular to the vector λ^α. Its contravariant components are given by

$$(25.19) \qquad \begin{aligned} \mu^\alpha &= g^{\alpha\beta}\mu_\beta = g^{\alpha\beta}\epsilon_{\gamma\beta}\lambda^\gamma = (g^{\alpha 2}\lambda^1 - g^{\alpha 1}\lambda^2)\sqrt{g} \\ &= (g^{\alpha 2}g^{\beta 1} - g^{\alpha 1}g^{\beta 2})\sqrt{g}\,\lambda_\beta = \epsilon^{\beta\alpha}\lambda_\beta. \end{aligned}$$

From (25.18), (25.19), and (25.15) we have

$$\mu^\alpha\mu_\alpha = \lambda^\alpha\lambda_\alpha = 1,$$

that is, μ^α is a unit vector, and

$$(25.20) \qquad \epsilon_{\alpha\beta}\lambda^\alpha\mu^\beta = \epsilon_{\alpha\beta}\lambda^\alpha\epsilon^{\gamma\beta}\lambda_\gamma = \lambda^\alpha\lambda_\alpha = 1.$$

Hence by theorem [25.5] we have

[25.6] *The quantities μ_α defined by (25.18) are the covariant components of the unit vector which makes a right angle with the unit vector λ^α.*

Consider next upon a surface the curvilinear quadrilateral formed by the coordinate lines, the vertices being the four points $P(u^1, u^2)$, $Q(u^1 + du^1, u^2)$, $R(u^1, u^2 + du^2)$, $S(u^1 + du^1, u^2 + du^2)$. To within terms of higher order the lengths of the sides PQ and PR are $\sqrt{g_{11}}\,du^1$ and $\sqrt{g_{22}}\,du^2$, and the sides opposite them have these respective lengths to within terms of higher order. Since the distances of Q and R from the tangent plane to the surface at P are of the second order in comparison with PQ and PR (see §11, Ex. 9), it follows that when Q, R, and S are projected orthogonally upon the tangent plane to within terms of higher order these projections are the vertices of a parallelogram of sides $\sqrt{g_{11}}\,du^1$ and $\sqrt{g_{22}}\,du^2$. The area of this parallelogram is equal to

$$\sin\omega\,\sqrt{g_{11}}\,du^1\,\sqrt{g_{22}}\,du^2 = \sqrt{g}\,du^1\,du^2,$$

the latter expression following from (25.5). Accordingly we have that the *element of area* $d\sigma$ of a surface is given by

$$(25.21) \qquad d\sigma = \sqrt{g}\,du^1\,du^2,$$

by means of which may be found by integration the area of any portion of a surface by evaluating the definite integral

$$\iint \sqrt{g}\, du^1\, du^2$$

for appropriate limits.

EXERCISES

1. Find the angle of the coordinate curves of a helicoid defined in §24, Ex. 6.

2. When the coordinate curves on a surface are the curves in which the surface is intersected by the planes $x^1 = $ const. and $x^2 = $ const., the parametric equations of the surface are

$$x^1 = x^1, \qquad x^2 = x^2, \qquad x^3 = f(x^1,\, x^2),$$

the linear element is

$$ds^2 = (1 + p_1^2)\, dx^{1^2} + 2p_1 p_2 dx^1 dx^2 + (1 + p_2^2)dx^{2^2},$$

where $p_\alpha \equiv \dfrac{\partial f}{\partial x^\alpha}$, and the angle ω of the coordinate curves is given by

$$\cos \omega = \frac{p_1 p_2}{\sqrt{(1 + p_1^2)(1 + p_2^2)}}.$$

3. When the coordinates of the enveloping space are general coordinates x^i, equations (24.7) are replaced by

$$g_{\alpha\beta} = a_{ij} \frac{\partial x^i}{\partial u^\alpha} \frac{\partial x^j}{\partial u^\beta},$$

where a_{ij} is the covariant metric tensor of space.

4. When the coordinates of the enveloping space are general coordinates x^i, the form of equations (24.23) is not changed.

5. Find the element of area of a sphere of radius a as defined in parametric form in §10, Ex. 1 (see §24, Ex. 1), and use the result to find the area of the sphere.

6. If θ_{10} and θ_{20} denote the angles at a point on a surface which the vectors $d_1 u^\alpha$ and $d_2 u^\alpha$ make with the vector $du^1 > 0$, $du^2 = 0$ at the point

$$\cos (\theta_{20} - \theta_{10}) = \cos \theta, \qquad \sin (\theta_{20} - \theta_{10}) = \sin \theta,$$

where $\cos \theta$ and $\sin \theta$ are given by (25.2) and (25.3). In particular,

$$\cos (\omega - \theta_0) = \frac{1}{\sqrt{g_{22}}} g_{\alpha 2} \frac{du^\alpha}{ds}, \qquad \sin (\omega - \theta_0) = \sqrt{\frac{g}{g_{22}}} \frac{du^1}{ds},$$

where ω and θ_0 are given by (25.5) and (25.6).

7. From (25.15) it follows that

$$\epsilon^{\alpha\beta} \epsilon_{\alpha\beta} = 2.$$

8. From (25.12) and (25.14) it follows that

(i) $\qquad$ $e^{\alpha\beta} g_{\alpha\gamma} g_{\beta\delta} = g e_{\gamma\delta}$, $\qquad$ $e_{\alpha\beta} g^{\alpha\gamma} g^{\beta\delta} = \dfrac{1}{g} e^{\gamma\delta}$;

(ii) $\qquad$ $\epsilon^{\alpha\beta} g_{\alpha\gamma} g_{\beta\delta} = \epsilon_{\gamma\delta}$, $\qquad$ $\epsilon_{\alpha\beta} g^{\alpha\gamma} g^{\beta\delta} = \epsilon^{\gamma\delta}$;

(iii) $\qquad$ $g^{\alpha\beta} \epsilon_{\gamma\alpha} \epsilon_{\delta\beta} = g_{\gamma\delta}$, $\qquad$ $g_{\alpha\beta} \epsilon^{\gamma\alpha} \epsilon^{\delta\beta} = g^{\gamma\delta}$;

(iv) $\qquad\qquad$ $\epsilon^{\gamma\alpha} g_{\gamma\beta} = \epsilon_{\beta\delta} g^{\alpha\delta}$.

26. FAMILIES OF CURVES IN A SURFACE. PRINCIPAL DIRECTIONS

Upon a surface defined in terms of coordinates u^{α} an equation

(26.1) $\qquad\qquad\qquad$ $f(u^1, u^2) = 0$

is an equation of a curve, as remarked in §10. If this curve is defined also by equations

$$u^{\alpha} = \varphi^{\alpha}(t),$$

in terms of a parameter t, and these expressions are substituted in (26.1) the resulting equation being an identity in t does not vary with t and consequently the derivative with respect to t is equal to zero; that is,

$$\frac{\partial f}{\partial u^{\alpha}} \frac{d\varphi^{\alpha}}{dt} = 0.$$

From this equation and (24.5) we have that differentials du^{α} determining the tangent at any point of the curve satisfy the equation

(26.2) $\qquad\qquad\qquad$ $\dfrac{\partial f}{\partial u^{\alpha}} du^{\alpha} = 0.$

Thus one of the differentials may be chosen arbitrarily ($\neq 0$), and the other is given by (26.2) when the values of u^1 and u^2 for the point are substituted in $\dfrac{\partial f}{\partial u^{\alpha}}$.

Consider an equation

(26.3) $\qquad\qquad\qquad$ $f(u^1, u^2) = c,$

where c is an arbitrary constant. For each value of c equation (26.3) is an equation in the coordinates u^{α} of a curve in a surface defined by equations of the form (24.1). Through each point of the surface for which f is single-valued there passes a curve of the family of curves (26.3). In fact, when the curvilinear coordinates u^1, u^2 of a point are substituted in (26.3), a value of c is uniquely determined, and evidently the equation (26.3) in which c is given this value is an equation of a curve through the point.

An equation

(26.4) $$f_1(u^1, u^2) = c_1$$

is an equation of the same family as (26.3), if f_1 is a function of f, that is, $f_1(u^1, u^2) = F(f(u^1, u^2))$. For, in this case a locus of points for which f is a constant is also a locus for which f_1 is a constant. Conversely, if every curve of the family (26.4) is a curve of the family (26.3), then f_1 must be a function of f. For, as follows from (26.2) the differentials du^α determining the tangent to the curve of each of the families (26.3) and (26.4) are given by

$$\frac{\partial f}{\partial u^\alpha} du^\alpha = 0, \qquad \frac{\partial f_1}{\partial u^\alpha} du^\alpha = 0.$$

If then each curve of one family is to coincide with a curve of the other family, we must have

$$\frac{\partial(f, f_1)}{\partial(u^1, u^2)} \equiv \begin{vmatrix} \dfrac{\partial f}{\partial u^1} & \dfrac{\partial f_1}{\partial u^1} \\ \dfrac{\partial f}{\partial u^2} & \dfrac{\partial f_1}{\partial u^2} \end{vmatrix} \equiv 0,$$

from which it follows that f and f_1 are functions of each other.

Consider next the differential equation

(26.5) $$M_\alpha du^\alpha = 0,$$

where M_1 and M_2 are functions of the u's. In accordance with the theory of such equations there exists a function $t(u^1, u^2)$, called an *integrating factor*,[*] such that a function f is defined by

(26.6) $$\frac{\partial f}{\partial u^\alpha} = tM_\alpha \qquad (\alpha = 1, 2),$$

and thus

$$tM_\alpha \, du^\alpha = df.$$

Consequently $f(u^1, u^2) = c$, where c is an arbitrary constant, is an integral of (26.5). If there is another integral of (26.5), say $f_1(u^1, u^2) = c_1$, corresponding to another integrating factor t_1, it follows from (26.6)

* Fine, 1927, 1, p. 292.

and similar equations in f_1 and t_1 that the jacobian of f and f_1 with respect to u^1 and u^2 is equal to zero, and consequently f_1 is a function of
f. Accordingly, and in view of the results of the preceding paragraph,
an equation (26.5) is an equation of one, and only one, family of curves
on the surface. We say that this family of curves is the family of
integral curves of the equation (26.5).

We seek now the differential equation whose integral curves are the
orthogonal trajectories of the integral curves of equation (26.5). If we
substitute in equations (26.5)

$$du^1 = \rho M_2, \qquad du^2 = -\rho M_1,$$

this equation is satisfied whatever be ρ. If then we substitute these
expressions in place of d_2u^1 and d_2u^2 in the equation

(26.7) $g_{\alpha\beta}\, d_2u^\alpha\, d_1u^\beta = 0,$

which from (25.2) is the condition that at a point of the surface the
vectors d_1u^α and d_2u^α are perpendicular, we obtain on discarding the
factor ρ

$$(g_{1\beta}M_2 - g_{2\beta}M_1)\, d_1u^\beta = 0.$$

Hence we have

[26.1] *The orthogonal trajectories of the integral curves of an equation*
$M_\alpha\, du^\alpha = 0$ *are the integral curves of the equation*

(26.8) $(g_{11}M_2 - g_{12}M_1)\, du^1 + (g_{12}M_2 - g_{22}M_1)\, du^2 = 0.$

As a corollary we have (setting $M_2 = 0$ and $M_1 = 0$ successively)

[26.2] *The orthogonal trajectories of the coordinate curves* $u^1 = $ const.
are the integral curves of the equation

(26.9) $g_{12}\, du^1 + g_{22}\, du^2 = 0;$

and of the curves $u^2 = $ const. *they are the integral curves of*

(26.10) $g_{11}\, du^1 + g_{12}\, du^2 = 0.$

If $f(u^1, u^2)$ is any function, for a change of coordinates in the surface
we have

$$\frac{\partial f}{\partial u^\alpha} = \frac{\partial f'}{\partial u'^\beta}\, \frac{\partial u'^\beta}{\partial u^\alpha},$$

where f' is the transform of f (see §15). From (24.25) it follows that
$\dfrac{\partial f}{\partial u^\alpha}$ are covariant components of a vector, the gradient of f(§17).

From this result and (26.6) one has that the quantities M_α in equation (26.5) are covariant components of a vector. Since its contravariant components are $g^{\alpha\beta}M_\alpha$, we have

$$g_{\alpha\beta}g^{\alpha\gamma}M_\gamma \, du^\beta = \delta^\gamma_\beta M_\gamma \, du^\beta = M_\beta \, du^\beta = 0.$$

Hence from (25.2) we have that M_α in equation (26.5) are the covariant components of the field of normals to the integral curves of the equations (26.5), these normals being tangent to the surface at the corresponding points of the integral curves.

Consider next a differential equation of the first order and second degree of the form

$$(26.11) \quad a_{\alpha\beta} \, du^\alpha \, du^\beta \equiv a_{11}(du^1)^2 + 2a_{12} \, du^1 \, du^2 + a_{22}(du^2)^2 = 0,$$

where $a_{11}a_{22} - a_{12}^2 \neq 0$. This equation is equivalent to the two equations

$$(26.12) \quad \begin{aligned} \left(a_{12} + \sqrt{a_{12}^2 - a_{11}a_{22}}\right) du^1 + a_{22}du^2 &= 0, \\ \left(a_{12} - \sqrt{a_{12}^2 - a_{11}a_{22}}\right) du^1 + a_{22}du^2 &= 0. \end{aligned}$$

We seek the condition that the integral curves of these equations, and consequently the two families of integral curves of equation (26.11), are orthogonal to one another, that is, form an *orthogonal net* (see §25). These integral curves are real or conjugate imaginary according as $a_{11}a_{22} - a_{12}^2$ is a negative or positive quantity. If we denote the differentials in the two equations (26.12) by d_1u^α and d_2u^α respectively, solve these equations for d_1u^2 and d_2u^2, and substitute in (26.7), we have

[26.3] *The two families of integral curves of an equation $a_{\alpha\beta} \, du^\alpha \, du^\beta = 0$ form an orthogonal net, if and only if*

$$(26.13) \qquad g_{11}a_{22} - 2g_{12}a_{12} + g_{22}a_{11} = 0.$$

Under a change of coordinates the equation (26.11) becomes

$$(26.14) \qquad a'_{\gamma\delta} \, du'^\gamma \, du'^\delta = 0,$$

where

$$(26.15) \qquad a'_{\gamma\delta} = a_{\alpha\beta} \frac{\partial u^\alpha}{\partial u'^\gamma} \frac{\partial u^\beta}{\partial u'^\delta}, \qquad a_{\alpha\beta} = a'_{\gamma\delta} \frac{\partial u'^\gamma}{\partial u^\alpha} \frac{\partial u'^\delta}{\partial u^\beta}.$$

Thus $a_{\alpha\beta}$ are the components of a covariant tensor of the second order, and there is no loss in generality in assuming that it is a symmetric tensor (see §19, Ex. 2). In consequence of (24.18) the condition (26.13) may be written

$$(26.16) \qquad g^{\alpha\beta}a_{\alpha\beta} = 0.$$

The left-hand member is a scalar quantity, whose vanishing is the condition that the integral curves of equation (26.11) form an orthogonal net.

We consider now the equations

$$(26.17) \qquad\qquad (a_{\alpha\beta} - r\, g_{\alpha\beta})\, \lambda^\beta = 0,$$

from which it follows that quantities λ^β are determined to within a common factor for each root r of the determinant equation

$$(26.18) \qquad\qquad | \, a_{\alpha\beta} - r\, g_{\alpha\beta} \, | = 0.$$

It is evident that equations (26.17) do not admit a common solution other than $\lambda^\beta = 0$ unless r is a root of equation (26.18). Since $a_{\alpha\beta}$ and $g_{\alpha\beta}$ are covariant tensors, it follows that in any other coordinate system the left-hand member of equation (26.18) is equal to

$$| \, a'_{\alpha\beta} - r\, g'_{\alpha\beta} \, | \left(\frac{\partial(u'^1, \, u'^2)}{\partial(u^1, \, u^2)} \right)^2 .$$

Since the jacobian is different from zero, we have in the coordinates u'^α an equation of the form (26.18), and since r is unaltered by a change of coordinates, it is a scalar. Moreover, the equations (26.17) transform into

$$(a'_{\gamma\delta} - r\, g'_{\gamma\delta})\, \lambda^\beta \frac{\partial u'^\gamma}{\partial u^\alpha} \frac{\partial u'^\delta}{\partial u^\beta} = 0,$$

which are equivalent to $(a'_{\gamma\delta} - rg'_{\gamma\delta})\lambda'^\delta = 0$, where

$$\lambda'^\delta = \lambda^\beta \frac{\partial u'^\delta}{\partial u^\beta},$$

that is, each set of quantities λ^β satisfying (26.17) for suitable values of r are the contravariant components of a vector.

When the determinant (26.18) is expanded, one obtains the quadratic equation

$$(26.19) \quad (g_{11}g_{22} - g_{12}^2)r^2 - (a_{11}g_{22} + a_{22}g_{11} - 2a_{12}g_{12})r + (a_{11}a_{22} - a_{12}^2) = 0.$$

The discriminant of this equation, namely

$$(a_{11}g_{22} + a_{22}g_{11} - 2a_{12}g_{12})^2 - 4(g_{11}g_{22} - g_{12}^2)(a_{11}a_{22} - a_{12}^2),$$

reduces to

$$(26.20) \quad (a_{11}g_{22} - a_{22}g_{11})^2 + 4(a_{12}g_{11} - a_{11}g_{12})(a_{12}g_{22} - a_{22}g_{12}).$$

Since the roots of equation (26.18) are scalars, if they are real and distinct or equal in one coordinate system, the same is true in any coordinate system. Suppose then that the coordinate curves form a real orthogonal net. For such a system the expression (26.20) reduces to

$$(a_{11}g_{22} - a_{22}g_{11})^2 + 4a_{12}^2 g_{11}g_{22},$$

which is non-negative since g_{11} and g_{22} are positive (§24). Hence the roots of (26.18) are real. They are distinct, unless the above expression is equal to zero. In this case

$$\frac{a_{11}}{g_{11}} = \frac{a_{22}}{g_{22}}, \qquad a_{12} = 0,$$

that is, $a_{\alpha\beta} = tg_{\alpha\beta}$, in which case $r = t$, and equations (26.17) are identities, and thus do not determine quantities λ^β.

We consider now the case when $a_{\alpha\beta}$ are not the same multiple of $g_{\alpha\beta}$, that is, the tensor $a_{\alpha\beta}$ is not proportional to the tensor $g_{\alpha\beta}$. We denote by r_1 and r_2 the roots of equation (26.18), and by $\lambda_{1|}^\beta$ and $\lambda_{2|}^\beta$ the corresponding vectors; thus we have

$$(26.21) \qquad (a_{\alpha\beta} - r_1 g_{\alpha\beta})\lambda_{1|}^\beta = 0, \qquad (a_{\alpha\beta} - r_2 g_{\alpha\beta})\lambda_{2|}^\beta = 0.$$

If we multiply these respective equations by $\lambda_{2|}^\alpha$ and $\lambda_{1|}^\alpha$, sum with respect to α in each case and subtract the resulting equations, on changing indices suitably we get

$$(r_2 - r_1)g_{\alpha\beta}\lambda_{1|}^\alpha \lambda_{2|}^\beta = 0.$$

Since $r_2 \neq r_1$ we have the first of the equations

$$(26.22) \qquad g_{\alpha\beta}\lambda_{1|}^\alpha \lambda_{2|}^\beta = 0, \qquad a_{\alpha\beta}\lambda_{1|}^\alpha \lambda_{2|}^\beta = 0,$$

the second being a consequence of the first and either of (26.21). From the first of (26.22) it follows from theorem [25.2] that the two vectors $\lambda_{1|}^\alpha$ and $\lambda_{2|}^\alpha$ are orthogonal, that is, the two vectors at each point are perpendicular.

These two vectors at each point are the tangents to the curves of an orthogonal net on the surface. In order to obtain the differential equation of which these curves are the integral curves, we replace λ^β by du^β in equations (26.17) and eliminate r from the two equations, as α takes the values 1, 2. The result is

$$(26.23) \qquad \begin{vmatrix} a_{1\beta}\,du^\beta & a_{2\beta}\,du^\beta \\ g_{1\beta}\,du^\beta & g_{2\beta}\,du^\beta \end{vmatrix} = 0,$$

which upon expansion is

$$(26.24) \qquad (a_{11}g_{12} - a_{12}g_{11})\,du^{1^2} + (a_{11}g_{22} - a_{22}g_{11})\,du^1\,du^2$$
$$+ (a_{12}g_{22} - a_{22}g_{12})\,du^{2^2} = 0.$$

Since these integral curves form an orthogonal net, when they are the coordinate curves we have in this coordinate system the first of the equations

$$(26.25) \qquad\qquad g_{12} = 0, \qquad a_{12} = 0,$$

and the second follows from the fact that in this coordinate system the coefficients of du^{1^2} and du^{2^2} in (26.24) are equal to zero, since (26.24) must be satisfied separately by $du^1 = 0$ and $du^2 = 0$, and g_{11} and g_{22} are different from zero.   In this coordinate system the directions of the integral curves of equation (26.11) are given by

$$\frac{du^2}{du^1} = \pm \sqrt{-\frac{a_{11}}{a_{22}}},$$

and consequently these directions are equally inclined to the curves $u^2 = $ const. as follows from (25.7).   As a result of the foregoing discussion we have

[26.4] *If $a_{\alpha\beta}$ is a tensor not proportional to the fundamental tensor $g_{\alpha\beta}$, the integral curves of the equation* (26.23) *form a real orthogonal net, and bisect the angles of the integral curves of the equation $a_{\alpha\beta}\,du^\alpha\,du^\beta = 0$.*

We call the directions determined by (26.17), that is, the directions of the integral curves of equation (26.23), the *principal directions for the tensor $a_{\alpha\beta}$*.   Another property of these directions and the significance of the roots of equation (26.18) follow from the consideration of the expression

$$(26.26) \qquad\qquad r = \frac{a_{\alpha\beta}\lambda^\alpha\lambda^\beta}{g_{\alpha\beta}\lambda^\alpha\lambda^\beta}.$$

At a point the value of r depends not only upon the point, but also upon the direction λ^α at the point.   The maximum and minimum values of r at a point are given by the directions for which $\dfrac{\partial r}{\partial \lambda^\alpha} = 0$, that is,

$$g_{\gamma\delta}\lambda^\gamma\lambda^\delta a_{\alpha\beta}\lambda^\beta - a_{\gamma\delta}\lambda^\gamma\lambda^\delta g_{\alpha\beta}\lambda^\beta = 0.$$

By means of (26.26) this is expressible in the form (26.17), and we have

[26.5] *At a point of a surface the maximum and minimum values of the expression $(a_{\alpha\beta}\lambda^\alpha\lambda^\beta)/(g_{\alpha\beta}\lambda^\alpha\lambda^\beta)$ are given by the principal directions for the tensor $a_{\alpha\beta}$, these directions being perpendicular.*

EXERCISES

1. Find upon a sphere the two families of curves which bisect the angles between the meridians and parallels, and then find the linear element of the sphere when these are the coordinate curves (see §10, Ex. 1 and §24, Ex. 1).

2. By definition a curve upon a surface of revolution which meets the meridians under constant angle is a *loxodromic curve*; the equation of loxodromic curves is (see §10, Ex. 2 and §24, Ex. 3) is

$$\int \frac{1}{u^1} \sqrt{1 + \varphi'^2}\, du^1 + bu^2 + c = 0,$$

where b, c are constants, $-b$ being the cotangent of the constant angle.

3. On the skew helicoid (see §24, Ex. 7)

$$x^1 = u^1 \cos u^2, \qquad x^2 = u^1 \sin u^2, \qquad x^3 = au^2$$

the integral curves of the equation

$$(du^1)^2 - (u^{1^2} + a^2)(du^2)^2 = 0$$

form an orthogonal net.

4. An angle θ between the integral curves of the equations

$$M_{1\alpha}\, du^\alpha = 0, \qquad M_{2\alpha}\, du^\alpha = 0$$

is given by

$$\cos \theta = \frac{g^{\alpha\beta} M_{1\alpha} M_{2\beta}}{\sqrt{(g^{\alpha\beta} M_{1\alpha} M_{1\beta})(g^{\gamma\delta} M_{2\gamma} M_{2\delta})}}.$$

5. By means of (24.18) and (25.14) equation (26.8) may be written

$$\epsilon_{\alpha\beta} g^{\alpha\gamma} M_\gamma\, du^\beta = 0.$$

6. Derive the result expressed by equation (26.16) by showing that this scalar is equal to zero when the integral curves of (26.11) form an orthogonal net and are the coordinate curves.

7. For the paraboloid

$$x^1 = au^1 \cos u^2, \qquad x^2 = bu^1 \sin u^2, \qquad x^3 = \tfrac{1}{2}u^{1^2} (a \cos^2 u^2 + b \sin^2 u^2),$$

where a and b are constants, find an equation of a curve on the surface such that the tangent planes along the curve make a constant angle with the $x^1 x^2$-plane; find the edge of regression of the developable surface which is the envelope of the tangent planes to the surface along one of these curves.

8. For a surface S with equations of the form

$$x^1 = e^{hu^2} f(u^1) \cos(u^1 + u^2), \qquad x^2 = e^{hu^2} f(u^1) \sin(u^1 + u^2), \qquad x^3 = e^{hu^2} \varphi(u^1),$$

where h is a constant, the curves $u^1 = $ const. lie on the quadric cones

$$\frac{1}{f^2} (x^{1^2} + x^{2^2}) - \frac{x^{3^2}}{\varphi^2} = 0,$$

and cut the generators of the cone under constant angle, that is, they are conical helices (see §3, Ex. 3). Such curves are also called *spirals*, and the surface S is called a *spiral surface*.

9. The first fundamental form of the surface S of Ex. 8 is

$$ds^2 = e^{2hu^2}[(f'^2 + f^2 + \varphi^2)\, du^{1^2} + 2(hff' + f^2 + h\varphi\varphi')\, du^1\, du^2$$

$$+ (f^2h^2 + f^2 + h^2\varphi^2)\, du^{2^2}],$$

where a prime indicates a derivative with respect to u^1; the orthogonal trajectories $u'^2 =$ const. of the curves $u^1 =$ const. can be found by quadratures and the linear element is reducible to the form

$$ds^2 = e^{2u'^2}(du'^{1^2} + \psi^2(u'^1)\, du'^{2^2}),$$

u'^1 being a suitable function of u^1.

27. THE INTRINSIC GEOMETRY OF A SURFACE. ISOMETRIC SURFACES

In §§24–26 it has been shown that the metric properties of a surface, that is, lengths of curves, angles between intersecting curves, and areas, are expressible completely by means of the first fundamental form of the surface. It is true that these results have been obtained by the consideration of these quantities from the standpoint of the enveloping euclidean space. In this sense the metric properties of the surface are *induced* by the euclidean metric of the enveloping space. However, once the formulas for the measurement of length, angle, and area have been found in terms of the first fundamental form of the surface, thereafter these metric formulas may be used without considering the surface as imbedded in space. For example, since the first fundamental form of a sphere of radius a is (see §24, Ex. 1)

$$(27.1) \qquad ds^2 = a^2((du^1)^2 + \sin^2 u^1(du^2)^2),$$

the coordinate curves being meridians and parallels of latitude, all the metric properties of the sphere are obtainable by the use of the formulas of §§24–26 applied to the form (27.1).

In the next chapter we consider geometric properties of a surface involving its shape as viewed from the enveloping space and we find that they are expressible in terms of the first fundamental form and another quadratic differential form, called the second fundamental form of the surface. In order to distinguish between the properties expressible entirely in terms of the first form from those expressible only in terms of the two forms, we say that the *intrinsic geometry of a surface* consists of the properties expressible in terms of the first form alone.

When two surfaces are such that there exists a coordinate system on each in terms of which the first fundamental forms of the two surfaces are identical, the intrinsic geometry of the two surfaces is the same. This means that so far as measurement on the two surfaces is concerned there is no difference in the two surfaces, no matter how different the surfaces may appear to be as viewed from the enveloping space. Two such surfaces are said to be *isometric*. From (24.12) it follows that, if the fundamental forms of two surfaces are identical for one coordinate system on each, they are identical when any and the same transformation of coordinates is applied to the two surfaces.

An example of isometric surfaces is afforded by the catenoid and the skew helicoid (see §24, Exs. 4, 7). From geometrical considerations it follows that a cylinder and a cone are isometric with the plane, since either may be rolled out upon a plane and thus brought into coincidence with a portion of the plane. In the case of a cylinder this is shown analytically in Ex. 1, the coordinate curves u'^α = const. on the surface corresponding to a rectangular cartesian system in the plane.

In §12 it was noted that in general the tangent planes to a surface constitute a two parameter family, but that there is a group of surfaces whose tangent planes constitute a one parameter family. Such a surface is called a developable surface, since it can be rolled out upon a plane, just as a cylinder or cone can be. In view of this property it is evident that a developable surface is isometric with a plane. It was shown also that with the exception of cones and cylinders a developable surface is the tangent surface of a curve, the tangent planes to the surface being the osculating planes of the curve.

In §24, Ex. 2 it was stated that the first fundamental form of the tangent surface of a curve is

$$\cdot \ (u^2 - u^1)^2 \kappa^2 \, du^{1^2} + du^{2^2},$$

u^1 being the arc of the curve. Since this expression does not involve the torsion τ of the curve, it follows that the tangent surfaces of the curves which have the same first intrinsic equation $\kappa = f_1(u^1)$ but different second intrinsic equations $\tau = f_2(u^1)$ (see (7.1)) are isometric. In particular, when $\tau = 0$ we have the tangent surface of the given twisted curve isometric with the plane, the correspondence being between points of the surface, and points of the tangents to a plane curve with the same intrinsic equation $\kappa = f_1(u^1)$ as the given twisted curve. Since the tangent planes to the surface are the osculating planes of the curve of which the surface is the tangent surface, when the developable surface is rolled out upon the plane, the tangents of the curve become the tangents to the plane curve of the same curvature κ and the principal normals of the former become the normals to the plane curve in its plane. With respect to a rectangular coordinate system in the plane the direction cosines of the tangent α^α and of the normal β^α are such that

$$(27.2) \qquad \alpha^\alpha = \frac{dx^\alpha}{ds}, \qquad \frac{d\alpha^\alpha}{ds} = \kappa\beta^\alpha, \qquad \frac{d\beta^\alpha}{ds} = -\kappa\alpha^\alpha,$$

as follows by processes for the plane which gave the Frenet formulas for a twisted curve; they follow also from (6.1) for $\tau = 0$.

In §12 we showed that any developable surface other than a cylinder or a cone is the tangent surface of some curve C, it being the envelope of the osculating planes of C. For a point x^i on C the coordinates of any point in the corresponding osculating plane are given by (see §6)

$$(27.3) \qquad\qquad X^i = x^i + u\alpha^i + v\beta^i,$$

for suitable values of u and v. When the developable surface is rolled out upon a plane, the curve C becomes a plane curve Γ. The tangent and principal normal to C at a point go into the tangent and normal to Γ at the corresponding point, and consequently the point X^i given by (27.3) goes into the point of coordinates

$$(27.4) \qquad\qquad \bar{X}^\alpha = \bar{x}^\alpha + u\alpha^\alpha + v\beta^\alpha,$$

where $\bar{X}^\alpha$ and $\bar{x}^\alpha$ are cartesian coordinates and where $\bar{x}^\alpha$, α^α, and β^α are appropriate functions of s, which is the same for both curves, and u and v are the same numbers as in (27.3). Differentiating equations (27.4) with respect to s, and making use of (27.2), we have

$$(27.5) \qquad \frac{d\bar{X}^\alpha}{ds} = \left(1 + \frac{du}{ds} - \kappa v\right)\alpha^\alpha + \left(\frac{dv}{ds} + \kappa u\right)\beta^\alpha.$$

If u and v are such that

$$(27.6) \qquad \frac{du}{ds} - \kappa v + 1 = 0, \qquad \frac{dv}{ds} + \kappa u = 0,$$

then $\bar{X}^\alpha$ are constants, that is, such values of u and v determine in each osculating plane of the curve a point such that all these points go into the same point in the plane when the surface is developed upon the plane. Equations (27.6) being the same as (6.10), whose solutions u and v determine an orthogonal trajectory of the osculating planes, we have

[27.1] *When a developable surface, other than a cone or cylinder, is developed upon a plane, all the points of an orthogonal trajectory of the tangent planes to the surface go into one and the same point of the plane.*

EXERCISES

1. When equations of any cylinder are given in the form

$$x^1 = f_1(u^1), \qquad x^2 = f_2(u^1), \qquad x^3 = u^2,$$

the first fundamental form is

$$(f'^2 + f_2'^2)(du^1)^2 + (du^2)^2.$$

In terms of parameters u'^1 and u'^2 defined by

$$u'^1 = \int \sqrt{f_1'^2 + f_2'^2}\, du^1, \qquad u'^2 = u^2$$

the form is $(du'^1)^2 + (du'^2)^2$. What are the coordinate curves in this case?

2. On a cylinder with equations as in Ex. 1 the helices, that is, the curves which meet the generators under constant angle, are defined by

$$a \int \sqrt{f_1'^2 + f_2'^2}\, du^1 + bu^2 + c = 0,$$

where a, b, c are constants.

3. Given a surface of revolution with equations as in §10, Ex. 2 and a right conoid with the equations (see §10, Ex. 4)

$$x^1 = u'^1 \cos u'^2, \qquad x^2 = u'^1 \sin u'^2, \qquad x^3 = \psi(u'^2);$$

in order that a surface of revolution and a right conoid be isometric with the meridians of the former corresponding to rulings on the latter, it is necessary and sufficient that $u'^2 = f(u^2)$, and

$$\sqrt{1 + \varphi'^2}\, du^1 = du'^1, \qquad u'^{1^2} + \psi'^2 = f'^2 u'^{1^2};$$

show that it follows that the surface of revolution is a catenoid (see §24, Ex. 4) and the right conoid is a skew helicoid (see §24, Ex. 7).

4. When the polar developable of a curve (see §12, Ex. 1) is developed upon a plane, the points of the curve go into one, and the same, point of the plane.

28. THE CHRISTOFFEL SYMBOLS FOR A SURFACE. THE RIEMANNIAN CURVATURE TENSOR. THE GAUSSIAN CURVATURE OF A SURFACE

From the definition (20.2) of Christoffel symbols of the second kind, and (24.18) it follows that the Christoffel symbols formed with respect to the first fundamental form of a surface are (no indices being summed)

$$(28.1) \quad \begin{cases} \begin{Bmatrix} \alpha \\ \alpha\alpha \end{Bmatrix} = \dfrac{1}{2g}\left(g_{\beta\beta}\dfrac{\partial g_{\alpha\alpha}}{\partial u^\alpha} + g_{\alpha\beta}\dfrac{\partial g_{\alpha\alpha}}{\partial u^\beta} - 2g_{\alpha\beta}\dfrac{\partial g_{\alpha\beta}}{\partial u^\alpha}\right), \\[2mm] \begin{Bmatrix} \beta \\ \alpha\alpha \end{Bmatrix} = \dfrac{1}{2g}\left(-g_{\alpha\beta}\dfrac{\partial g_{\alpha\alpha}}{\partial u^\alpha} - g_{\alpha\alpha}\dfrac{\partial g_{\alpha\alpha}}{\partial u^\beta} + 2g_{\alpha\alpha}\dfrac{\partial g_{\alpha\beta}}{\partial u^\alpha}\right), \\[2mm] \begin{Bmatrix} \alpha \\ \alpha\beta \end{Bmatrix} = \dfrac{1}{2g}\left(g_{\beta\beta}\dfrac{\partial g_{\alpha\alpha}}{\partial u^\beta} - g_{\alpha\beta}\dfrac{\partial g_{\beta\beta}}{\partial u^\alpha}\right) \qquad (\alpha, \beta = 1, 2; \alpha \neq \beta). \end{cases}$$

From (20.5), (20.6) and (20.7) we have

(28.2)
$$\frac{\partial g_{\alpha\beta}}{\partial u^\gamma} = g_{\alpha\delta} \begin{Bmatrix} \delta \\ \beta\gamma \end{Bmatrix} + g_{\delta\beta} \begin{Bmatrix} \delta \\ \alpha\gamma \end{Bmatrix},$$

$$\frac{\partial g^{\alpha\beta}}{\partial u^\gamma} = -g^{\alpha\delta} \begin{Bmatrix} \beta \\ \delta\gamma \end{Bmatrix} - g^{\delta\beta} \begin{Bmatrix} \alpha \\ \delta\gamma \end{Bmatrix},$$

$$\frac{\partial \log \sqrt{g}}{\partial u^\alpha} = \begin{Bmatrix} \beta \\ \beta\alpha \end{Bmatrix},$$

in all of which the summation convention applies.

From the expression in §20, Ex. 7 for the covariant components of the Riemann tensor it follows that the tensor is skew-symmetric in the first two indices and also in the last two indices. Consequently for a surface

(28.3) $R_{\alpha\alpha\beta\gamma} = R_{\alpha\beta\gamma\gamma} = 0,$ $R_{1212} = R_{2121} = -R_{2112} = -R_{1221}.$

Hence every non-zero component is equal to R_{1212}, or to its negative. From the form (ii) of Ex. 7 in §20, we have

(28.4)
$$R_{1212} = \frac{1}{2} \left(2 \frac{\partial^2 g_{12}}{\partial u^1 \partial u^2} - \frac{\partial^2 g_{11}}{\partial u^{2^2}} - \frac{\partial^2 g_{22}}{\partial u^{1^2}} \right)$$
$$+ g_{\alpha\beta} \left[\begin{Bmatrix} \alpha \\ 12 \end{Bmatrix} \begin{Bmatrix} \beta \\ 12 \end{Bmatrix} - \begin{Bmatrix} \alpha \\ 11 \end{Bmatrix} \begin{Bmatrix} \beta \\ 22 \end{Bmatrix} \right].$$

When a surface is isometric with the plane, there necessarily exist upon it orthogonal nets with respect to which as coordinate curves

$$g_{11} = g_{22} = 1, \qquad g_{12} = 0,$$

the net in the plane consisting of lines parallel to the coordinate axes. Hence $R_{1212} = 0$ in this coordinate system and consequently in every coordinate system by theorem [18.1]. Conversely, if the Riemann tensor for a surface is a zero tensor, it follows from theorem [23.3], which applies to any n-space for $n > 1$, that there exist coordinate systems on the surface with respect to which the above equations hold. Hence we have

[28.1] *A surface is isometric with the plane, if and only if the Riemann tensor is a zero tensor.*

The quantity K defined by

(28.5)
$$K = \frac{R_{1212}}{g}$$

is called the *Gaussian curvature* of a surface, and also the *total curvature of a surface*. As thus defined it is an intrinsic property of the surface as was shown by Gauss.* The geometric significance of the Gaussian curvature of a surface as viewed from the enveloping space is treated in §46.

From (28.5), (28.3) and (25.14) we have

$$(28.6) \qquad\qquad R_{\alpha\beta\gamma\delta} = K\epsilon_{\alpha\beta}\epsilon_{\gamma\delta},$$

and consequently (see §25, Ex. 7)

$$(28.7) \qquad\qquad K = \tfrac{1}{4}R_{\alpha\beta\gamma\delta}\epsilon^{\alpha\beta}\epsilon^{\gamma\delta}.$$

Hence from theorems [25.4] and [19.2] it follows that (see Ex. 8)

[28.2] *The Gaussian curvature of a surface is a scalar.*

From (28.2) and analogously to theorem [22.1] we have

[28.3] *The covariant derivatives of $g_{\alpha\beta}$ and of $g^{\alpha\beta}$ are equal to zero.*

We leave it as an exercise to show that in consequence of the third of equations (28.2)

[28.4] *The covariant derivatives of $\epsilon_{\alpha\beta}$ and of $\epsilon^{\alpha\beta}$ are equal to zero.*

Although we have defined K to be the Gaussian curvature of a surface with the first fundamental form $g_{\alpha\beta} du^{\alpha} du^{\beta}$, it is advisable also to speak of K as the curvature of the quadratic form $g_{\alpha\beta} du^{\alpha} du^{\beta}$. In §26 we considered the integral curves of the equation $a_{\alpha\beta} du^{\alpha} du^{\beta} = 0$ and pointed out that these curves are real only in case the determinant $a \equiv a_{11}a_{22} - a_{12}^2$ is non-positive. When this determinant is negative, which is the condition that there be two distinct families of real curves, we say that the form $a_{\alpha\beta} du^{\alpha} du^{\beta}$ is *indefinite* just as we say that the first fundamental form of a surface is *definite* since $g > 0$. When a is negative, it is possible to find real coordinates $\bar{u}^{\alpha}$ such that

$$a_{\alpha\beta} du^{\alpha} du^{\beta} = t \, d\bar{u}^1 \, d\bar{u}^2,$$

where t is some factor. In this case $\bar{u}^{\alpha} = $ constant are the integral curves. We wish to consider particularly the case when $t = 1$. In this case the curvature of the right-hand member is zero, and consequently this case can arise only when the curvature of the left-hand number is zero since curvature is a scalar.

* 1827, 1, p. 236.

When $t = 1$, we factor the left-hand member of the above equation and write the equation as follows

$$\left(\sqrt{a_{11}}\, du^1 + \frac{a_{12} + \sqrt{-a}}{\sqrt{a_{11}}}\, du^2\right)\left(\sqrt{a_{11}}\, du^1 + \frac{a_{12} - \sqrt{-a}}{\sqrt{a_{11}}}\, du^2\right)$$
$$= d\bar{u}^1\, d\bar{u}^2.$$

We replace this equation by the two

(28.8)
$$\begin{cases} e^{\mu}\left(\sqrt{a_{11}}\, du^1 + \dfrac{a_{12} + \sqrt{-a}}{\sqrt{a_{11}}}\, du^2\right) = d\bar{u}^1, \\[2mm] e^{-\mu}\left(\sqrt{a_{11}}\, du^1 + \dfrac{a_{12} - \sqrt{-a}}{\sqrt{a_{11}}}\, du^2\right) = d\bar{u}^2, \end{cases}$$

where e^{μ} and $e^{-\mu}$ are to be determined.   Evidently they are integrating factors of their respective equations.   Consequently μ is such that

$$\frac{\partial}{\partial u^2}\left(e^{\mu}\sqrt{a_{11}}\right) - \frac{\partial}{\partial u^1}\left(e^{\mu}\frac{a_{12} + \sqrt{-a}}{\sqrt{a_{11}}}\right) = 0,$$

$$\frac{\partial}{\partial u^2}\left(e^{-\mu}\sqrt{a_{11}}\right) - \frac{\partial}{\partial u^1}\left(e^{-\mu}\frac{a_{12} - \sqrt{-a}}{\sqrt{a_{11}}}\right) = 0.$$

From these two equations one can obtain $\dfrac{\partial\mu}{\partial u^1}$ and $\dfrac{\partial\mu}{\partial u^2}$.   Their condition of integrability is necessarily satisfied by the fact that the curvature of the above form is zero.   Hence μ can be found by a quadrature and then $\bar{u}^1$ and $\bar{u}^2$ by further quadratures from (28.8).   We accordingly have

[28.5] *When a quadratic form $a_{\alpha\beta}\, du^{\alpha}\, du^{\beta}$ is indefinite and its curvature is zero, real coordinates $\bar{u}^1$ and $\bar{u}^2$ can be found by quadratures in terms of which the form is equal to $d\bar{u}^1\, d\bar{u}^2$.*

When the form is definite, $\bar{u}^1$ and $\bar{u}^2$ as derived by the above processes are conjugate imaginary.   If then $\bar{u}^1$ and $\bar{u}^2$ are replaced by $\bar{u}^1 + i\bar{u}^2$ and $\bar{u}^1 - i\bar{u}^2$ respectively we have the following theorem:

[28.6] *When a quadratic form $a_{\alpha\beta}\, du^{\alpha}\, du^{\beta}$ is definite and its curvature is zero, real coordinates $\bar{u}^1$ and $\bar{u}^2$ can be found by quadratures in terms of which the form is $(d\bar{u}^1)^2 + (d\bar{u}^2)^2$.*

As a corollary of this theorem we have

[28.7] *Upon a developable surface cartesian coordinates can be found by quadratures, that is, the coordinates are cartesian when the surface is developed upon a plane.*

EXERCISES

1. When the coordinate curves on a surface form an orthogonal net, the corresponding Christoffel symbols are (see §20, Ex. 2)

$$\begin{Bmatrix} \alpha \\ \alpha\alpha \end{Bmatrix} = \frac{\partial \log \sqrt{g_{\alpha\alpha}}}{\partial u^\alpha}, \qquad \begin{Bmatrix} \beta \\ \alpha\alpha \end{Bmatrix} = -\frac{1}{2g_{\beta\beta}} \frac{\partial g_{\alpha\alpha}}{\partial u^\beta},$$

$$(\beta \neq \alpha),$$

$$\begin{Bmatrix} \alpha \\ \alpha\beta \end{Bmatrix} = \frac{\partial \log \sqrt{g_{\alpha\alpha}}}{\partial u^\beta},$$

from which and (28.4) it follows that

$$R_{1212} = -\frac{1}{2}\left(\frac{\partial^2 g_{11}}{\partial u^{2^2}} + \frac{\partial^2 g_{22}}{\partial u^{1^2}}\right) + \frac{1}{4g_{11}}\left[\left(\frac{\partial g_{11}}{\partial u^2}\right)^2 + \frac{\partial g_{11}}{\partial u^1}\frac{\partial g_{22}}{\partial u^1}\right] + \frac{1}{4g_{22}}\left[\left(\frac{\partial g_{22}}{\partial u^1}\right)^2 + \frac{\partial g_{11}}{\partial u^2}\frac{\partial g_{22}}{\partial u^2}\right]$$

$$= -\frac{1}{2}\sqrt{g_{11}g_{22}}\left[\frac{\partial}{\partial u^1}\left(\frac{1}{\sqrt{g_{11}g_{22}}}\frac{\partial g_{22}}{\partial u^1}\right) + \frac{\partial}{\partial u^2}\left(\frac{1}{\sqrt{g_{11}g_{22}}}\frac{\partial g_{11}}{\partial u^2}\right)\right].$$

2. The components of the Ricci tensor (see §20, Ex. 10) for a surface are given by

$$R_{\alpha\beta} = g^{\gamma\delta}R_{\gamma\alpha\beta\delta} = -\frac{g_{\alpha\beta}}{g}R_{1212},$$

and the scalar curvature K of the surface by

$$-\frac{1}{2}R = -\frac{1}{2}g^{\alpha\beta}R_{\alpha\beta} = \frac{R_{1212}}{g}.$$

3. For a sphere of radius a with the equations of the form in §10, Ex. 1 it follows from Ex. 1 and §24, Ex. 1 that

$$K = \frac{R_{1212}}{g_{11}g_{22}} = \frac{1}{a^2}.$$

4. For a surface of revolution with the equations

$$x^1 = u^1 \cos u^2, \qquad x^2 = u^1 \sin u^2, \qquad x^3 = \varphi(u^1),$$

it follows from Ex. 1 and §24, Ex. 3 that

$$K = \frac{R_{1212}}{g_{11}g_{22}} = \frac{\varphi'\varphi''}{u^1(1+\varphi'^2)}.$$

5. For a surface with the first fundamental form

$$a^2[\cos^2\omega(du^1)^2 + \sin^2\omega(du^2)^2],$$

where ω is a function of u^1 and u^2

$$R_{1212} = a^2 \sin\omega \cos\omega \left(\frac{\partial^2\omega}{\partial u^{2^2}} - \frac{\partial^2\omega}{\partial u^{1^2}}\right).$$

6. The integral curves of the equation

$$g_{\alpha\beta} \, du^{\alpha} \, du^{\beta} = 0$$

are conjugate imaginary curves of length zero, that is, minimal curves (see §3).

7. The equations

$$x^1 = a \frac{u^1 + u^2}{1 + u^1 u^2}, \qquad x^2 = a \frac{i(u^2 - u^1)}{1 + u^1 u^2}, \qquad x^3 = a \frac{u^1 u^2 - 1}{1 + u^1 u^2},$$

where a is a constant, $i = \sqrt{-1}$, and u^1 and u^2 are conjugate imaginary, are parametric equations of a sphere of radius a; the coordinate curves are minimal lines.

8. In consequence of the remark following (28.3)

$$\frac{R_{1212}}{g} = \frac{R'_{\alpha\beta\gamma\delta} \dfrac{\partial u'^{\alpha}}{\partial u^1} \dfrac{\partial u'^{\beta}}{\partial u^2} \dfrac{\partial u'^{\gamma}}{\partial u^1} \dfrac{\partial u'^{\delta}}{\partial u^2}}{g' \left[\dfrac{\partial (u'^1, \, u'^2)}{\partial (u^1, \, u^2)} \right]^2} = \frac{R'_{1212}}{g'};$$

consequently K is a scalar.

9. From equations analogous to (20.15) and from (28.3) and (28.5) it follows that

$$R^{\alpha}{}_{\beta\gamma\delta} = K(\delta^{\alpha}_{\gamma} g_{\beta\delta} - \delta^{\alpha}_{\delta} g_{\beta\gamma}).$$

10. From §20, Ex. 7 and (28.5) it follows that

(i)
$$K = \frac{1}{2\sqrt{g}} \left[\frac{\partial}{\partial u^1} \left(\frac{g_{12}}{g_{11} \sqrt{g}} \frac{\partial g_{11}}{\partial u^2} - \frac{1}{\sqrt{g}} \frac{\partial g_{22}}{\partial u^1} \right) \right.$$
$$\left. + \frac{\partial}{\partial u^2} \left(\frac{2}{\sqrt{g}} \frac{\partial g_{12}}{\partial u^1} - \frac{1}{\sqrt{g}} \frac{\partial g_{11}}{\partial u^2} - \frac{g_{12}}{g_{11} \sqrt{g}} \frac{\partial g_{11}}{\partial u^1} \right) \right],$$

and in consequence of (25.5)

(ii)
$$K = - \frac{1}{\sqrt{g}} \left[\frac{\partial^2 \omega}{\partial u^1 \partial u^2} + \frac{\partial}{\partial u^1} \left(\frac{\dfrac{\partial \sqrt{g_{22}}}{\partial u^1} - \dfrac{\partial \sqrt{g_{11}}}{\partial u^2} \cos \omega}{\sqrt{g_{11}} \sin \omega} \right) \right.$$
$$\left. + \frac{\partial}{\partial u^2} \left(\frac{\dfrac{\partial \sqrt{g_{11}}}{\partial u^2} - \dfrac{\partial \sqrt{g_{22}}}{\partial u^1} \cos \omega}{\sqrt{g_{22}} \sin \omega} \right) \right].$$

11. When the equations of a surface are of the form in §25, Ex. 2 the Christoffel symbols have the values

$$\left\{ \begin{matrix} \alpha \\ \beta\gamma \end{matrix} \right\} = \frac{1}{\sqrt{g}} f_{\alpha} f_{\beta\gamma},$$

where $g = 1 + f_1^2 + f_2^2$, and $f_{\beta\gamma} = \dfrac{\partial^2 f}{\partial x^{\beta} \partial x^{\gamma}}$.

12. From the definition (§22) of the covariant derivative based upon $g_{\alpha\beta}$ of the contravariant components λ^α of a vector, namely

(i) $$\lambda^\alpha{}_{,\beta} = \frac{\partial \lambda^\alpha}{\partial u^\beta} + \lambda^\gamma \left\{ \begin{matrix} \alpha \\ \beta\gamma \end{matrix} \right\},$$

and from (28.2) one obtains

(ii) $$\frac{\partial}{\partial u^\alpha} \left(\sqrt{g}\, \lambda^\alpha \right) = \sqrt{g}\, \lambda^\alpha{}_{,\alpha}.$$

The scalar $\lambda^\alpha{}_{,\alpha}$ is called the *divergence* of the vector λ^α (see §22, Ex. 2).

13. For the covariant derivative $\lambda_{\alpha,\beta}$ of the covariant components λ_α of a vector one has that

$$\epsilon^{\alpha\beta} \lambda_{\alpha,\beta} = \frac{1}{\sqrt{g}} \left(\frac{\partial \lambda_1}{\partial u^2} - \frac{\partial \lambda_2}{\partial u^1} \right)$$

is a scalar, called the *curl* of the vector λ_α ; when the curl is equal to zero, the vector is a gradient (see §17).

14. If λ^α and λ_α are the contravariant and covariant components of a unit vector, one has $\lambda^\alpha{}_{,\beta}\lambda_\alpha = \lambda_{\alpha,\beta}\lambda^\alpha = 0$, from which it follows that

$$\lambda^\alpha{}_{,\beta} = \mu^\alpha \nu_\beta, \qquad \lambda_{\alpha,\beta} = \mu_\alpha \nu_\beta,$$

where μ^α is the vector perpendicular to the given vector, and ν_β is some vector.

15. When the coordinates are chosen so that an indefinite quadratic form reduces to $2a_{12}\, du^1\, du^2$ the curvature of the form is given by

$$K = -\frac{1}{a_{12}} \frac{\partial^2}{\partial u^1 \partial u^2} \log a_{12}.$$

29. DIFFERENTIAL PARAMETERS

If φ is any function of u^1 and u^2, $\dfrac{\partial \varphi}{\partial u^\alpha}$ are the covariant components of a vector, the gradient of φ, as observed in §26. Consequently the quantity $\Delta_1\varphi$ defined by

(29.1) $$\Delta_1 \varphi \equiv g^{\alpha\beta} \frac{\partial \varphi}{\partial u^\alpha} \frac{\partial \varphi}{\partial u^\beta}$$

is a scalar (see theorem [19.2]). It is the square of the length of the gradient $\dfrac{\partial \varphi}{\partial u^\alpha}$ by theorem [24.1]. Also

(29.2) $$\Delta_1(\varphi_1, \varphi_2) \equiv g^{\alpha\beta} \frac{\partial \varphi_1}{\partial u^\alpha} \frac{\partial \varphi_2}{\partial u^\beta},$$

where φ_1 and φ_2 are any functions of u^1 and u^2, is a scalar.

Consider also the quantity

$$(29.3) \qquad \Theta_1(\varphi_1, \varphi_2) \equiv \frac{1}{\sqrt{g}} \frac{\partial(\varphi_1, \varphi_2)}{\partial(u^1, u^2)} = \epsilon^{\alpha\beta} \frac{\partial \varphi_1}{\partial u^\alpha} \frac{\partial \varphi_2}{\partial u^\beta}.$$

In consequence of theorem [25.4] this quantity is a scalar for a positive transformation.

The scalars $\Delta_1\varphi$, $\Delta_1(\varphi_1, \varphi_2)$, $\Theta_1(\varphi_1, \varphi_2)$ are called *differential parameters of the first order*, involving as they do derivatives of the first order.

In order to give a geometric interpretation to these differential parameters, we note that for any function φ of u^1 and u^2 one has

$$(29.4) \qquad \frac{\partial \varphi}{\partial u^\alpha} du^\alpha = 0,$$

where du^α are the contravariant components of the tangent vector at a point of each curve of the family $\varphi(u^1, u^2) = $ const. From (29.4) and theorem [25.2] it follows that this tangent vector and the vector $\frac{\partial \varphi}{\partial u^\alpha}$ at each point are perpendicular.

For two families of curves $\varphi_1(u^1, u^2) = c_1$ and $\varphi_2(u^1, u^2) = c_2$ the angle θ which the normal vector $\frac{\partial \varphi_2}{\partial u^\alpha}$ at a point makes with the normal vector $\frac{\partial \varphi_1}{\partial u^\alpha}$ at the point, and consequently an angle between the curves at the point, is given by

$$(29.5) \qquad \cos \theta = \frac{\Delta_1(\varphi_1, \varphi_2)}{\sqrt{\Delta_1\varphi_1 \cdot \Delta_1\varphi_2}}, \qquad \sin \theta = \frac{\Theta_1(\varphi_1, \varphi_2)}{\sqrt{\Delta_1\varphi_1 \cdot \Delta_1\varphi_2}},$$

as follows from (25.10), (29.2), and (29.3) (see Ex. 5).

From (29.5) we have

[29.1] *Two families of curves $\varphi_1(u^1, u^2) = $ const. and $\varphi_2(u^1, u^2) = $ const. form an orthogonal net, if and only if $\Delta_1(\varphi_1, \varphi_2) = 0$.*

From (29.5) it follows that the following identity holds between the differential parameters of the first order:

$$(29.6) \qquad (\Delta_1(\varphi_1, \varphi_2))^2 + (\Theta_1(\varphi_1, \varphi_2))^2 = \Delta_1\varphi_1 \cdot \Delta_1\varphi_2.$$

When, in particular, we take $\varphi_1 = u^1$, $\varphi_2 = u^2$, we have from (29.1), (29.2), and (24.18) in the coordinate system u^α

$$(29.7) \qquad \Delta_1 u^1 = g^{11} = \frac{g_{22}}{g}, \qquad \Delta_1 u^2 = g^{22} = \frac{g_{11}}{g},$$

$$\Delta_1(u^1, u^2) = g^{12} = -\frac{g_{12}}{g}.$$

Since differential parameters are scalars, when $\Delta_1 u^1$, $\Delta_1 u^2$, $\Delta_1(u^1, u^2)$ are written out in another coordinate system u'^α, we have

(29.8)
$$g^{\alpha\beta} = g'^{\gamma\delta} \frac{\partial u^\alpha}{\partial u'^\gamma} \frac{\partial u^\beta}{\partial u'^\delta},$$

that is, equations (24.19). Since g_{11}, g_{22} and g are positive quantities when u^1 and u^2 are real coordinates (see §24), it follows from (29.7) that $\Delta_1\varphi$ is a positive scalar when φ is a real function.

Let φ be a solution of the differential equation

(29.9)
$$\Delta_1\varphi = 1,$$

and ψ a solution of the equation

(29.10)
$$\Delta_1(\varphi, \psi) = 0,$$

in which φ is the given solution of (29.9). By theorem [29.1] the curves $\varphi = $ const. and $\psi = $ const. form an orthogonal net. If they are taken as the parametric curves $u'^1 = $ const. and $u'^2 = $ const. respectively, it follows from (29.7) that $g'_{12} = 0$, $g' = g'_{11} g'_{22}$, and

$$\frac{g'_{22}}{g'_{11} g'_{22}} = \Delta_1 u'^1 = \Delta_1\varphi = 1,$$

and consequently the linear element is

(29.11)
$$ds^2 = (du'^1)^2 + g'_{22}(du'^2)^2.$$

From this result it follows that along a curve $\psi \equiv u'^2 = $ const., $ds = du'^1$, and thus the distance along this curve from a curve $\varphi = c_1$ to a curve $\varphi = c_2$ is equal to $c_2 - c_1$. Since this distance is the same along all curves $\psi = $ const., we say that the curves $\varphi = $ const. are *parallel curves*.

Since $\Delta_1\varphi$ has been shown to be positive for a real function φ, if $\Delta_1\varphi$ is a positive function of φ, say

(29.12)
$$\Delta_1\varphi = F(\varphi),$$

a real function $f(\varphi)$ is defined by

$$f(\varphi) = \int \frac{d\varphi}{\sqrt{F(\varphi)}}.$$

From (29.1) one has for any function $f(\varphi)$

$$\Delta_1 f(\varphi) = \left(\frac{df}{d\varphi}\right)^2 \Delta_1\varphi.$$

Hence for the above function $f(\varphi)$ one has $\Delta_1 f(\varphi) = 1$. Accordingly $f(\varphi) = $ const. are parallel curves, and therefore $\varphi = $ const. are. Hence we have

[29.2] *If φ is any real function of real coordinates u^α and $\Delta_1 \varphi = F(\varphi)$, the curves $\varphi = $ const. are parallel curves.*

For example, when the linear element is

$$ds^2 = \varphi^2(u^1)(du^1)^2 + g_{22}(du^2)^2,$$

$\Delta_1 u^1 = \dfrac{1}{\varphi^2(u^1)}$, as follows from (29.7) and consequently the curves $u^1 = $ const. are parallel.

The quantities $\varphi_{,\alpha\beta}$, defined by

$$(29.13) \qquad \varphi_{,\alpha\beta} \equiv \frac{\partial^2 \varphi}{\partial u^\alpha \, \partial u^\beta} - \frac{\partial \varphi}{\partial u^\gamma} \left\{ \begin{matrix} \gamma \\ \alpha\beta \end{matrix} \right\},$$

being the second covariant derivatives of φ, are the components of a symmetric covariant tensor of the second order. Consequently the quantity $\Delta_2 \varphi$ defined by

$$(29.14) \qquad \Delta_2 \varphi \equiv g^{\alpha\beta} \varphi_{,\alpha\beta}$$

is a scalar. It is called the *fundamental differential parameter of φ of the second order*; it involves derivatives of the second and first orders. The differential parameters $\Delta_1 \varphi$ and $\Delta_2 \varphi$ were introduced by Lamé[*] for space referred to general coordinates in his study of physical problems. They were introduced in the study of the geometry of a surface by Beltrami.[†]

In order to write $\Delta_2 \varphi$ in another form, we note from the definition (22.4) of the covariant derivative of a contravariant vector that the covariant derivative of the vector $g^{\alpha\gamma} \dfrac{\partial \varphi}{\partial u^\alpha}$ $(\equiv g^{\alpha\gamma} \varphi_{,\alpha})$ is given by

$$(29.15) \qquad (g^{\alpha\gamma} \varphi_{,\alpha})_{,\beta} = \frac{\partial}{\partial u^\beta} (g^{\alpha\gamma} \varphi_{,\alpha}) + g^{\alpha\delta} \varphi_{,\alpha} \left\{ \begin{matrix} \gamma \\ \delta\beta \end{matrix} \right\}.$$

By theorem [28.3] the left-hand member of the above equation is equal to $g^{\alpha\gamma} \varphi_{,\alpha\beta}$. From this result and (29.15) it follows that (29.14) can be written

$$\Delta_2 \varphi = \frac{\partial}{\partial u^\beta} (g^{\alpha\beta} \varphi_{,\alpha}) + g^{\alpha\delta} \varphi_{,\alpha} \left\{ \begin{matrix} \beta \\ \delta\beta \end{matrix} \right\}.$$

* 1859, 1, pp. 5, 17.
† 1864, 1, pp. 359, 365.

In consequence of this result and (28.2) we have

$$(29.16) \qquad \Delta_2 \varphi = \frac{1}{\sqrt{g}} \frac{\partial}{\partial u^\beta} \left(\sqrt{g} \, g^{\alpha\beta} \, \varphi_{,\alpha} \right),$$

which because of (24.18) may be given the form

$$(29.17) \quad \Delta_2 \varphi = \frac{1}{\sqrt{g}} \left[\frac{\partial}{\partial u^1} \left(\frac{g_{22} \dfrac{\partial\varphi}{\partial u^1} - g_{12} \dfrac{\partial\varphi}{\partial u^2}}{\sqrt{g}} \right) + \frac{\partial}{\partial u^2} \left(\frac{g_{11} \dfrac{\partial\varphi}{\partial u^2} - g_{12} \dfrac{\partial\varphi}{\partial u^1}}{\sqrt{g}} \right) \right].$$

EXERCISES

1. Show that

$$\frac{\partial}{\partial u^\gamma} \Delta_1 \varphi = 2 g^{\alpha\beta} \varphi_{,\alpha} \varphi_{,\beta\gamma}.$$

2. Show that

$$\Delta_2 u^\alpha = - g^{\beta\gamma} \begin{Bmatrix} \alpha \\ \beta\gamma \end{Bmatrix} \qquad\qquad (\alpha = 1, 2).$$

3. The following are differential parameters of the second order:

$$\Delta_1 \Delta_1 \varphi, \ \Delta_1(\varphi, \Delta_1\varphi), \ \Theta_1(\varphi, \Delta_1\varphi),$$

$$\Delta_1\Delta_1(\varphi, \psi), \ \Delta_1(\Delta_1\varphi, \Delta_1\psi), \ \Theta_1(\Delta_1\varphi, \Delta_1\psi).$$

4. If f and g are any functions of u^1 and u^2,

$$\Delta_1 f = \left(\frac{\partial f}{\partial u^1} \right)^2 \Delta_1 u^1 + 2 \frac{\partial f}{\partial u^1} \frac{\partial f}{\partial u^2} \Delta_1(u^1, u^2) + \left(\frac{\partial f}{\partial u^2} \right)^2 \Delta_1 u^2,$$

$$\Delta_1(f, g) = \frac{\partial f}{\partial u^1} \frac{\partial g}{\partial u^1} \Delta_1 u^1 + \left(\frac{\partial f}{\partial u^1} \frac{\partial g}{\partial u^2} + \frac{\partial f}{\partial u^2} \frac{\partial g}{\partial u^1} \right) \Delta_1(u^1, u^2) + \frac{\partial f}{\partial u^2} \frac{\partial g}{\partial u^2} \Delta_1 u^2,$$

$$\Delta_2 f = \frac{\partial f}{\partial u^\alpha} \Delta_2 u^\alpha + \frac{\partial^2 f}{\partial u^{1^2}} \Delta_1 u^1 + 2 \frac{\partial^2 f}{\partial u^1 \partial u^2} \Delta_1(u^1, u^2) + \frac{\partial^2 f}{\partial u^{2^2}} \Delta_1 u^2.$$

5. At a point of a curve $\varphi(u^1, u^2) = 0$ the unit normal vector $\dfrac{\partial\varphi}{\partial u^\alpha} / \sqrt{\Delta_1 \varphi}$ makes a right-angle with the unit tangent vector $\dfrac{du^\alpha}{ds}$ when the latter is chosen in direction so that by theorem [25.6]

$$\frac{\partial\varphi}{\partial u^\alpha} = \sqrt{\Delta_1 \varphi} \, \epsilon_{\beta\alpha} \frac{du^\beta}{ds}.$$

If for two intersecting curves $\varphi(u^1, u^2) = 0$ and $\psi(u^1, u^2) = 0$ one has respectively

$$\frac{\partial\varphi}{\partial u^\alpha} = \sqrt{\Delta_1 \varphi} \, \epsilon_{\beta\alpha} \frac{d_1 u^\beta}{d_1 s}, \qquad \frac{\partial\psi}{\partial u^\alpha} = \sqrt{\Delta_1 \psi} \, \epsilon_{\beta\alpha} \frac{d_2 u^\beta}{d_2 s},$$

the expression (29.5) for $\sin\theta$ is equal to (25.3) (see (25.15)).

6. On a surface of revolution (see §10, Ex. 2) the curves u^1 = const., that is, the orthogonal trajectories of the meridian curves, are parallel curves (see §24, Ex. 3).

7. For a helicoid, as defined in §24, Ex. 6, the helices are parallel curves, and their orthogonal trajectories are the integral curves of the equation

$$\frac{a\varphi'}{u^{1^2} + a^2} du^1 + du^2 = 0;$$

find the linear element of the surface when the helices and their orthogonal trajectories are the coordinate curves, and show that any given helicoid is isometric with some surface of revolution.

8. When there exists upon a surface an orthogonal coordinate net such that the quantities g_{11} and g_{22} are functions of u^1 alone or u^2 alone, the surface is isometric with a surface of revolution.

9. If φ is a solution of the equation $\Delta_1\varphi = 0$, the curves φ = const. are imaginary, as follows from (29.7).

10. Of the differential parameters $\Delta_1\varphi$, $\Delta_1(\varphi, \psi)$, $\Delta_2\varphi$, $\Theta_1(\varphi, \psi)$ the last is the only one which changes sign (but not magnitude) when a transformation is negative.

11. When the coordinate net is orthogonal,

$$\Delta_2 u^\alpha = \frac{1}{\sqrt{g_{11} g_{22}}} \frac{\partial}{\partial u^\alpha} \sqrt{\frac{g_{\beta\beta}}{g_{\alpha\alpha}}} \qquad (\beta \ne \alpha);$$

from this result and (29.7) it follows that if u^α are solutions of the equation

(i) $$\Delta_1(\varphi, \Delta_1\varphi) = 2\Delta_2\varphi(\Delta_1\varphi - 1)$$

other than those for which $\Delta_1\varphi = 1$, then

$$\frac{\partial}{\partial u^\alpha} \left(\frac{1 - g_{\alpha\alpha}}{g_{\beta\beta}} \right) = 0 \qquad (\beta \ne \alpha).$$

In this case the fundamental form may be written

(ii) $$ds^2 = \cos^2 \theta \, du^{1^2} + \sin^2 \theta \, du^2 \; ;$$

if φ is any solution of (i) such that $\Delta_1\varphi \ne 1$, the function ψ of the orthogonal trajectories ψ = const. of the curves φ = const. can be chosen so that the fundamental form is

$$ds^2 = \cos^2 \bar{\theta} \, d\varphi^2 + \sin^2 \bar{\theta} \, d\psi^2.$$

12. When the linear element is the form (ii) of Ex. 11 and one effects the transformation of coordinates

$$u^1 = u'^1 + u'^2, \qquad u^2 = u'^1 - u'^2,$$

the linear element becomes

(i) $$ds^2 = (du'^1)^2 + 2 \cos \omega \, du'^1 \, du'^2 + (du'^2)^2,$$

where $\omega(= 2\theta)$ is the angle of the new coordinate curves. A net with respect to which as coordinate the linear element is of the form (i) is called a *Tchebychef net*; a necessary and sufficient condition that the coordinate curves form such a net is

$$\left\{ \begin{matrix} \alpha \\ \alpha\beta \end{matrix} \right\} = 0 \qquad\qquad (\alpha \neq \beta).$$

When the linear element is $g_{\alpha\beta}\, du^\alpha\, du^\beta$ and one effects the transformation

$$u'^1 = \int \sqrt{g_{11}}\, du^1, \qquad u'^2 = \int \sqrt{g_{22}}\, du^2,$$

the new coordinate curves form a Tchebychef net; the determination of all such nets is equivalent to the solution of equation (i) of Ex. 11.

30. ISOMETRIC ORTHOGONAL NETS. ISOMETRIC COORDINATES

When the coordinate curves on a surface form an orthogonal net such that $g_{11} = g_{22} = t^2$, in which case the linear element of the surface is

$$(30.1) \qquad\qquad ds^2 = t^2(du^{1^2} + du^{2^2}),$$

the elements of length of the u^1- and u^2-coordinate curves are $t\, du^1$ and $t\, du^2$ respectively. Hence the coordinate curves divide the surface into small squares to a first approximation. Such a net is called an *isometric orthogonal net* and u^α *isometric coordinates*.

From (29.7) it follows that a necessary and sufficient condition that two families of curves $\varphi(u^1, u^2) = $ const. and $\psi(u^1, u^2) = $ const. form an isometric net and that φ and ψ be isometric coordinates is that

$$(30.2) \qquad\qquad \Delta_1\varphi = \Delta_1\psi, \qquad \Delta_1(\varphi, \psi) = 0.$$

In this case the linear element of the surface is

$$(30.3) \qquad\qquad ds^2 = t^2(d\varphi^2 + d\psi^2),$$

where t^2 is the reciprocal of the common value of the two members of the first of equations (30.2).

When the linear element is in the form (30.3) it follows from (29.17) that

$$\Delta_2 u^\alpha = 0 \qquad\qquad (\alpha = 1, 2).$$

It will now be shown conversely that each real solution of the equation $\Delta_2\theta = 0$ determines an isometric orthogonal net.

If φ is a solution of the differential equation of the second order $\Delta_2\theta = 0$, we have from (29.16)

$$(30.4) \qquad \frac{\partial}{\partial u^1}\left(\sqrt{g}\, g^{\alpha 1}\, \varphi_{,\alpha}\right) + \frac{\partial}{\partial u^2}\left(\sqrt{g}\, g^{\alpha 2}\, \varphi_{,\alpha}\right) = 0.$$

From this equation it follows that a function ψ is defined by

(30.5) $\sqrt{g}\, g^{\alpha 1}\varphi_{,\alpha} = \dfrac{\partial \psi}{\partial u^2} \equiv \psi_{,2}\,,$ $\sqrt{g}\, g^{\alpha 2}\varphi_{,\alpha} = -\dfrac{\partial \psi}{\partial u^1} \equiv -\psi_{,1}\,,$

since (30.4) expresses the condition of integrability of equations (30.5). If these equations are multiplied by $g_{\gamma 1}$ and $g_{\gamma 2}$ respectively and the resulting equations are added, we have

$$g_{\gamma\beta}g^{\alpha\beta}\varphi_{,\alpha} = \frac{1}{\sqrt{g}}\,(g_{\gamma 1}\psi_{,2} - g_{\gamma 2}\psi_{,1}).$$

The left-hand member of this equation reduces to $\varphi_{,\gamma}$. For the values $\gamma = 1$ and $\gamma = 2$ these equations are reducible, in consequence of (24.18), to

(30.6) $\varphi_{,1} = \sqrt{g}\, g^{\beta 2}\psi_{,\beta}\,,$ $\varphi_{,2} = -\sqrt{g}\, g^{\beta 1}\psi_{,\beta}\,.$

Expressing the condition of integrability of these equations, we have

$$\frac{\partial}{\partial u^\alpha}\,(\sqrt{g}\, g^{\beta\alpha}\psi_{,\beta}) = 0,$$

which by (29.16) is equivalent to the condition $\Delta_2\psi = 0$.

If equations (30.5) are multiplied by $\varphi_{,1}$ and $\varphi_{,2}$, and the resulting equations added, we have

$$\sqrt{g}\, g^{\alpha\beta}\varphi_{,\alpha}\varphi_{,\beta} = \varphi_{,1}\psi_{,2} - \varphi_{,2}\psi_{,1}\,,$$

which in consequence of (29.1) and (25.14) may be written

(30.7) $\Delta_1\varphi = \epsilon^{\alpha\beta}\varphi_{,\alpha}\psi_{,\beta}\,.$

Likewise, if equations (30.6) are multiplied by $\psi_{,2}$ and $\psi_{,1}$ respectively and the resulting equations are subtracted, we have

(30.8) $\epsilon^{\alpha\beta}\varphi_{,\alpha}\psi_{,\beta} = \Delta_1\psi.$

From these two equations we have the first of equations (30.2).

Again, if equations (30.5) are multiplied by $\psi_{,1}$ and $\psi_{,2}$, and the resulting equations are added, we have the second of equations (30.2). Hence we have

[30.1] *Any real solution φ of the equation $\Delta_2\theta = 0$ and the function ψ obtained by quadrature from the corresponding equations (30.5) are isometric coordinates of an isometric orthogonal net.*

If now we define coordinates u^1 and u^2 by

(30.9) $\varphi = f_1(u^1),$ $\psi = f_2(u^2),$

the linear element (30.3) becomes

(30.10) $$ds^2 = t^2(f_1'^2 du^{1^2} + f_2'^2 du^{2^2}).$$

Thus the linear element is no longer of the form (30.1), but the coordinate curves are the same as before (see (10.14)). For the form (30.10)

(30.11) $$\frac{g_{11}}{g_{22}} = \frac{U_1^2}{U_2^2},$$

where U_1 and U_2 are functions of u^1 and u^2 respectively.

Conversely, if for an orthogonal net on a surface the quantities g_{11} and g_{22} are in the relation (30.11), the linear element is of the form

(30.12) $$ds^2 = t^2(U_1^2 du^{1^2} + U_2^2 du^{2^2}).$$

If then we define coordinates u'^α by

$$u'^1 = \int U_1 du^1, \qquad u'^2 = \int U_2 du^2,$$

the coordinate curves are the same as before, but the linear element is

$$ds^2 = t^2(du'^{1^2} + du'^{2^2}).$$

Hence we have

[30.2] *When for an orthogonal coordinate net, the condition* (30.11) *is satisfied, the net is isometric, and the isometric coordinates can be obtained by quadratures.*

If any function of φ, say $f(\varphi)$ is a solution of the equation $\Delta_2 \theta = 0$, in which case $f(\varphi)$ is one of a pair of isometric coordinates, we have from (29.16)

(30.13) $$\Delta_2\varphi f'(\varphi) + \Delta_1\varphi f''(\varphi) = 0,$$

where the primes indicate derivatives with respect to φ. When this equation is written in the form

$$\frac{\Delta_2 \varphi}{\Delta_1 \varphi} = -\frac{f''(\varphi)}{f'(\varphi)},$$

we have that, if $\Delta_2\varphi/\Delta_1\varphi$ is a function of φ, say $F(\varphi)$, then the function $f(\varphi)$ obtained by two quadratures from

(30.14) $$f'(\varphi) = e^{-\int F(\varphi)d\varphi} = e^{-\int (\Delta_2\varphi/\Delta_1\varphi)d\varphi}$$

is such that $f(\varphi)$ is an isometric coordinate. Hence we have

[30.3] *A necessary and sufficient condition that a family of curves $\varphi = $ const. and their orthogonal trajectories form an isometric net is that $\Delta_2\varphi = 0$ or that the ratio of $\Delta_2\varphi$ and $\Delta_1\varphi$ be a function of φ.*

From (30.4) it follows that (30.13) may be written in the form

$$\frac{\partial}{\partial u^1} (f'(\varphi)\sqrt{g}\, g^{\alpha 1} \varphi_{,\alpha}) + \frac{\partial}{\partial u^2} (f'(\varphi)\sqrt{g}\, g^{\alpha 2} \varphi_{,\alpha}) = 0.$$

Hence a function ψ is defined by

$$(30.15) \quad f'(\varphi)\sqrt{g}\, g^{\alpha 1} \varphi_{,\alpha} = \psi_{,2}, \qquad f'(\varphi)\sqrt{g}\, g^{\alpha 2} \varphi_{,\alpha} = -\psi_{,1},$$

and ψ is such that the second of (30.2) is satisfied, that is, the curves $\psi = $ const. are the orthogonal trajectories of the curves $\varphi = $ const. and form with the latter an isometric net.

Equations (30.15) follow also from (30.5) when φ is replaced by $f(\varphi)$. Consequently in place of (30.6) we have

$$(30.16) \quad f'(\varphi)\varphi_{,1} = \sqrt{g}\, g^{\beta 2}\psi_{,\beta}, \qquad f'(\varphi)\varphi_{,2} = -\sqrt{g}\, g^{\beta 1}\psi_{,\beta}.$$

From (30.15) and (30.16) we have similarly to (30.7) and (30.8)

$$f'(\varphi)\Delta_1 \varphi = \epsilon^{\alpha\beta} \varphi_{,\alpha}\psi_{,\beta} = \frac{1}{f'(\varphi)} \Delta_1 \psi,$$

that is, in place of the first of (30.2) we have

$$(30.17) \qquad\qquad \Delta_1\psi = f'^2(\varphi)\Delta_1\varphi.$$

From this result, equations (29.7), and (30.14) we find that the linear element is

$$(30.18) \qquad ds^2 = \frac{1}{\Delta_1\varphi} (d\varphi^2 + e^{2\int (\Delta_2\varphi/\Delta_1\varphi)d\varphi}\, d\psi^2).$$

If u^1 and u^2 are isometric coordinates of an isometric orthogonal net, that is, if the linear element is of the form (30.1) and functions φ and ψ are defined by

$$(30.19) \quad \varphi + i\psi = f(u^1 \pm iu^2), \qquad \varphi - i\psi = f_0(u^1 \mp iu^2),$$

where f and f_0 are conjugate functions, we have

$$d\varphi^2 + d\psi^2 = f'f_0'(du^{1^2} + du^{2^2}).$$

Consequently the linear element is

$$ds^2 = \frac{t^2}{f'f_0'} (d\varphi^2 + d\psi^2),$$

and hence φ and ψ are isometric coordinates of a real isometric orthogonal net, different from the given one since φ and ψ are not functions of u^1 and u^2 respectively, or of u^2 and u^1 respectively.

When the linear element is of the form (30.1), equations (30.5) become

$$\frac{\partial \varphi}{\partial u^1} = \frac{\partial \psi}{\partial u^2}, \qquad \frac{\partial \varphi}{\partial u^2} = -\frac{\partial \psi}{\partial u^1},$$

that is, the Cauchy-Riemann equations,* the integrals of which are given by (30.19). Hence we have

[30.4] *When one isometric orthogonal net is known for a surface, all other such nets may be obtained directly by equations of the form (30.19).*

EXERCISES

1. The meridians and their orthogonal trajectories on a surface of revolution (see §24, Ex. 3) form an isometric net.

2. The rulings on a skew helicoid and their orthogonal trajectories form an isometric orthogonal net (see §24, Ex. 7).

3. When the coordinate curves of a surface form an isometric orthogonal net, the curves

$$u^1 + u^2 = \text{const.}, \qquad u^1 - u^2 = \text{const.}$$

are the bisectors of the angles of the coordinate net, and form an isometric orthogonal net.

4. When on a surface two families of curves $\varphi = \text{const.}$ and $\psi = \text{const.}$ form an isometric orthogonal net such that

$$\Delta_1 \varphi = \Delta_1 \psi = f(\varphi)$$

the surface is isometric with a surface of revolution.

5. For a plane referred to cartesian coordinates the equations

$$u^1 + iu^2 = \frac{1}{x^1 - ix^2}$$

define an isometric orthogonal net consisting of two families of circles.

6. For a central quadric with parametric equations of §10, Ex. 3

$$g_{\alpha\alpha} = \frac{u^\alpha(u^\alpha - u^\beta)}{4(a_1 - u^\alpha)(a_2 - u^\alpha)(a_3 - u^\alpha)}, \qquad g_{\alpha\beta} = 0 \qquad (\alpha, \beta = 1, 2; \alpha \neq \beta),$$

hence the coordinate curves form an isometric orthogonal net.

7. The equations

$$x^1 = \pm \sqrt{\frac{a_1 - a_2}{a_2} u^1 u^2}, \qquad x^2 = \pm \frac{1}{a_2} \sqrt{\frac{a_2 - a_1}{a_1} (1 + a_1 u^1)(1 + a_2 u^2)},$$

$$x^3 = \frac{1}{2} \cdot \frac{a_2 - a_1}{a_1 a_2} (1 + a_1 u^1 + a_2 u^2),$$

---

* Fine, 1927, 1, pp. 408, 409.

in which the a's are constants are equations of a paraboloid. For the coordinate system u^α

$$g_{\alpha\alpha} = \frac{a_1 - a_2}{4a_2^2}(u^\alpha - u^\beta)\frac{a_1(a_1 - a_2)u^\alpha - a_2}{u^\alpha(1 + a_1 u^\alpha)}, \qquad g_{\alpha\beta} = 0 \qquad (\alpha, \beta = 1, 2; \alpha \neq \beta);$$

and consequently the coordinate curves form an isometric orthogonal net.

31. ISOMETRIC SURFACES

In this section there is derived a necessary and sufficient condition that two surfaces S and S' be isometric (see §27), their respective first fundamental forms being

$$(31.1) \qquad g_{\alpha\beta}du^\alpha du^\beta, \qquad g'_{\alpha\beta}du'^\alpha du'^\beta.$$

The surfaces are isometric if there exist two independent equations

$$(31.2) \qquad \varphi(u^1, u^2) = \varphi'(u'^1, u'^2), \qquad \psi(u^1, u^2) = \psi'(u'^1, u'^2)$$

establishing a one-to-one correspondence between points of S and S' such that by means of (31.2) either of the quadratic forms (31.1) is transformed into the other. Since differential parameters are scalars, it follows that a necessary condition that S and S' be isometric is that

$$(31.3) \quad \Delta_1\varphi = \Delta_1'\varphi', \qquad \Delta_1(\varphi, \psi) = \Delta_1'(\varphi', \psi'), \qquad \Delta_1\psi = \Delta_1'\psi',$$

where the differential parameters on the left and right are formed with respect to the respective forms (31.1). Conversely, the conditions (31.3) are sufficient conditions that S and S' be isometric. In fact, if the curves $\varphi = $ const., $\psi = $ const. are taken as coordinate on S and $\varphi' = $ const., $\psi' = $ const. on S', in consequence of (29.7) the respective quadratic forms may be written in the form

$$\frac{\Delta_1\psi \, d\varphi^2 - 2\Delta_1(\varphi, \psi) \, d\varphi \, d\psi + \Delta_1\varphi \, d\psi^2}{\Delta_1\varphi\Delta_1\psi - \Delta_1^2(\varphi, \psi)},$$

$$\frac{\Delta_1'\psi' \, d\varphi'^2 - 2\Delta_1'(\varphi', \psi') \, d\varphi' \, d\psi' + \Delta_1'\varphi' \, d\psi'^2}{\Delta_1'\varphi'\Delta_1'\psi' - \Delta_1'^2(\varphi', \psi')}.$$

Hence when there exist two independent equations (31.2) such that (31.3) are satisfied, the surfaces S and S' are isometric.

Since by theorem [28.2] the Gaussian curvature is a scalar, a necessary condition that surfaces S and S' with the fundamental forms (31.1) be isometric is that

$$(31.4) \qquad K(u^1, u^2) = K'(u'^1, u'^2),$$

where K and K' are the Gaussian curvatures of S and S' respectively. We now seek a necessary and sufficient condition that S and S' be isometric.

We consider first the case when K and K' are equal constants, that is,

$$(31.5) \qquad R_{1212} = ag, \qquad R'_{1212} = ag'.$$

If $a = 0$, each surface is isometric with the plane by theorem [28.1], and consequently with the other.

In order to discuss the case $a \neq 0$ we remark that equations (31.5) are equivalent to

$$(31.6) \quad R_{\alpha\beta\gamma\delta} = a(g_{\alpha\gamma}g_{\beta\delta} - g_{\alpha\delta}g_{\beta\gamma}), \qquad R'_{\alpha\beta\gamma\delta} = a(g'_{\alpha\gamma}g'_{\beta\delta} - g'_{\alpha\delta}g'_{\beta\gamma}).$$

We apply to this case the discussion of equations (23.12) and (23.13) for the functions u^α of u'^1, u'^2 and the functions $p^\alpha_\beta = \dfrac{\partial u^\alpha}{\partial u'^\beta}$. Because of (31.6) the corresponding equations (23.16) are satisfied in consequence of

$$g'_{\mu\nu} = g_{\alpha\beta}p^\alpha_\mu p^\beta_\nu,$$

that is, equations (24.12). These equations, three in number, are the set E_0, and all the sets $E_1, \cdots$ are satisfied because of the set E_0. Since there are six functions u^α, p^α_μ and three equations in the set E_0, we have in consequence of theorem [23.2]

[31.1] *Two surfaces of equal constant Gaussian curvature are isometric, the equations giving the one-to-one correspondence involving three arbitrary constants.*

In this connection it must be remarked that such correspondence is limited to domains for which theorem [23.2] applies.

When K is not a constant, we consider in addition to (31.4), the equation

$$(31.7) \qquad \Delta_1 K = \Delta'_1 K'.$$

We note that $\Delta_1 K \neq 0$ for K real (see §29). If then equations (31.4) and (31.7) are independent, these equations establish a correspondence between S and S' which is isometric, if and only if

$$(31.8) \quad \Delta_1(K, \Delta_1 K) = \Delta'_1(K', \Delta'_1 K'), \qquad \Delta_1\Delta_1 K = \Delta'_1\Delta'_1 K'.$$

If equations (31.4) and (31.7) are not independent, and they are to be satisfied, we must have

$$(31.9) \qquad \Delta_1 K = f(K), \qquad \Delta'_1 K' = f(K'),$$

where f is some positive function of K and K' respectively.   In this case we may take for equations (31.2) equations (31.4) and

$$(31.10) \qquad\qquad \Delta_2 K = \Delta_2' K',$$

unless

$$(31.11) \qquad\qquad \Delta_2 K = f_1(K), \qquad \Delta_2' K' = f_1(K').$$

If equations of the form (31.11) do not exist, then equations (31.4), (31.9), (31.10) and

$$(31.12) \quad \Delta_1(K, \Delta_2 K) = \Delta_1'(K', \Delta_2' K'), \qquad \Delta_1 \Delta_2 K = \Delta_1' \Delta_2' K'$$

constitute a necessary and sufficient condition that S and S' be isometric.

We consider finally the case when both (31.9) and (31.11) are satisfied, including the possibility $f_1(K) = f_1(K') = 0$.   If $f_1(K) \neq 0$, the ratio of $\Delta_2 K$ and $\Delta_1 K$ is a function of K, and by theorem [30.3] the curves $K =$ const. and their orthogonal trajectories $\psi =$ const. form an isometric net, the function ψ being obtained by a quadrature (see §30).   Furthermore, the respective quadratic forms are by (30.18)

$$(31.13) \quad \frac{1}{f(K)} (dK^2 + e^{2\int (f_1/f) dK} d\psi^2), \qquad \frac{1}{f(K')} (dK'^2 + e^{2\int (f_1/f) dK'} d\psi'^2),$$

and when $\Delta_2 K = 0$, $\Delta_2' K' = 0$, by (30.3), they are

$$(31.14) \qquad\qquad \frac{1}{f(K)} (dK^2 + d\psi^2), \qquad \frac{1}{f(K')} (dK'^2 + d\psi'^2).$$

In either case it is seen from (31.13) and (31.14) that the equations

$$K = K', \qquad \psi = \pm \psi' + a,$$

where a is an arbitrary constant, define the isometric correspondence of the two surfaces.

We have thus treated all possible cases and as the result have

[31.2] *Given two surfaces whose Gaussian curvatures are not constant; it can be determined directly, that is, without quadratures, whether the surfaces are isometric.*

When (31.9) and (31.11) are not both satisfied, the equations determining the correspondence are given directly, whereas in the case when (31.9) and (31.11) are both satisfied, the determination involves a quadrature.   In this case the correspondence can be effected in an infinity of ways.   What this means geometrically follows from the observation that from the forms (31.13) and (31.14) it is seen that S and

S' are isometric with a surface of revolution (see §24, Ex. 3).   When one considers a surface of revolution from the standpoint of the enveloping space, one sees that a surface of revolution is isometric with itself in an infinity of ways, each such correspondence being given by a suitable rotation of the surface about its axis.

It should be remarked that for any two surfaces in isometric correspondence such correspondence is established only for the domains for which the equations given above, such as (31.4), (31.7), and (31.10), are independent in each of the cases considered.

EXERCISES

1. When the linear element of a surface is written in the form

(i) $$ds^2 = du^{1^2} + g_{22}\,du^{2^2},$$

one has

(ii) $\Delta_1 u^1 = 1, \qquad \Delta_2 u^1 = \dfrac{\partial \log \sqrt{g_{22}}}{\partial u^1}, \qquad K = -\dfrac{1}{\sqrt{g_{22}}}\dfrac{\partial^2 \sqrt{g_{22}}}{\partial u^{1^2}},$

from which it follows that

$$\Delta(u^1, \Delta_2 u^1) = -K - (\Delta_2 u^1)^2.$$

Hence for any function φ the equation

$$\Delta_1(\varphi, \Delta_2\varphi) = -K - (\Delta_2\varphi)^2$$

is an identity.

2. If equations (31.4) and (31.9) are satisfied, for the functions σ and σ' defined by

$$\sigma = \int \frac{dK}{f(K)}, \qquad \sigma' = \int \frac{dK'}{f(K')},$$

the equation (31.10) reduces to $\Delta_2\sigma = \Delta_2'\sigma'$, and the equation $\Delta_1(\sigma, \Delta_2\sigma) = \Delta_1'(\sigma', \Delta_2'\sigma')$ is a consequence of (31.10) and (31.4) (see Ex. 1).

3. The equation (i) of Ex. 1 is the linear element of a surface isometric with a surface of revolution if g_{22} is a function of u^1 alone (see §24, Ex. 3).   In order that the surface have constant Gaussian curvature g_{22} must be a solution of the third of equations (ii) of Ex. 1, in which K is a constant.   If $K \neq 0$, there are the two cases

(i) $K = \dfrac{1}{a^2}, \qquad \sqrt{g_{22}} = b \cos \dfrac{u^1}{a} + c \sin \dfrac{u^1}{a},$

(ii) $K = -\dfrac{1}{a^2}, \qquad \sqrt{g_{22}} = b \cosh \dfrac{u^1}{a} + c \sinh \dfrac{u^1}{a},$

where b and c are arbitrary constants.

4. For a surface with the linear element

$$ds^2 = \frac{du^{1^2} + du^{2^2}}{[1 + a(u^{1^2} + u^{2^2})]^2},$$

where a is a constant, the Gaussian curvature is equal to $4a$.

32. GEODESICS

The Christoffel symbols $\left\{ \begin{matrix} \alpha \\ \beta\gamma \end{matrix} \right\}$ and $\left\{ \begin{matrix} \lambda \\ \mu\nu \end{matrix} \right\}'$ in two coordinate systems u^α and u'^α respectively in a surface are in the relation

(32.1) $$\frac{\partial^2 u^\alpha}{\partial u'^\mu \, \partial u'^\nu} + \left\{ \begin{matrix} \alpha \\ \beta\gamma \end{matrix} \right\} \frac{\partial u^\beta}{\partial u'^\mu} \frac{\partial u^\gamma}{\partial u'^\nu} = \left\{ \begin{matrix} \lambda \\ \mu\nu \end{matrix} \right\}' \frac{\partial u^\alpha}{\partial u'^\lambda}.$$

This result follows from (20.11), which result is general and applies to any transformation of coordinates in any number of variables, provided only that the determinant of the covariant tensor with respect to which the Christoffel symbols are formed, in the present case $g_{\alpha\beta}$, is not equal to zero.

For any curve on the surface defined by u^α as functions of s, and for any transformation of coordinates, we have

(32.2) $$\frac{du'^\mu}{ds} = \frac{\partial u'^\mu}{\partial u^\alpha} \frac{du^\alpha}{ds}, \qquad \frac{du^\alpha}{ds} = \frac{\partial u^\alpha}{\partial u'^\mu} \frac{du'^\mu}{ds}.$$

Thus λ^α defined by

(32.3) $$\lambda^\alpha = \frac{du^\alpha}{ds}$$

are contravariant components of the tangent vector to the curve, and it is a unit vector since from (24.6)

(32.4) $$g_{\alpha\beta} \frac{du^\alpha}{ds} \frac{du^\beta}{ds} = 1.$$

Differentiating the second set of equations (32.2) with respect to s and making use of (32.1), we have

$$\frac{d^2 u^\alpha}{ds^2} = \frac{\partial u^\alpha}{\partial u'^\mu} \frac{d^2 u'^\mu}{ds^2} + \left(\left\{ \begin{matrix} \lambda \\ \mu\nu \end{matrix} \right\}' \frac{\partial u^\alpha}{\partial u'^\lambda} - \left\{ \begin{matrix} \alpha \\ \beta\gamma \end{matrix} \right\} \frac{\partial u^\beta}{\partial u'^\mu} \frac{\partial u^\gamma}{\partial u'^\nu} \right) \frac{du'^\mu}{ds} \frac{du'^\nu}{ds},$$

from which it follows that

(32.5) $$\frac{d^2 u^\alpha}{ds^2} + \left\{ \begin{matrix} \alpha \\ \beta\gamma \end{matrix} \right\} \frac{du^\beta}{ds} \frac{du^\gamma}{ds} = \frac{\partial u^\alpha}{\partial u'^\lambda} \left(\frac{d^2 u'^\lambda}{ds^2} + \left\{ \begin{matrix} \lambda \\ \mu\nu \end{matrix} \right\}' \frac{du'^\mu}{ds} \frac{du'^\nu}{ds} \right).$$

If we differentiate (32.4) with respect to s, we obtain

$$2g_{\alpha\beta}\frac{d^2 u^\alpha}{ds^2}\frac{du^\beta}{ds} + \frac{\partial g_{\alpha\beta}}{\partial u^\gamma}\frac{du^\alpha}{ds}\frac{du^\beta}{ds}\frac{du^\gamma}{ds} = 0.$$

From this result and (28.2) we have

(32.6) $$g_{\alpha\beta}\left(\frac{d^2 u^\gamma}{ds^2} + \left\{\begin{matrix}\alpha\\ \delta\gamma\end{matrix}\right\}\frac{du^\delta}{ds}\frac{du^\gamma}{ds}\right)\frac{du^\beta}{ds} = 0.$$

If then we put (changing dummy indices)

(32.7) $$\frac{d^2 u^\alpha}{ds^2} + \left\{\begin{matrix}\alpha\\ \beta\gamma\end{matrix}\right\}\frac{du^\beta}{ds}\frac{du^\gamma}{ds} = \mu^\alpha,$$

it follows from (32.5) that μ^α are the contravariant components of a vector at each point of the curve, which may be a zero vector (that is, $\mu^\alpha = 0$). From (32.6) it follows that if the vectors μ^α at points of a curve are not zero vectors they are perpendicular to the tangents at the corresponding points; this case is considered in §34.

We consider now the curves at each point of which the vector μ^α is a zero vector, that is, the curves for which u^α as functions of s are solutions of the equations

(32.8) $$\frac{d^2 u^\alpha}{ds^2} + \left\{\begin{matrix}\alpha\\ \beta\gamma\end{matrix}\right\}\frac{du^\beta}{ds}\frac{du^\gamma}{ds} = 0.$$

These curves are called *geodesics*.

Before considering geodesics on a general surface, we observe that, if the surface is a plane and the coordinates are cartesian, equations (32.8) reduce to $\frac{d^2 u^\alpha}{ds^2} = 0$, the integral of which is

$$u^\alpha = a^\alpha + b^\alpha s,$$

when the a's and b's are constants. Hence the geodesics of a plane are straight lines, and conversely any straight line is a geodesic. The reader should compare the results in this section concerning geodesics on any surface with the properties of straight lines in the plane. From the form of equation (32.8) it follows that isometric surfaces (§27) have the same equations of geodesics. In particular, the geodesics on a developable surface are such that they become straight lines when the surface is rolled out upon a plane. A characteristic property of geodesics on a surface as viewed from enveloping space is shown in §44.

A solution of equations (32.8) is determined by initial values of u^α and $\dfrac{du^\alpha}{ds}$, that is, the values when $s = 0$. For such values we obtain from (32.8) the corresponding initial values of $\dfrac{d^2 u^\alpha}{ds^2}$, and from the result of differentiating (32.8) corresponding initial values of all higher derivatives of u^α. In accordance with the theory of the existence of integrals of ordinary differential equations,* the corresponding integral of equations (32.8) is given by

$$(32.9) \quad u^\alpha = u_0^\alpha + \left(\frac{du^\alpha}{ds}\right)_0 s + \frac{1}{2}\left(\frac{d^2 u^\alpha}{ds^2}\right)_0 s^2 + \frac{1}{\underline{3}}\left(\frac{d^3 u^\alpha}{ds^3}\right)_0 s^3 + \cdots ,$$

for values of s for which the series converge, the subscript 0 indicating initial values. If we differentiate the equation

$$(32.10) \qquad\qquad g_{\alpha\beta} \frac{du^\alpha}{ds} \frac{du^\beta}{ds} = \text{const.,}$$

we obtain (32.6), and consequently any integral of equations (32.8) satisfies (32.10). Hence, in order that s in a solution (32.9) shall be the arc, it is necessary and sufficient to choose the initial values $\left(\dfrac{du^\alpha}{ds}\right)_0$ so that (32.4) be satisfied.

Since the values $\left(\dfrac{du^\alpha}{ds}\right)_0$ determine the direction of the geodesic at the initial point u_0^α, we have the following fundamental theorem:

[32.1] *Through each point in a surface and in any given direction there passes a unique geodesic.*

From (32.8) and (32.6) we have

[32.2] *The coordinate curves $u^\alpha = $ const. for $\alpha = 1$ or 2 are geodesics, if and only if*

$$(32.11) \qquad\qquad \left\{ \begin{matrix} \alpha \\ \beta\beta \end{matrix} \right\} = 0 \qquad (\alpha = 1 \text{ or } 2; \beta = 2 \text{ or } 1).$$

---

* If we put $\dfrac{du^\alpha}{ds} = p^\alpha$, these equations and $\dfrac{dp^\alpha}{ds} + \left\{ \begin{matrix} \alpha \\ \beta\gamma \end{matrix} \right\} p^\beta p^\gamma = 0$, which are equations (32.8), are of the form discussed by Darboux (see §23).

When the coordinate curves form an orthogonal net, equations (32.11) reduce to $\frac{\partial g_{\beta\beta}}{\partial u^\alpha} = 0$ (see §28, Ex. 1). Hence we have

[32.3] *When the coordinate curves on a surface form an orthogonal net, a necessary and sufficient condition that the curves $u^\alpha = $ const. be geodesics is that $g_{\beta\beta}$, where $\beta \neq \alpha$, be a function of u^β alone.*

Thus for example, the meridians on a surface of revolution are geodesics (see §24, Ex. 3), as are the rulings on a right conoid (see §24, Ex. 5).

Returning to the consideration of the series (32.9), and the remarks preceding (32.9), we see that if we put

$$(32.12) \qquad\qquad \bar{u}^\alpha = s\left(\frac{du^\alpha}{ds}\right)_0,$$

we have

$$u^\alpha - u_0^\alpha = \bar{u}^\alpha + a^\alpha \bar{u}^{1^2} + b^\alpha \bar{u}^1 \bar{u}^2 + c^\alpha \bar{u}^{2^2} + \cdots,$$

where $a, b, c, \cdots$, are functions of u_0^1, and u_0^2. These series are convergent for values of $\bar{u}^1$ and $\bar{u}^2$ in absolute value less than some fixed quantity. Since the jacobian of u^α with respect to $\bar{u}^\beta (\alpha, \beta = 1, 2)$ for $\bar{u}^\alpha = 0$ is equal to $+1$, these series may be inverted giving $\bar{u}^\alpha$ as power series in $u^1 - u_0^1$ and $u^2 - u_0^2$, which are convergent so long as $u^\alpha - u_0^\alpha$ in absolute value are less than some fixed quantity.* For such values of $u^\alpha - u_0^\alpha$ the values of $\bar{u}^\alpha$ are uniquely determined, and consequently there passes only one geodesic through the points u_0^α and u^α. Moreover, from (32.4) and (32.12), it follows that the length of the arc of the geodesics between these points is given by

$$s^2 = (g_{\alpha\beta})_0 \bar{u}^\alpha \bar{u}^\beta.$$

Hence we have

[32.4] *Through two sufficiently near points on a surface there passes one and only one geodesic.†*

In order to find the rate of change with respect to s along a geodesic of the angle θ_0 which the geodesic makes with the curves $u^2 = $ const., we differentiate with respect to s equation (25.7) written in the form

$$\theta_0 = \tan^{-1} \frac{\sqrt{g}\,\dfrac{du^2}{ds}}{g_{1\alpha}\,\dfrac{du^\alpha}{ds}}.$$

---

* See Goursat, 1924, 1, vol. 1, p. 474.
† See Darboux, 1889, 1, p. 408.

Making use of (28.2) and (24.6), the resulting equation is reducible to

$$
\begin{aligned}
(32.13) \quad \frac{d\theta_0}{ds} = \sqrt{g}\Bigg[& \frac{du^1}{ds}\frac{d^2u^2}{ds^2} - \frac{du^2}{ds}\frac{d^2u^1}{ds^2} \\
& + \frac{1}{g_{11}}\left(g_{1\alpha}\left\{\begin{matrix}2\\\beta 2\end{matrix}\right\} - g_{2\alpha}\left\{\begin{matrix}2\\\beta 1\end{matrix}\right\} - g_{1\gamma}\left\{\begin{matrix}\gamma\\\alpha\beta\end{matrix}\right\}\right)\frac{du^2}{ds}\frac{du^\alpha}{ds}\frac{du^\beta}{ds}\Bigg] \\
= \; & \epsilon_{\alpha\beta}\frac{du^\alpha}{ds}\left(\frac{d^2u^\beta}{ds^2}+\left\{\begin{matrix}\beta\\\gamma\delta\end{matrix}\right\}\frac{du^\gamma}{ds}\frac{du^\delta}{ds}\right) - \frac{\sqrt{g}}{g_{11}}\left\{\begin{matrix}2\\1\alpha\end{matrix}\right\}\frac{du^\alpha}{ds},
\end{aligned}
$$

where $\epsilon_{\alpha\beta}$ are defined in (25.14). This result holds for any curve on a surface. When the curve is a geodesic, the above equation reduces in consequence of equations (32.8) to

$$
(32.14) \qquad \frac{d\theta_0}{ds} + \frac{\sqrt{g}}{g_{11}}\left\{\begin{matrix}2\\1\alpha\end{matrix}\right\}\frac{du^\alpha}{ds} = 0.
$$

Consider now in a surface an orthogonal net of coordinate curves for which the curves $u^2 = $ const. are geodesics. By theorem [32.3] the coordinate u^1 can be chosen so that the linear element is

$$
(32.15) \qquad ds^2 = du^{1^2} + g_{22}du^{2^2}.
$$

From this result it follows that the length of the segment of a curve $u^2 = $ const. between the curves $u^1 = c_1$ and $u^1 = c_2$ is given by

$$
\int du^1 = c_2 - c_1.
$$

Since this length does not depend upon u^2, the lengths of the segments of all the geodesics $u^2 = $ const. between any two orthogonal trajectories are equal. In consequence of this result and theorem [32.1] we have the following theorem of Gauss:*

[32.5] *Given any curve C upon a surface and the geodesics orthogonal to C; when equal lengths are measured from C along these geodesics, the locus of their end points is an orthogonal trajectory of the geodesics.*

The curves thus defined are called *geodesic parallels* to the curve C. From the discussion of equations (29.9) and (29.11) it follows that the curves there called parallel are geodesic parallels. From theorem [29.2], and the above discussion we have

[32.6] *A necessary and sufficient condition that a family of curves $\varphi = $ const. be geodesic parallels is that $\Delta_1\varphi = F(\varphi)$, where $F(\varphi)$ is any positive*

* 1827, 1, p. 241.

function of φ; φ is the length of the geodesics measured from one of the curves
$\varphi = const.$, if and only if $\Delta_1\varphi = 1$.

Suppose then that P_1 and P_2 are points of a surface through which there passes only one geodesic, and that we take this for the curve $u^2 = 0$ of a system of geodesic parallels $u^2 = const.$ and for $u^1 = const.$ their orthogonal trajectories, so that we have the linear element (32.15). An equation of any other curve which passes through P_1 and P_2 is of the form $u^2 = \varphi(u^1)$, and the length of the arc P_1P_2 of this curve is given by

$$ s = \int_{u_1^1}^{u_2^1} \sqrt{1 + g_{22}\varphi'^2}\, du^1, $$

where u_1^1 and u_2^1 $(> u_1^1)$ are the values of u^1 at the points P_1 and P_2 respectively. Since the arc P_1P_2 of the given geodesic is $u_2^1 - u_1^1$, and g_{22} being positive the quantity $1 + g_{22}\varphi'^2$ is greater than one, we have the following fundamental theorem:

[32.7] *If two points in a surface are such that only one geodesic passes through them, the length of the segment of the geodesic is the shortest distance in the surface between the two points.*

If one has a solution $\varphi(u^1, u^2, a)$ of the equation

(32.16) $$\Delta_1\varphi = 1$$

such that $\dfrac{\partial\varphi}{\partial a}$ involves the constant a, when the solution is substituted in (32.16) and the resulting identity is differentiated with respect to a, we have

(32.17) $$\Delta_1\left(\varphi, \frac{\partial\varphi}{\partial a}\right) = 0.$$

Consequently for each value of a the curves

(32.18) $$\frac{\partial\varphi}{\partial a} = b,$$

where b is a constant, are geodesics, being the orthogonal trajectories of the curves $\varphi = const.$, which by theorem [32.6] are geodesic parallels. For a particular point u_0^α the components du^α of the tangent to a geodesic through the point are given by $\dfrac{\partial^2\varphi}{\partial a\,\partial u^\alpha}\, du^\alpha = 0$ for each value of a. Conversely, if du^α are given, the corresponding value of a is determined by this equation since a geodesic is uniquely determined by a point and a direction.   Then b is determined by the equation

$$\frac{\partial\varphi(u_0^1,\, u_0^2,\, a)}{\partial a} = b.$$

Hence all of the geodesics in a surface are given by (32.18), and we have

[32.8] *If* $\varphi(u^1,\, u^2,\, a)$ *is a solution of the equation* $\Delta_1\varphi = 1$ *such that* $\dfrac{\partial\varphi}{\partial a}$ *involves the constant* a, *the equation*

$$\frac{\partial\varphi}{\partial a} = b$$

for all values of the constant b *is the finite equation of the geodesics of the surface, and the arc of the geodesics is measured by* φ.

A surface possessing an orthogonal parametric net with respect to which the linear element is

(32.19) $$ds^2 = (A_1 + A_2)(B_1^2 du^{1^2} + B_2^2 du^{2^2}),$$

where A_α and B_α are functions of u^α alone is called a *surface of Liouville*. For such a surface equation (32.16) is reducible to

$$\frac{1}{B_1^2}\left(\frac{\partial\varphi}{\partial u^1}\right)^2 - A_1 = -\frac{1}{B_2^2}\left(\frac{\partial\varphi}{\partial u^2}\right)^2 + A_2.$$

We seek the solution of this equation such that each member of this equation is equal to a constant a, and find that φ is given by the two quadratures

$$\varphi = \int B_1\sqrt{A_1 + a}\, du^1 + \int B_2\sqrt{A_2 - a}\, du^2.$$

Hence by theorem [32.8] an equation of the geodesics is

(32.20) $$\int \frac{B_1}{\sqrt{A_1 + a}}\, du^1 - \int \frac{B_2}{\sqrt{A_2 - a}}\, du^2 = 2b,$$

and we have

[32.9] *The geodesics on a surface of Liouville, when its linear element is given in the form* (32.19), *can be found by two quadratures.*

We close this section with the derivation anew of the equations of geodesics, using the property of geodesics in theorem [32.7]. To this end we consider a curve C with the equations

(32.21) $$u^\alpha = f^\alpha(t),$$

and two points P_1 and P_2 on the curve with the respective values t_1 and t_2 of the parameter t. Any curve in the neighborhood of C and passing through the points P_1 and P_2 is defined by

(32.22) $$u^\alpha = f^\alpha(t) + \epsilon\omega^\alpha(t)$$

for a sufficiently small absolute value of the constant ϵ, and for functions $\omega^\alpha(t)$ such that

(32.23) $$\omega^\alpha(t_1) = \omega^\alpha(t_2) = 0.$$

The arc s of C between P_1P_2 is given by

(32.24) $$s = \int_{t_1}^{t_2} \sqrt{g_{\alpha\beta} f^{\alpha\prime} f^{\beta\prime}}\, dt \equiv \int_{t_1}^{t_2} \varphi(f^1, f^2, f^{1\prime}, f^{2\prime})\, dt,$$

and the arc s_1 of C_1 by

(32.25) $$s_1 = \int_{t_1}^{t_2} \varphi(f^1 + \epsilon\omega^1, f^2 + \epsilon\omega^2, f^{1\prime} + \epsilon\omega^{1\prime}, f^{2\prime} + \epsilon\omega^{2\prime})\, dt,$$

where the primes indicate differentiation with respect to t. In order that the arc s of C be the minimum of the arcs of the curves through P_1 and P_2 in the neighborhood of C, it is necessary that the derivative of s_1 with respect to ϵ be zero for $\epsilon = 0$, that is,

$$\int_{t_1}^{t_2} \left(\frac{\partial\varphi}{\partial f^\alpha} \omega^\alpha + \frac{\partial\varphi}{\partial f^{\alpha\prime}} \omega^{\alpha\prime} \right) dt = 0.$$

Since it is understood that the derivatives involved are continuous in the interval t_1, t_2, on integrating by parts the second term in the integrand, we obtain in consequence of (32.23)

$$\int_{t_1}^{t_2} \omega^\alpha \left(\frac{\partial\varphi}{\partial f^\alpha} - \frac{d}{dt}\frac{\partial\varphi}{\partial f^{\alpha\prime}} \right) dt = 0.$$

Since this equation must hold for arbitrary functions ω^α such that (32.23) is satisfied, we have the equations of Euler

(32.26) $$\frac{d}{dt}\frac{\partial\varphi}{\partial f^{\alpha\prime}} - \frac{\partial\varphi}{\partial f^\alpha} = 0.$$

From the definition (32.24) of φ we have, noting that f^α is u^α,

$$\frac{\partial \varphi}{\partial f^{\alpha'}} = \frac{g_{\alpha\beta} f^{\beta'}}{\sqrt{g_{\alpha\beta} f^{\alpha'} f^{\beta'}}} = \frac{g_{\alpha\beta} \dfrac{du^\beta}{dt}}{\dfrac{ds}{dt}}, \qquad \frac{\partial \varphi}{\partial f^\alpha} = \frac{\dfrac{1}{2} \dfrac{\partial g_{\beta\gamma}}{\partial u^\alpha} \dfrac{du^\beta}{dt} \dfrac{du^\gamma}{dt}}{\dfrac{ds}{dt}}.$$

Substituting these expressions in (32.26), we obtain

$$g_{\alpha\beta} \frac{d^2 u^\beta}{dt^2} + \frac{\partial g_{\alpha\beta}}{\partial u^\gamma} \frac{du^\gamma}{dt} \frac{du^\beta}{dt} - \frac{1}{2} \frac{\partial g_{\beta\gamma}}{\partial u^\alpha} \frac{du^\beta}{dt} \frac{du^\gamma}{dt} - g_{\alpha\beta} \frac{du^\beta}{dt} \frac{\dfrac{d^2 s}{dt^2}}{\dfrac{ds}{dt}} = 0,$$

which is reducible by means of (28.2) to

$$g_{\alpha\beta} \left(\frac{d^2 u^\beta}{dt^2} + \begin{Bmatrix} \beta \\ \gamma\delta \end{Bmatrix} \frac{du^\gamma}{dt} \frac{du^\delta}{dt} - \frac{du^\beta}{dt} \frac{\dfrac{d^2 s}{dt^2}}{\dfrac{ds}{dt}} \right) = 0.$$

Since $g \neq 0$, it follows that

$$(32.27) \qquad \frac{d^2 u^\beta}{dt^2} + \begin{Bmatrix} \beta \\ \gamma\delta \end{Bmatrix} \frac{du^\gamma}{dt} \frac{du^\delta}{dt} - \frac{du^\beta}{dt} \frac{\dfrac{d^2 s}{dt^2}}{\dfrac{ds}{dt}} = 0.$$

When the parameter t is the arc these equations reduce to (32.8).

EXERCISES

1. The great circles on a sphere are its geodesics.

2. A necessary and sufficient condition that there exist upon a surface a family of geodesics whose orthogonal trajectories also are geodesics is that the surface be isometric with the plane.

3. The geodesics on a cylinder are helices (see §27, Ex. 2).

4. When the linear element of a surface of revolution is written in the form (see §24, Ex. 3)

$$(i) \qquad ds^2 = (1 + \varphi'^2)\, du^{1^2} + u^{1^2}\, du^{2^2},$$

from the second of equations (32.8) one obtains by integration $u^{1^2} \dfrac{du^2}{ds} = c$, where c is an arbitrary constant. From this result and (i) it follows that

$$\frac{du^1}{ds} = \pm \frac{\sqrt{u^{1^2} - c^2}}{u^1 \sqrt{1 + \varphi'^2}}, \qquad \frac{du^2}{ds} = \frac{c}{u^{1^2}}.$$

From this result one obtains that $u^2 = \pm c \int \dfrac{\sqrt{1+\varphi'^2}}{u^1\sqrt{u^{1^2}-c^2}}\, du^1 + d$, where d is a constant, is an equation of the geodesics.

5. When the linear element of a surface of revolution is written in the form $ds^2 = du^{1^2} + \psi^2(u^1)\, du^{2^2}$,

$$u^2 = \pm c \int \frac{du^1}{\psi\sqrt{\psi^2-c^2}} + d,$$

where c and d are constants, is an equation of the geodesics in the surface, and consequently in any surface isometric to it.

6. If a family of geodesics and their orthogonal trajectories on a surface form an isometric net, the surface is isometric with a surface of revolution.

7. A necessary and sufficient condition that $u^2 = \varphi(u^1)$ be an equation of a geodesic is that

$$\varphi'' - \left\{\begin{matrix}1\\22\end{matrix}\right\}\varphi'^3 + \left(\left\{\begin{matrix}2\\22\end{matrix}\right\} - 2\left\{\begin{matrix}1\\12\end{matrix}\right\}\right)\varphi'^2 + \left(2\left\{\begin{matrix}2\\12\end{matrix}\right\} - \left\{\begin{matrix}1\\11\end{matrix}\right\}\right)\varphi' + \left\{\begin{matrix}2\\11\end{matrix}\right\} = 0,$$

where the primes indicate differentiation with respect to u^1.

8. A necessary and sufficient condition that φ be a solution of the equation $\Delta_1\varphi = 1$ is that $ds^2 - d\varphi^2$ be a perfect square.

9. If $\varphi = a\theta_1 + \theta_2$, where θ_1 and θ_2 are functions of u^1 and u^2, are solutions of the equation $\Delta_1\varphi = 1$ for all values of the constant a, the curves $\theta_1 = $ const. are minimal curves, and the curves $\theta_2 = $ const. are geodesic parallels.

10. When the linear element of a spiral surface is written in the form (see §26, Ex. 9)

$$ds^2 = e^{2u^2}[du^{1^2} + U_1(u^1)\, du^{2^2}],$$

the equation $\Delta_1\varphi = 1$ admits the solution $e^{u^2}\psi(u^1)$, where ψ is any solution of the equation $\psi'^2 + \dfrac{\psi^2}{U_1} = 1$; by the integration of this equation one obtains all the geodesics in the surface.

11. The orthogonal trajectories of the curves $\theta(u^1, u^2) = $ const. are integral curves of the equation (see theorem [26.1])

$$g^{2\alpha}\theta_{,\alpha}\, du^1 - g^{1\beta}\theta_{,\beta}\, du^2 = 0.$$

The integral $\varphi(u^1, u^2) = $ const. of this equation is given by

$$\frac{\partial\varphi}{\partial u^1} = \frac{t}{\sqrt{g}}\,\epsilon_{1\gamma}g^{\gamma\alpha}\theta_{,\alpha}, \qquad \frac{\partial\varphi}{\partial u^2} = \frac{t}{\sqrt{g}}\,\epsilon_{2\gamma}g^{\gamma\alpha}\theta_{,\alpha},$$

where t is an integrating factor. If the curves $\theta = $ const. are geodesics, by a suitable choice of t the function φ is such that $\Delta_1\varphi = 1$. Hence t is given by

$$1 = \frac{t^2}{g}\,g^{\sigma\tau}\epsilon_{\sigma\gamma}g^{\gamma\alpha}\theta_{,\alpha}\,\epsilon_{\tau\delta}g^{\delta\beta}\theta_{,\beta}.$$

By means of §25, Ex. 8 the right-hand member of this equation reduces to $\dfrac{t^2}{g}\Delta_1\theta$

and consequently φ is given by the quadrature

$$\varphi = \int \frac{\sqrt{g}}{\sqrt{\Delta_1 \theta}} \left(g^{\alpha 2} \frac{\partial \theta}{\partial u^\alpha} du^1 - g^{\beta 1} \frac{\partial \theta}{\partial u^\beta} du^2 \right) \equiv \int \frac{1}{\sqrt{\Delta_1 \theta}} \epsilon_{\gamma \delta} g^{\alpha \delta} \theta_{,\alpha} du^\gamma.$$

12. If the integral curves of an equation $M_\alpha du^\alpha = 0$ are geodesics, their orthogonal trajectories are given by the quadrature

$$\varphi = \int \sqrt{g} \frac{(g^{\alpha 2} M_\alpha du^1 - g^{\beta 1} M_\beta du^2)}{\sqrt{g^{\alpha \beta} M_\alpha M_\beta}} = \int \frac{\epsilon_{\gamma \delta} g^{\alpha \delta} M_\alpha du^\gamma}{\sqrt{g^{\alpha \beta} M_\alpha M_\beta}}.$$

13. If $\dfrac{d\varphi}{du^1} = \psi(u^1, u^2, a)$, where a is a constant, is a first integral of the equation of Ex. 7, it follows from theorem [32.8] and Ex. 12 that the finite equations of the geodesics is

$$\frac{\partial}{\partial a} \int \frac{(g_{11} + g_{12}\psi) du^1 + (g_{12} + g_{22}\psi) du^2}{\sqrt{g_{11} + 2g_{12}\psi + g_{22}\psi^2}} = b.$$

14. Surfaces isometric with surfaces of revolution and the quadric surfaces (see §30, Exs. 6 and 7) are surfaces of Liouville.

15. A surface of constant Gaussian curvature is a surface of Liouville, in consequence of theorem [31.1] and §31, Ex. 3.

16. For a surface of Liouville with the linear element (32.19) the angle θ_0 which a geodesic makes with the curves $u^2 =$ const. is given by

$$A_1 \sin^2 \theta_0 - A'_2 \cos^2 \theta_0 + a = 0,$$

where a is the constant appearing in (32.20).

33. GEODESIC POLAR COORDINATES. GEODESIC TRIANGLES

In accordance with theorem [32.1] through a point P of a surface there passes a geodesic in each direction. Consider a domain about P such that no two geodesics through P meet again within the domain. When then the geodesics through P are taken for the parametric curves $u^2 =$ const. and their orthogonal trajectories the curves $u^1 =$ const. the linear element of the surface for the domain under consideration is of the form (32.15) by a suitable choice of the parameter u^1. If u^1 is replaced by u^1 plus a constant, the form (32.15) is unaltered. It follows from the discussion following (32.15) that u^1 may be chosen so that it is the distance along each geodesic from the point P. Thus each curve $u^1 =$ const. is the locus of a point at a constant distance from P, this distance being measured along the geodesics through P. In this sense the curves $u^1 =$ const. are *geodesic circles*.

Since at P $\dfrac{\partial x^i}{\partial u^2} = 0$, it follows from (24.7) that $g_{22} = 0$ at P, as is also $g_{12} = 0$. The former result follows also from the fact that the arc of a curve $u^1 =$ const. between two geodesics $u^2 = 0$ and $u^2 = c_1$ is given by $\int_0^{c_1} \sqrt{g_{22}}\, du^2$, and this quantity approaches zero in the limit as u^1 approaches zero.   In like manner the angle at P between these geodesics is given by

$$\lim_{u^1 \to 0} \frac{\int_0^{c_1} \sqrt{g_{22}}\, du^2}{u^1} = \lim_{u^1 \to 0} \frac{\int_0^{c_1} \frac{\partial \sqrt{g_{22}}}{\partial u^1}\, du^2}{1},$$

the right-hand member being obtained by the use of differentiation to evaluate the indeterminate form of the left-hand member.   Consequently a necessary and sufficient condition that the coordinate u^2 be the angle made at P with the geodesic $u^2 = 0$ by the coordinate curves $u^2 = \text{const.}$ is that $\left(\frac{\partial \sqrt{g_{22}}}{\partial u^1} \right)_{u^1=0} = 1$.   In this case the coordinates are called *geodesic polar coordinates*, because they are analogous to polar coordinates in the plane.   Hence we have

[33.1] *A necessary and sufficient condition that the coordinates in terms of which the linear element is*

$$(33.1) \qquad ds^2 = du^{1^2} + g_{22}\, du^{2^2}$$

be geodesic polar coordinates is that

$$(33.2) \qquad (g_{22})_{u^1=0} = 0, \qquad \left(\frac{\partial \sqrt{g_{22}}}{\partial u^1} \right)_{u^1=0} = 1.$$

Consider, for example, the sphere with the equations

$$x^1 = a \sin \frac{u^1}{a} \cos u^2, \qquad x^2 = a \sin \frac{u^1}{a} \sin u^2, \qquad x^3 = a \cos \frac{u^1}{a}.$$

The curves $u^2 = \text{const.}$ are the great circles through the point $(0, 0, a)$. Now $g_{11} = 1$, $g_{12} = 0$, $g_{22} = a^2 \sin^2 \frac{u^1}{a}$, which satisfy the conditions of theorem [33.1].

From §28, Ex. 1 we have that for the linear element in the form (33.1) the Gaussian curvature (28.5) of the surface is given by

$$(33.3) \qquad K = -\frac{1}{\sqrt{g_{22}}} \frac{\partial^2 \sqrt{g_{22}}}{\partial u^{1^2}}.$$

If $K = 0$, $\sqrt{g_{22}} = au^1 + b$, where a and b do not involve u^1.   In order that the conditions (33.2) be satisfied, we must have $a = 1$, $b = 0$, that is,

$$(33.4) \qquad ds^2 = du^{1^2} + u^{1^2} du^{2^2}.$$

This is the linear element of the plane in polar coordinates, and consequently, the surface is isometric with the plane. Conversely, if a surface is isometric with the plane, its linear element can be given the form (33.4). Thus we have another proof of theorem [28.1].

If $K \neq 0$ for a surface referred to polar geodesic coordinates, it follows from (33.3) and (33.2) that

$$\left(\frac{\partial^2 \sqrt{g_{22}}}{\partial u^{1^2}} \right)_{u^1=0} = 0,$$

(33.5)

$$K_0 = \frac{-\left(\dfrac{\partial^3 \sqrt{g_{22}}}{\partial u^{1^3}} \right)_{u^1=0}}{\left(\dfrac{\partial \sqrt{g_{22}}}{\partial u^1} \right)_{u^1=0}} = -\left(\frac{\partial^3 \sqrt{g_{22}}}{\partial u^{1^3}} \right)_{u^1=0},$$

where K_0 is the value of K at the pole of the coordinate system.

From (33.2) and (33.5) we have

$$\sqrt{g_{22}} = u^1 - \frac{1}{\lfloor 3} K_0 u^{1^3} + \cdots .$$

Hence the perimeter and area respectively of a geodesic circle of radius u^1 are given by

(33.6) $$\int_0^{2\pi} \sqrt{g_{22}} \, du^2 = 2\pi \left(u^1 - \frac{1}{\lfloor 3} K_0 u^{1^3} + \cdots \right),$$

(33.7) $$\int_0^{u^1} \int_0^{2\pi} \sqrt{g_{22}} \, du^1 \, du^2 = \pi \left(u^{1^2} - \frac{K_0 u^{1^4}}{12} + \cdots \right).$$

We consider next the integral of the absolute value of the Gaussian curvature over a geodesic triangle, that is, a triangle whose three sides are geodesics, the triangle being of such size that no two geodesics through a vertex meet again within or on the triangle, and such that that K has the same sign within and on the triangle. Such a triangle is shown in Fig. 9. We choose a geodesic polar coordinate system with pole at P. and with the sides PP_1 and PP_2 as the geodesics $u^2 = 0$ and $u^2 = \alpha$. In consequence of (33.3) the integral is

(33.8) $$I = e \iint K \sqrt{g} \, du^1 \, du^2 = -e \iint \frac{\partial^2 \sqrt{g_{22}}}{\partial u^{1^2}} \, du^1 \, du^2,$$

where e is $+1$ or -1 according as K is positive or negative. Integrating with respect to u^1 between the limits 0 and u^1, we have, in consequence of (33.2),

$$I = e \int \left(1 - \frac{\partial \sqrt{g_{22}}}{\partial u^1} \right) du^2.$$

From (32.14) we have along the geodesic P_1P_2

$$d\theta = -\frac{\partial \sqrt{g_{22}}}{\partial u^1} du^2,$$

and consequently

(33.9) $I = e\left(\int_0^\alpha du^2 + \int_{\pi-\beta}^\gamma d\theta\right) = e(\alpha + \beta + \gamma - \pi).$

We say that I defined by (33.8) measures the *total curvature of the triangle*.

From (33.9) it follows that for a geodesic triangle such that no two geodesics through at least one of the vertices meet again within or on the triangle, and such that the Gaussian curvature has the same sign at all points of the same, the total curvature of the triangle is equal to the excess over 180° of the sum of the angles of the triangle or to the deficit

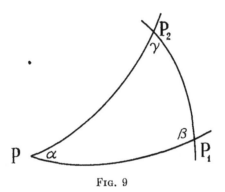

Fig. 9

from 180° according as the Gaussian curvature is positive or negative. If for a geodesic triangle geodesics through each vertex meet again within the triangle and thus there does not exist a polar geodesic system with one of the vertices as pole which system applies to the whole triangle, it is possible to subdivide the given triangle into a number n of smaller geodesic triangles, for each of which a polar geodesic system holds, in consequence of theorem [32.4]. Equation (33.9) applies to each of these triangles, and when the corresponding equations are added one obtains e times the sum of all the angles of the n triangles minus $n\pi$. The situation at all the vertices of the smaller triangles except at the vertices of the given triangle is the same as when a plane triangle is subdivided into n small triangles. In this case the sum of all the angles of $n\pi$, and since the sum of those at the vertices is π, it follows that the sum of all the others is $(n - 1)\pi$. Hence in the case of a geodesic triangle divided into n geodesic triangles when we subtract $(n - 1)\pi$

which is the sum of the angles not at the vertices of the given triangle we obtain the right-hand member of equation (33.9). Hence we have the celebrated theorem of Gauss*

[33.2] *The total curvature of a geodesic triangle such that at all points K has the same sign is equal to the excess over 180° of the sum of the angles of the triangle or the deficit from 180° according as K is positive or negative.*

As a consequence of this theorem we prove the following:

[33.3] *Two geodesics on a surface of negative curvature cannot meet in two points and enclose a simply connected area.*

For suppose two geodesics through a point A meet again in a point B, and enclose a simply connected area, as in Fig. 10. In consequence of theorem [32.4] it is possible to find two points C and D on the two geodesics sufficiently near to A so that there is a unique geodesic through C and D, thus forming two geodesic triangles ACD and BCD. The total curvature of these two triangles is equal to 2π minus the angles

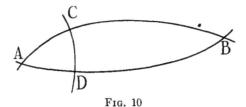

Fig. 10

at A, B, C and D. But the sum of the angles at C and D is 2π, and thus the total curvature of the two triangles is a negative quantity, whereas it is understood to be positive, as defined in (33.8). Hence the theorem is proved.

It follows from theorem [32.6] that if θ is a solution of $\Delta_1\theta = 1$, the curves $\theta = $ const. are geodesic parallels, and θ is the arc of the orthogonal geodesics measured from one of the parallels. If then we take two curves C_1 and C_2 not geodesically parallel, their geodesic parallels form a net on the surface. If u^1 and u^2 denote the geodesic distances of these parallels from C_1 and C_2 respectively, they are solutions of $\Delta_1\theta = 1$, and in the coordinate system u^α we have from (29.7)

(33.10) $$\frac{g_{22}}{g} = \frac{g_{11}}{g} = 1.$$

From these equations and (25.5) we have

$$g_{11} = g_{22} = \frac{1}{\sin^2\omega}, \qquad g_{12} = \frac{\cos\omega}{\sin^2\omega},$$

* 1827, 1, p. 246.

where ω is the angle of the two coordinate families of geodesic parallels. Hence the linear element of the surface is

$$(33.11) \qquad ds^2 = \frac{du^{1^2} + 2\cos\omega\, du^1\, du^2 + du^{2^2}}{\sin^2\omega}.$$

A similar result follows, if we take for coordinate curves geodesic circles with centers at two points F_1 and F_2 on the surface.

Conversely, equations (33.10) follow from (33.11), and consequently when the linear element of a surface is of the form (33.11), the coordinate curves are geodesic parallels which may be two families of geodesic circles.

If we put

$$(33.12) \qquad u'^1 = \tfrac{1}{2}(u^1 + u^2), \qquad u'^2 = \tfrac{1}{2}(u^1 - u^2),$$

a curve $u'^1 = $ const. or $u'^2 = $ const. is the locus of a point such that the sum or difference of the geodesic distances from C_1 and C_2, or the points F_1 and F_2 as the case may be, is a constant. In the latter case these curves are analogous to ellipses and hyperbolas in the plane. Accordingly they are called *geodesic ellipses* and *hyperbolas* not only in this case, but also when the distances are measured from curves C_1 and C_2. From (33.12) and (33.11) we find that in terms of the coordinates u'^α the linear element of the surface is

$$(33.13) \qquad ds^2 = \frac{(du'^1)^2}{\sin^2\dfrac{\omega}{2}} + \frac{(du'^2)^2}{\cos^2\dfrac{\omega}{2}}.$$

Conversely, by means of (33.12) the linear element (33.13) is transformed into (33.11), and consequently when the linear element of a surface is expressible in the form (33.13), that is, when $\dfrac{1}{g_{11}} + \dfrac{1}{g_{22}} = 1$, the coordinate curves are geodesic ellipses and hyperbolas. As in the case of confocal ellipses and hyperbolas in the plane, we have

[33.4] *A set of geodesic ellipses and hyperbolas form an orthogonal net.*

EXERCISES

1. The angles of any two families of geodesic parallels on a surface are bisected by the corresponding geodesic ellipses and hyperbolas.

2. In order that a coordinate system of geodesic ellipses and hyperbolas on a surface form an isometric orthogonal net, it is necessary and sufficient that the linear element be of the form

$$ds^2 = (U_1 + U_2)(du^{1^2} + du^{2^2}),$$

where U_1 and U_2 are positive functions of u^1 and u^2 respectively.

3. When the plane is referred ⟨ a system of confocal ellipses and hyperbolas, whose foci are at the distance $2a$ apart, the linear element can be written

$$ds^2 = (u^{1^2} - u^{2^2}) \left(\frac{du^{1^2}}{u^{1^2} - a^2} + \frac{du^{2^2}}{a^2 - u^{2^2}} \right).$$

4. A necessary and sufficient condition that an orthogonal coordinate net on a surface be a system of geodesic ellipses and hyperbolas is that

$$\frac{U_1}{g_{11}} + \frac{U_2}{g_{22}} = 1,$$

where U_1 and U_2 are positive functions of u^1 and u^2 respectively.

5. If an orthogonal coordinate net is to be a system of geodesic ellipses and hyperbolas in two ways it is necessary and sufficient that there be two sets of positive functions $U_{\alpha\beta}(\alpha = 1, 2)$, where $U_{\alpha 1}$ and $U_{\alpha 2}$ are functions of u^1 and u^2 respectively such that

(i)
$$\frac{U_{\alpha 1}}{g_{11}} + \frac{U_{\alpha 2}}{g_{22}} = 1;$$

in this case the linear element of the surface is reducible to the form in Ex. 2.

6. If the conditions (i) of Ex. 6 are satisfied, it follows that

$$\frac{tU_{11} + (1 - t) U_{21}}{g_{11}} + \frac{tU_{12} + (1 - t)U_{22}}{g_{22}} = 1,$$

which for t a constant is of the form (i); consequently, if an orthogonal net consists of a system of geodesic ellipses and hyperbolas in two ways, it is such a system in an endless number of ways.

7. When two families of geodesics on a surface meet under constant angle and their orthogonal trajectories are the curves $u^\alpha = $ const., the angle ω in (33.11) is a constant, and consequently the surface is isometric with the plane.

8. When g_{22} in theorem [33.1] is a function of u^1 alone, it follows from (33.2) and (33.3) that for a surface of constant Gaussian curvature one has (see §31, Ex. 3)

(i)
$$K = \frac{1}{a^2}, \qquad g_{22} = a^2 \sin^2 \frac{u^1}{a};$$

(ii)
$$K = -\frac{1}{a^2}, \qquad g_{22} = a^2 \sinh^2 \frac{u^1}{a}.$$

9. If C denotes the circumference of a geodesic circle of radius a, the Gaussian curvature K at the center of the circle is given by $\lim\limits_{a \to 0} \dfrac{3(2\pi a - C)}{\pi a^3}$, as follows from (33.6).

34. GEODESIC CURVATURE

We return to the consideration of equations (32.7) when μ^α is not a zero vector, that is, when the curve is not a geodesic. From (32.6) it follows that this vector is perpendicular to the tangent vector to the curve, that is, to the unit vector $\dfrac{du^\alpha}{ds}$. We replace μ^α in (32.7) by $\kappa_g \mu^\alpha$ and obtain

(34.1)
$$\frac{d^2 u^\alpha}{ds^2} + \left\{ \begin{matrix} \alpha \\ \beta\gamma \end{matrix} \right\} \frac{du^\beta}{ds} \frac{du^\gamma}{ds} = \kappa_g \mu^\alpha ,$$

where now it is understood that μ^α is the unit vector which makes a right angle with the unit vector $\dfrac{du^\alpha}{ds}$, that is, (see theorem [25.5]),

(34.2)
$$\frac{du^1}{ds} \mu^2 - \frac{du^2}{ds} \mu^1 = \frac{1}{\sqrt{g}}.$$

Since μ^α in (34.1) is a unit vector, the absolute value of κ_g is the length of the vector whose components are the left-hand members of equations (34.1). We call the vector μ^α the *curvature vector* of the curve at a point, and κ_g the *geodesic curvature* of the curve.

From (34.1) and (34.2) we have

(34.3)
$$\begin{aligned}
\kappa_g = \sqrt{g} &\left[\frac{du^1}{ds} \left(\frac{d^2 u^2}{ds^2} + \left\{ \begin{matrix} 2 \\ \beta\gamma \end{matrix} \right\} \frac{du^\beta}{ds} \frac{du^\gamma}{ds} \right) \right. \\
&\left. - \frac{du^2}{ds} \left(\frac{d^2 u^1}{ds^2} + \left\{ \begin{matrix} 1 \\ \beta\gamma \end{matrix} \right\} \frac{du^\beta}{ds} \frac{du^\gamma}{ds} \right) \right] \\
&= \epsilon_{\alpha\beta} \frac{du^\alpha}{ds} \left(\frac{d^2 u^\beta}{ds^2} + \left\{ \begin{matrix} \beta \\ \gamma\delta \end{matrix} \right\} \frac{du^\gamma}{ds} \frac{du^\delta}{ds} \right).
\end{aligned}$$

For a geodesic $\kappa_g = 0$ as follows from (34.1) and (32.8). Conversely, if $\kappa_g = 0$, it follows from the first form of (34.3) that

$$\frac{d^2 u^\alpha}{ds^2} + \left\{ \begin{matrix} \alpha \\ \beta\gamma \end{matrix} \right\} \frac{du^\beta}{ds} \frac{du^\gamma}{ds} = t \frac{du^\alpha}{ds} \qquad (\alpha = 1, 2),$$

where t is a factor to be determined. In consequence of (32.6) we have that $t = 0$, since $g_{\alpha\beta} \dfrac{du^\alpha}{ds} \dfrac{du^\beta}{ds} = 1$. Hence we have

[34.1] *A necessary and sufficient condition that a curve be a geodesic is that the geodesic curvature of the curve be zero.*

When the coordinate curves form an orthogonal net, for a curve $u^2 =$ const. we denote the geodesic curvature by κ_{g1} and take $\dfrac{du^1}{ds} = \dfrac{1}{\sqrt{g_{11}}}$, that is, we take the tangent vector in the direction of u^1 increasing. By means of §28, Ex. 1 we have from (34.3)

(34.4)
$$\kappa_{g1} = - \frac{1}{\sqrt{g_{22}}} \frac{\partial \log \sqrt{g_{11}}}{\partial u^2}.$$

In like manner for a curve $u^1 = $ const., if we take $\dfrac{du^2}{ds} = -\dfrac{1}{\sqrt{g_{22}}}$, the geodesic curvature $\kappa_{\varrho 2}$ is given by

$$(34.5) \qquad\qquad \kappa_{\varrho 2} = -\frac{1}{\sqrt{g_{11}}}\frac{\partial \log \sqrt{g_{22}}}{\partial u^1}.$$

From (34.2) and the choice $\dfrac{du^1}{ds} = \dfrac{1}{\sqrt{g_{11}}}$ in the first case and $\dfrac{du^2}{ds} = -\dfrac{1}{\sqrt{g_{22}}}$ in the second case it follows that in the first case μ^2 is positive, and in the second case μ^1 is positive. Hence in the first case the normal μ^α is tangent to the curve $u^1 = $ const. and in the direction in which u^2 is in-

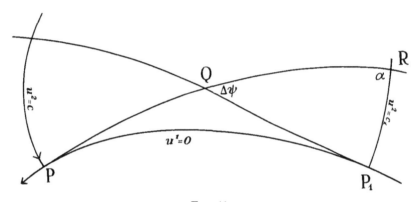

Fig. 11

creasing, and in the second case it is tangent to the curve $u^2 = $ const. and in the direction in which u^1 is increasing.

Consider now any curve C on the surface, and two nearby points P and P_1 on C. At P and P_1 draw the unique geodesics g and g_1 tangent to C, and denote by Q their point of intersection and by $\Delta\psi$ the angle under which they meet, as shown in Fig. 11. We shall prove the following theorem:

[34.2] *The limit of the ratio* $\dfrac{\Delta\psi}{\Delta s}$ *as P_1 approaches P along the curve is the geodesic curvature of C at P.*

In order to prove this theorem we take the curve C as the curve $u^1 = 0$ of an orthogonal coordinate net. The tangent geodesic at P makes with the curve $u^2 = c \ (>c_1)$ the angle $3\pi/2$ and with the curve $u^2 = c_1$ the angle $\pi + \alpha$, hence $\Delta\theta_0 = \alpha - \pi/2$. From (33.9) applied to the geodesic triangle QP_1R we have

$$I = e\left(\Delta\psi + \frac{\pi}{2} + \alpha - \pi\right) = e(\Delta\psi + \Delta\theta_0).$$

As P_1 approachs P along the curve, $\dfrac{I}{\Delta s}$ approaches zero and consequently $\dfrac{\Delta\psi}{\Delta s}$ approaches $-\dfrac{d\theta_0}{ds}$, the latter being given by (32.14). Hence we have

$$\lim \frac{\Delta\psi}{\Delta s} = -\frac{d\theta_0}{ds} = \frac{\sqrt{g}}{g_{11}}\begin{Bmatrix}2\\12\end{Bmatrix}\frac{du^2}{ds} = -\frac{1}{\sqrt{g_{11}}}\frac{\partial\log\sqrt{g_{22}}}{\partial u^1},$$

and thus in consequence of (34.5) the theorem is proved.

In view of this result geodesic curvature of a curve on a surface is a generalization of curvature of a plane curve, and from theorem [34.1] we have that a geodesic is a generalization of a straight line in the plane. The geometric significance of geodesic curvature as viewed from the enveloping euclidean space is established in §44.

We shall show in what manner the quantity κ_g for a curve $\varphi(u^1, u^2) = $ const., is expressible in terms of differential parameters of φ. To this end we observe that for the coordinate system used in establishing equation (34.4) we have from (29.7), (29.2), and (29.16)

$$\Delta_1 u^2 = \frac{1}{g_{22}}, \quad \Delta_1(u^2, \sqrt{g_{22}}) = \frac{1}{g_{22}}\frac{\partial\sqrt{g_{22}}}{\partial u^2}, \quad \Delta_2 u^2 = \frac{1}{\sqrt{g_{11}g_{22}}}\frac{\partial}{\partial u^2}\sqrt{\frac{g_{11}}{g_{22}}}.$$

By means of these expressions equation (34.4) can be written in the form

$$\kappa_{g1} = -\left[\frac{\Delta_2 u^2}{\sqrt{\Delta_1 u^2}} + \Delta_1\left(u^2, \frac{1}{\sqrt{\Delta_1 u^2}}\right)\right].$$

In any other coordinate system the curves $u^2 = $ const. are defined by an equation $\varphi(u^1, u^2) = $ const., and consequently we have the theorem of Beltrami:[*]

[34.3] *The geodesic curvature of a curve $\varphi = $ const. is given by*

$$(34.6) \qquad \kappa_g = -\left[\frac{\Delta_2\varphi}{\sqrt{\Delta_1\varphi}} + \Delta_1\left(\varphi, \frac{1}{\sqrt{\Delta_1\varphi}}\right)\right].$$

By means of (29.16) the quantity in brackets is equal to

$$\frac{\dfrac{\partial}{\partial u^\beta}(\sqrt{g}\,g^{\alpha\beta}\varphi_{,\alpha})}{\sqrt{g}\,\sqrt{g^{\alpha\beta}\varphi_{,\alpha}\varphi_{,\beta}}} + g^{\alpha\beta}\varphi_{,\alpha}\frac{\partial}{\partial u^\beta}\left(\frac{1}{\sqrt{g^{\gamma\delta}\varphi_{,\gamma}\varphi_{,\delta}}}\right),$$

* 1865, 1, p. 83.

and consequently (34.6) may be written in the following form due to Bonnet:*

$$(34.7) \qquad \kappa_g = - \frac{1}{\sqrt{g}} \frac{\partial}{\partial u^\beta} \left(\frac{\sqrt{g}\, g^{\alpha\beta}\, \varphi_{,\alpha}}{\sqrt{g^{\gamma\delta}\, \varphi_{,\gamma}\, \varphi_{,\delta}}} \right).$$

In particular, if κ_{g1} and κ_{g2} denote the geodesic curvatures of the curves $u^2 = $ const. and $u^1 = $ const. respectively of any coordinate system, we have (see (24.18))

$$(34.8) \qquad
\begin{aligned}
\kappa_{g\alpha} &= - \frac{1}{\sqrt{g}} \frac{\partial}{\partial u^\gamma} \left(\frac{\sqrt{g}\, g^{\beta\gamma}}{\sqrt{g^{\beta\beta}}} \right) \qquad (\beta \neq \alpha) \\
&= \frac{1}{\sqrt{g}} \left(\frac{\partial}{\partial u^\alpha} \frac{g_{\alpha\beta}}{\sqrt{g_{\alpha\alpha}}} - \frac{\partial}{\partial u^\beta} \sqrt{g_{\alpha\alpha}} \right),
\end{aligned}$$

of which (34.4) and (34.5) are particular cases when the coordinate curves form an orthogonal net.

Equation (32.13) gives the rate of change with s of the angle θ_0 which a curve makes with the curves $u^2 = $ const. By means of this result the geodesic curvature of the curve, as given by (34.3), is expressible in the form

$$(34.9) \qquad \kappa_g = \frac{d\theta_0}{ds} + \frac{\sqrt{g}}{g_{11}} \begin{Bmatrix} 2 \\ \alpha 1 \end{Bmatrix} \frac{du^\alpha}{ds}.$$

With the aid of this equation we establish an important result due to Bonnet. To this end we use the formula of Green, namely

$$(34.10) \qquad \iint_S \left(\frac{\partial \mu_2}{\partial u^1} - \frac{\partial \mu_1}{\partial u^2} \right) du^1\, du^2 = \int_C \mu_1\, du^1 + \mu_2\, du^2.$$

In this formula the surface integral is applied to a simply connected portion of the surface, and the curvilinear integral to the contour C, the positive sense of the latter being such that as a point describes C in this sense the portion of the surface is on the left.† It is understood that μ_1, μ_2 and their first derivatives are finite and continuous within and on the contour.

From (34.10) and (34.9) we have

$$(34.11) \qquad
\begin{aligned}
\int_C d\theta_0 - \int_C \kappa_g\, ds &= - \int_C \frac{\sqrt{g}}{g_{11}} \left(\begin{Bmatrix} 2 \\ 11 \end{Bmatrix} du^1 + \begin{Bmatrix} 2 \\ 12 \end{Bmatrix} du^2 \right) \\
&= \iint_S \left[\frac{\partial}{\partial u^2} \left(\frac{\sqrt{g}}{g_{11}} \begin{Bmatrix} 2 \\ 11 \end{Bmatrix} \right) - \frac{\partial}{\partial u^1} \left(\frac{\sqrt{g}}{g_{11}} \begin{Bmatrix} 2 \\ 12 \end{Bmatrix} \right) \right] du^1\, du^2.
\end{aligned}$$

* 1860, 1, p. 166.
† Fine, 1927, 1, p. 337.

Since the left-hand member of this equation is independent of the co-ordinate system u^α, the same is true of the right-hand member. By (25.21) the element of area in any coordinate system is $d\sigma = \sqrt{g}\, du^1\, du^2$. If then the coordinate system is such that the linear element is of the form (33.1), one finds by means of §28, Ex. 1 that the quantity in brackets in (34.11) multiplied by $1/\sqrt{g}$ reduces to $-\dfrac{1}{\sqrt{g_{22}}}\dfrac{\partial^2 \sqrt{g_{22}}}{\partial u^{1^2}}$, as the reader should verify. From this result and (33.3) it follows that in any coordinate system the integrand of the right-hand member of (34.11) is $K\, d\sigma$.

If the contour C consists of arcs of curves forming a curved polygon with exterior angles $\theta_1, \cdots, \theta_p$, the first integral in (34.11) is equal to $2\pi - \overset{1\cdots p}{\underset{i}{\sum}} \theta_i$. Hence we have the *Gauss-Bonnet theorem:**

[34.4] *For a simply connected portion S of a surface for which the Gaussian curvature K is finite and continuous and the geodesic curvature κ_g of the contour C is finite and continuous*

$$(34.12) \qquad \int_C \kappa_g\, ds + \iint_S K \sqrt{g}\, du^1\, du^2 = 2\pi - \overset{1\cdots p}{\underset{i}{\sum}} \theta_i,$$

where $\theta_1, \cdots, \theta_p$ are the exterior angles at the vertices of the contour C if any.

When the coordinates u^α undergo any transformation, the quantities μ_α in the right-hand member of (34.10) are seen to be the covariant components of a vector. The integrand in the left-hand member of (34.10) is equal to $(\mu_{2,1} - \mu_{1,2})du^1 du^2$. In terms of the quantities $\epsilon^{\alpha\beta}$ defined in (25.14) equation (34.10) may be written

$$(34.13) \qquad \int_C \mu_\alpha\, du^\alpha = - \iint_S \epsilon^{\alpha\beta} \mu_{\alpha,\beta} \sqrt{g}\, du^1\, du^2.$$

If λ^α is any vector, then μ_α defined by

$$(34.14) \qquad \mu_\alpha = \epsilon_{\beta\alpha} \lambda^\beta$$

are the covariant components of a vector since $\epsilon_{\beta\alpha}$ are the covariant components of a tensor (see theorem [25.4]). When the expressions from (34.14) are substituted in the left-hand member of (34.10), we

* See 1848, 1, p. 131.

obtain in consequence of (25.14), (28.2) and §28, Ex. 12

$$\iint_S \left[\frac{\partial}{\partial u^1} (\epsilon_{\beta 2} \lambda^\beta) - \frac{\partial}{\partial u^2} (\epsilon_{\beta 1} \lambda^\beta) \right] du^1 \, du^2 = \iint_S \frac{\partial}{\partial u^\alpha} (\lambda^\alpha \sqrt{g}) \, du^1 \, du^2$$

$$= \iint_S \lambda^\alpha_{,\alpha} \sqrt{g} \, du^1 \, du^2 .$$

Hence another form of the formula of Green is

$$(34.15) \qquad \iint_S \lambda^\alpha_{,\alpha} \sqrt{g} \, du^1 \, du^2 = \int_C \epsilon_{\beta\alpha} \lambda^\beta \, du^\alpha = - \int_C \nu_\beta \lambda^\beta \, ds,$$

where

$$(34.16) \qquad\qquad \nu_\beta = \epsilon_{\alpha\beta} \frac{du^\alpha}{ds} ;$$

thus ν_β are the covariant components of the unit vector which makes a right-angle with the unit vector $\dfrac{du^\alpha}{ds}$ by theorem [25.6], and whose direction depends upon the sense of rotation (§25).

EXERCISES

1. The parallels on a surface of revolution are curves of constant geodesic curvature.

2. A small circle on a sphere has constant geodesic curvature.

3. When a surface has an orthogonal net such that the curves of one family are geodesics, and those of the other family have constant ($\neq 0$) geodesic curvature, the surface is isometric with a surface of revolution.

4. For a family of loxodromic curves upon a surface of revolution (see §26, Ex. 2), that is, curves which make the same constant angle with the meridians, the geodesic curvature of all these curves is the same at their points of intersection with a parallel.

5. For a surface with the linear element

$$ds^2 = \frac{du^{1^2} + du^{2^2}}{(U_1 + U_2)^2},$$

where U_α is a function of u^α alone, $\kappa_{g1} = U'_2$, $\kappa_{g2} = U'_1$, where the primes indicate differentiation with respect to the argument, that is, the coordinate curves have constant geodesic curvature. Conversely, when the curves of a coordinate orthogonal net have constant geodesic curvature the linear element is reducible to the above form.

6. If the curves of one family of an isometric orthogonal net have constant geodesic curvature, the curves of the other family have the same property.

7. The geodesic curvature of the integral curves of the equation $M_\alpha \, du^\alpha = 0$ is given by

$$\kappa_g = -\frac{1}{\sqrt{g}} \frac{\partial}{\partial u^\beta} \left(\frac{\sqrt{g} \, g^{\alpha\beta} M_\alpha}{\sqrt{g^{\gamma\delta} M_\gamma M_\delta}} \right).$$

8. Derive the theorem of Gauss [33.2] from theorem [34.4].

9. From (34.8) and (25.5) one has

$$\kappa_{g\alpha} = \frac{1}{\sqrt{g}} \left[\frac{\partial}{\partial u^\alpha} (\cos \omega \sqrt{g_{\beta\beta}}) - \frac{\partial \sqrt{g_{\alpha\alpha}}}{\partial u^\beta} \right] \qquad (\beta \neq \alpha).$$

10. From the discussion of (34.11) it follows that

$$K = \frac{1}{\sqrt{g}} \left[\frac{\partial}{\partial u^2} \left(\frac{\sqrt{g}}{g_{11}} \begin{Bmatrix} 2 \\ 11 \end{Bmatrix} \right) - \frac{\partial}{\partial u^1} \left(\frac{\sqrt{g}}{g_{11}} \begin{Bmatrix} 2 \\ 12 \end{Bmatrix} \right) \right];$$

this should be verified by means of (28.5), (28.2) and (20.13).

11. From §28, Ex. 10 and (34.8) one obtains the *formula of Liouville*

$$K = \frac{1}{\sqrt{g}} \left[\frac{\partial^2 \omega}{\partial u^1 \partial u^2} + \frac{\partial}{\partial u^1} (\sqrt{g_{22}} \, \kappa_{g2}) + \frac{\partial}{\partial u^2} (\sqrt{g_{11}} \, \kappa_{g1}) \right].$$

12. When the coordinate curves form an orthogonal net, the formula of Liouville may be written

$$K = \frac{1}{\sqrt{g_{11}}} \frac{\partial}{\partial u^1} \kappa_{g2} + \frac{1}{\sqrt{g_{22}}} \frac{\partial}{\partial u^2} \kappa_{g1} - \kappa_{g1}^2 - \kappa_{g2}^2,$$

and equation (34.9) may be written (see §28, Ex. 1)

$$\kappa_g = \frac{d\theta_0}{ds} + \cos\theta_0 \kappa_{g1} - \sin\theta_0 \kappa_{g2}.$$

13. From (29.2) and (29.14) one has

$$\iint_S \Delta_1(\varphi, \psi) \, d\sigma = \iint_S [(g^{\alpha\beta} \varphi_{,\alpha} \psi)_{,\beta} - \psi \Delta_2 \varphi] \, d\sigma,$$

from which and (34.15) it follows that

$$\iint_S \Delta_1(\varphi, \psi) \, d\sigma + \iint_S \psi \Delta_2 \varphi \, d\sigma = - \int_C \psi g^{\alpha\beta} \varphi_{,\alpha} \nu_\beta \, ds.$$

14. From (29.16) and §28, Ex. 12 one has

$$\iint_S \Delta_2 \varphi \sqrt{g} \, du^1 \, du^2 = \iint_S \frac{\partial}{\partial u^\alpha} (\sqrt{g} \, g^{\alpha\beta}, \varphi_\beta) \, du^1 \, du^2 = \iint_S (g^{\alpha\beta} \varphi_{,\beta})_{,\alpha} \sqrt{g} \, du^1 \, du^2,$$

and from this result and (34.15)

$$\iint_S \Delta_2 \varphi \, d\sigma = - \int_C \nu_\alpha g^{\alpha\beta} \varphi_{,\beta} \, ds.$$

35. THE VECTOR ASSOCIATE TO A GIVEN VECTOR WITH RESPECT TO A CURVE. PARALLELISM OF VECTORS

Let C be any curve upon a surface defined by $u^\alpha = f^\alpha(s)$, where s is the arc of C, and let λ^α be the components of a family of unit vectors one at each point of C, λ^α being functions of s. Since the vectors are unit vectors,

$$(35.1) \qquad\qquad g_{\alpha\beta}\lambda^\alpha\lambda^\beta = 1.$$

Differentiating this equation with respect to s and making use of (28.2), we obtain

$$(35.2) \qquad\qquad g_{\alpha\beta}\lambda^\beta\left(\frac{d\lambda^\alpha}{ds} + \lambda^\gamma\begin{Bmatrix}\alpha\\\gamma\delta\end{Bmatrix}\frac{du^\delta}{ds}\right) = 0.$$

The components λ'^α of these vectors in any other coordinate system u'^α are given by

$$(35.3) \qquad\qquad \lambda^\alpha = \lambda'^\mu\frac{\partial u^\alpha}{\partial u'^\mu}, \qquad \lambda'^\mu = \lambda^\alpha\frac{\partial u'^\mu}{\partial u^\alpha}.$$

Differentiating the first of these equations with respect to s and making use of (32.1), we may write the resulting equation in the form

$$(35.4) \qquad \frac{d\lambda^\alpha}{ds} + \lambda^\gamma\begin{Bmatrix}\alpha\\\gamma\beta\end{Bmatrix}\frac{du^\beta}{ds} = \frac{\partial u^\alpha}{\partial u'^\mu}\left(\frac{d\lambda'^\mu}{ds} + \lambda'^\nu\begin{Bmatrix}\mu\\\nu\sigma\end{Bmatrix}'\frac{du'^\sigma}{ds}\right).$$

Hence, if we put

$$(35.5) \qquad\qquad \frac{d\lambda^\alpha}{ds} + \lambda^\gamma\begin{Bmatrix}\alpha\\\gamma\beta\end{Bmatrix}\frac{du^\beta}{ds} = \nu^\alpha,$$

it follows from (35.4) that ν^α are the contravariant components of a vector, and from (35.2) that the vector ν^α at a point is perpendicular to the vector λ^α at the point. Following Bianchi* we call ν^α the *vector associate to* λ^α *with respect to the curve* C.

When in particular the vector λ^α is the tangent vector $\dfrac{du^\alpha}{ds}$, ν^α is the vector $\kappa_g\mu^\alpha$ in equation (34.1), and we have

[35.1] *The vector associate to the tangent vector with respect to a curve is the geodesic curvature vector of the curve.*

Following McConnell† we call the left-hand member of (35.5) the

---

* 1922, 1, p. 161.
† 1931, 1, p. 179–181; see also Synge, 1926, 2.

intrinsic derivative of λ^α for the curve C and denote it by $\dfrac{\delta\lambda^\alpha}{\delta s}$, that is,

$$(35.6) \qquad \frac{\delta\lambda^\alpha}{\delta s} \equiv \frac{d\lambda^\alpha}{ds} + \lambda^\gamma \begin{Bmatrix} \alpha \\ \gamma\beta \end{Bmatrix} \frac{du^\beta}{ds}.$$

Thus the intrinsic derivative of a contravariant vector for a curve is a contravariant vector. In like manner by means of (32.1) it can be shown that the intrinsic derivative of a covariant vector λ_α defined by

$$(35.7) \qquad \frac{\delta\lambda_\alpha}{\delta s} \equiv \frac{d\lambda_\alpha}{ds} - \lambda_\gamma \begin{Bmatrix} \gamma \\ \alpha\beta \end{Bmatrix} \frac{du^\beta}{ds}$$

is a covariant vector.

If λ^α and λ_α are the contravariant and covariant components of a vector defined at all points of a surface, then for a curve one has

$$(35.8) \qquad \frac{\delta\lambda^\alpha}{\delta s} = \lambda^\alpha_{,\beta} \frac{du^\beta}{ds}, \qquad \frac{\delta\lambda_\alpha}{\delta s} = \lambda_{\alpha,\beta} \frac{du^\beta}{ds}.$$

In like manner if one has a tensor $a^{\alpha_1\cdots\alpha_r}_{\beta_1\cdots\beta_s}$ defined at all points of a surface, at points of a curve the quantities

$$(35.9) \qquad \frac{\delta a^{\alpha_1\cdots\alpha_r}_{\beta_1\cdots\beta_s}}{\delta s} \equiv a^{\alpha_1\cdots\alpha_r}_{\beta_1\cdots\beta_s,\gamma} \frac{du^\gamma}{ds}$$

are the components of a tensor of the same order. The results concerning intrinsic differentiation follow from equations (32.1), just as the results concerning covariant differentiation in §22 are a consequence of the corresponding equations (20.11). Hence for intrinsic differentiation the rules of the ordinary calculus as regards the differentiation of the sum, difference and product of quantities apply (see §22). Also just as the covariant derivative of a scalar is the ordinary derivative of the scalar, so the intrinsic derivative of a scalar is the ordinary derivative.

From the above results and theorems [28.3] and [28.4] we have

[35.2] *For any curve*

$$(35.10) \qquad \frac{\delta g_{\alpha\beta}}{\delta s} = 0, \qquad \frac{\delta g^{\alpha\beta}}{\delta s} = 0, \qquad \frac{\delta \epsilon_{\alpha\beta}}{\delta s} = 0, \qquad \frac{\delta \epsilon^{\alpha\beta}}{\delta s} = 0.$$

From (25.16) we have that the angle θ made with the unit vector $\dfrac{du^\alpha}{ds}$ tangent to a curve C by a unit vector λ^α at all points of C and not tangent to C is given by

$$(35.11) \qquad \sin\theta = \epsilon_{\alpha\beta} \frac{du^\alpha}{ds} \lambda^\beta.$$

In consequence of (35.10), (34.1) and (35.5) the intrinsic derivative of equation (35.11) is reducible to

$$(35.12) \qquad \cos \theta \, \frac{\delta \theta}{\delta s} = \epsilon_{\alpha\beta} \left(\kappa_g \mu^\alpha \lambda^\beta + \frac{du^\alpha}{ds} \nu^\beta \right).$$

When C is a geodesic, that is, when $\kappa_g = 0$, the condition that the angle θ be a constant is

$$\epsilon_{\alpha\beta} \frac{du^\alpha}{ds} \nu^\beta = 0,$$

that is, either ν^β is a zero vector ($\nu^\beta = 0$), or the vector ν^β is tangent to C. In the latter case we should have from (35.2) $g_{\alpha\beta}\lambda^\alpha \dfrac{du^\beta}{ds} = 0$, the intrinsic derivative of which is $g_{\alpha\beta}\nu^\alpha \dfrac{du^\beta}{ds}$; that is, ν^α is normal to C. Hence $\nu^\alpha = 0$ and we have

[35.3] *The unit vectors* λ^α *at points of a geodesic make equal angles with the geodesic, if and only if*

$$(35.13) \qquad \frac{\delta \lambda^\alpha}{\delta s} \equiv \frac{d\lambda^\alpha}{ds} + \lambda^\gamma \begin{Bmatrix} \alpha \\ \gamma\beta \end{Bmatrix} \frac{du^\beta}{ds} = 0.$$

When the surface is a plane, in which case a geodesic is a straight line, the vectors λ^α are parallel in the euclidean sense. Accordingly it is a natural generalization to say that the family of vectors λ^α satisfying (35.13) are parallel with respect to the geodesic involved.

Although we have introduced the notion of the parallelism of vectors at points of a geodesic, we make a further generalization and say that given any curve, where u^α are functions of the arc, the functions λ^α satisfying (35.13) are the components of a family of *parallel vectors with respect to the curve*; this concept is due to Levi-Civita (see §45). From the theory of differential equations of the form (35.13) it follows that any such family of parallel vectors is completely determined by the values of the components at some point of the curve, that is, by initial values of λ^α for the value of s at the point. Since equations (35.13) involve the equations of the curve, it follows that if one has two curves intersecting at two points P_1 and P_2, and finds with respect to each of the curves the families of parallel vectors having the same values at P_1, in general the vectors at P_2 will be different for the two families. Thus parallelism as just defined is *relative to a curve*. When the surface is a plane referred to cartesian coordinates, equations (35.13) reduce to $\dfrac{d\lambda^\alpha}{ds} = 0$, that is, λ^α are constants, and parallelism is *absolute*, that is, it is not relative to a curve. In this case the vectors are parallel in the euclidean sense.

Since for $\lambda^\alpha = \dfrac{du^\alpha}{ds}$ equations (35.13) are equations of geodesics, we have that the tangents to a geodesic are parallel with respect to the curve, and in this sense geodesics may be called the straightest lines on the surface. In this connection it is interesting to observe that for the domain about a point P within which no two geodesics through P meet again coordinates u^α can be chosen in terms of which equations of the geodesics through P are given by (32.12), that is,

$$\bar{u}^\alpha = s\left(\frac{du^\alpha}{ds}\right)_0,$$

where u^α is a given set of coordinates on the surface. Thus in this coordinate system equations of the geodesics through P are of the form of equations of straight lines in the plane referred to cartesian coordinates. The equations of geodesics not through P do not have this simple form in this coordinate system.

Since equations (35.13) involve only the first fundamental form of a surface, parallelism is an intrinsic property of a surface. As a consequence we have

[35.4] *If λ^α are the components of a family of parallel vectors with respect to a curve on a surface, they are the components of vectors parallel with respect to the corresponding curve on a surface isometric with the given surface.*

If λ^α are the contravariant components of a unit vector-field such that

$$(35.14) \qquad\qquad \lambda^\alpha_{,\beta} = 0,$$

the vectors of the field at two points P_1 and P_2 are parallel with respect to every curve through the two points and thus are absolutely parallel. In particular, for the plane referred to cartesian coordinates equations (35.14) reduce to $\dfrac{\partial \lambda^\alpha}{\partial u^\beta} = 0$, that is, $\lambda^\alpha = $ const. The conditions of integrability of equations (35.14), namely the Ricci identities (see (22.22))

$$(35.15) \qquad \lambda^\alpha_{,\beta\gamma} - \lambda^\alpha_{,\gamma\beta} = -\lambda^\delta R^\alpha_{\delta\beta\gamma} = -\lambda^\delta g^{\alpha\epsilon} R_{\epsilon\delta\beta\gamma},$$

are satisfied identically for a plane, or any surface isometric with the plane. In this case equations (35.14) are completely integrable, that is, a field of absolutely parallel vectors is determined in any coordinate system by initial values of λ^α.

Conversely, in order that there be a field of absolutely parallel unit vectors in a surface, its contravariant components must be solutions of equations (35.14), and consequently from (35.15) one has that $\lambda^\delta R_{\epsilon\delta\beta\gamma} = 0$, which equations are equivalent to (see (28.3))

$$\lambda^1 R_{2112} = 0, \qquad \lambda^2 R_{1212} = 0.$$

Hence $R_{1212} = 0$ and we have

[35.5] *A necessary and sufficient condition that there be a field of absolutely parallel unit vectors in a surface is that the surface be isometric with the plane.*

In the plane the angle of two vectors not at the same points is by definition the angle between either vector at its point of application and a vector at this point parallel to the other vector. In consequence of theorem [35.5] this definition does not apply to a general surface. However, we can speak of the angle of two vectors relative to a curve, this angle being the angle between either vector at its point of application and a vector at this point parallel to the other vector with respect to the curve. The angles thus formed at each point are equal, these being the angles referred to in Ex. 2.

We return to the consideration of equations (35.5) when $\nu^\alpha \neq 0$, and recall that the vector ν^α is perpendicular to the vector λ^α. We write equations (35.5) in the form

$$(35.16) \qquad \frac{\delta\lambda^\alpha}{\delta s} \equiv \frac{d\lambda^\alpha}{ds} + \lambda^\gamma \begin{Bmatrix} \alpha \\ \gamma\beta \end{Bmatrix} \frac{du^\beta}{ds} = r\nu^\alpha,$$

where now ν^α is the unit vector which makes a right angle with the vector λ^α. In this case the absolute value of r is the length of the associate vector.

Differentiating intrinsically the equation

$$g_{\alpha\beta}\lambda^\alpha\nu^\beta = 0,$$

we have in consequence of (35.10) and (35.16)

$$g_{\alpha\beta}\lambda^\alpha \frac{\delta\nu^\beta}{\delta s} + r = 0.$$

Since ν^α is a unit vector and $-\lambda^\alpha$ makes a right angle with it, we have analogously to (35.16) $\dfrac{\delta\nu^\beta}{\delta s} = -r'\lambda^\beta$. On substitution in the above equation we find that $r' = r$. Hence we have

(35.17) $$\frac{\delta \nu^\alpha}{\delta s} \equiv \frac{d\nu^\alpha}{ds} + \nu^\gamma \begin{Bmatrix} \alpha \\ \gamma\beta \end{Bmatrix} \frac{du^\beta}{ds} = - r\lambda^\alpha.$$

When, in particular, λ^α is the unit tangent vector to a curve, ν^α is the curvature vector of the curve, and r is the geodesic curvature. Consequently we have

[35.6] *If λ^α and μ^α are the contravariant components of the unit tangent vector and unit curvature vector of a curve on a surface, then*

(35.18)
$$\frac{\delta \lambda^\alpha}{\delta s} \equiv \frac{d\lambda^\alpha}{ds} + \lambda^\gamma \begin{Bmatrix} \alpha \\ \gamma\beta \end{Bmatrix} \frac{du^\beta}{ds} = \kappa_g \mu^\alpha,$$

$$\frac{\delta \mu^\alpha}{\delta s} \equiv \frac{d\mu^\alpha}{ds} + \mu^\gamma \begin{Bmatrix} \alpha \\ \gamma\beta \end{Bmatrix} \frac{du^\beta}{ds} = - \kappa_g \lambda^\alpha.$$

These are the Frenet formulas of a curve in a surface (see §21).

If the unit vectors $\lambda^\alpha(s)$ are parallel with respect to a curve C, the angle θ which λ^α makes with a unit vector $\mu^\alpha(s)$ at each point of C is given by (see (25.16))

$$\sin \theta = \epsilon_{\alpha\beta} \mu^\alpha \lambda^\beta.$$

Taking the intrinsic derivative of this equation and making use of the first of (25.9) and (35.13), one obtains

$$\lambda^\beta \left(\mu_\beta \frac{d\theta}{ds} - \epsilon_{\alpha\beta} \frac{\delta \mu^\alpha}{\delta s} \right) = 0.$$

In consequence of Ex. 2 the expression in parentheses is independent of the choice of the parallel vectors λ^α, and consequently

$$\frac{d\theta}{ds} = \epsilon_{\alpha\beta} \frac{\delta \mu^\alpha}{\delta s} \mu^\beta.$$

When, in particular, μ^α is the unit tangent vector to C, this equation becomes, in consequence of theorems [35.1] and [25.5],

(35.19) $$\frac{d\theta}{ds} = - \kappa_g,$$

where κ_g is the geodesic curvature of C. Hence we have*

[35.7] *If a vector undergoes a parallel displacement along a curve C the arc-rate of change of the angle which the vector makes with the curve is the negative of the geodesic curvature of C.*

* See 1927, 2, p. 136.

Consider a smooth closed curve C in a surface, enclosing a simply connected region, and a vector λ at a point P of the curve. Let this vector be displaced parallel to itself around C in the positive sense, and denote by $\bar\lambda$ its final position at P, and by φ the angle $\bar\lambda$ makes with λ. This angle is equal to its rotation relative to the tangent to C plus the rotation of the tangent about C, namely 2π. In consequence of (35.19) we have

$$(35.20) \qquad \varphi = 2\pi - \int_C \kappa_g \, ds = \iint_S K\sqrt{g} \, du^1 \, du^2,$$

the last expression being a consequence of theorem [34.4]. We observe that this result is independent of the position of the point P on C and of the vector λ.

EXERCISES

1. The covariant components λ_α of a set of unit vectors parallel with respect to a curve are solutions of the equation

$$\frac{d\lambda_\alpha}{ds} - \lambda_\gamma \begin{Bmatrix} \gamma \\ \alpha\beta \end{Bmatrix} \frac{du^\beta}{ds} = 0.$$

2. For two families of unit vectors λ_1^α and λ_2^α parallel with respect to a curve the angle between the vectors at a point is the same for all points of the curve.

3. A necessary and sufficient condition that the unit tangent vectors to the curves $u^\alpha = $ const. for $\alpha = 1$ or 2 be parallel with respect to a curve C is that the latter be an integral curve of the equation

$$\begin{Bmatrix} \alpha \\ \beta\gamma \end{Bmatrix} du^\gamma = 0 \qquad\qquad\qquad (\beta \neq \alpha).$$

4. The unit tangent vectors to a family of geodesic parallels (§32) at points of intersection with each of the orthogonal geodesics are parallel with respect to the geodesic.

5. When the coordinate curves of a surface form a Tchebychef net (§29, Ex. 12), in which case the linear element is $du^{1^2} + 2\cos\omega \, du^1 \, du^2 + du^{2^2}$, the tangents to the parametric curves of either family at their points of meeting with a curve of the other family are parallel with respect to the latter.

6. Let P_1, P_2, P_3 be the vertices of a geodesic triangle on a surface, and θ_1, θ_2, θ_3 the interior angles at these respective points; when the tangent vector at P_1 to the geodesic P_1P_2 is transported parallel to itself around the triangle in the direction $P_1P_2P_3P_1$ it makes the angle $\pi - \theta_1 - \theta_2 - \theta_3$ with its original direction at P_1.

7. If $\lambda^\alpha(s)$ is a solution of equations (35.13) and one puts $\lambda^\alpha = \mu^\alpha\varphi(s)$, where $\varphi(s)$ is any function of s, then

$$\mu^2 \left(\frac{d\mu^1}{ds} + \mu^\beta \begin{Bmatrix} 1 \\ \beta\alpha \end{Bmatrix} \frac{du^\alpha}{ds} \right) - \mu^1 \left(\frac{d\mu^2}{ds} + \mu^\beta \begin{Bmatrix} 2 \\ \beta\alpha \end{Bmatrix} \frac{du^\alpha}{ds} \right) = 0$$

is independent of $\varphi(s)$, and this is the condition that a family of non-unit vectors be parallel with respect to the curve.

8. Show that theorem [34.2] is a consequence of equation (35.19).

9. Given a field of unit vectors λ^α and the curves C to which these vectors are tangent; from (35.18) one has

(i) $$\lambda^\alpha_{,\beta}\lambda^\beta = \kappa_g\mu^\alpha, \qquad \mu^\alpha_{,\beta}\mu^\beta = -\kappa_g\lambda^\alpha,$$

where μ^α is the unit vector at each point P of a curve C normal to C and κ_g is the geodesic curvature of C at P. Since $\mu^\alpha\mu_\alpha = 1$, $\lambda^\alpha\mu_\alpha = 0$, one has (see §28, Ex. 14)

$$\mu^\alpha_{,\beta} = \nu_\beta\lambda^\alpha,$$

from which and the second of equations (i) it follows that $\lambda^\beta\nu_\beta = -\kappa_g$, and consequently $\mu^\alpha_{,\sigma} = -\kappa_g$.

36. CONFORMAL CORRESPONDENCE OF TWO SURFACES

In §27 we defined as isometric two surfaces such that there exists on each a coordinate system in terms of which the first fundamental forms of the two surfaces are identical. For two isometric surfaces corresponding lengths are equal, and corresponding angles are equal as follows from (25.2) and (25.3). The converse is not true, that is, the equality of lengths is not a consequence of the equality of angles. We consider now the case when two surfaces are in one-to-one correspondence such that corresponding angles are equal. In this case the correspondence is said to be *conformal*.

Given two surfaces S and $\bar{S}$ expressed in terms of coordinates u^α and $\bar{u}^\alpha$, a one-to-one correspondence is established between points of the two surfaces by equations of the form

$$u^\alpha = \varphi^\alpha(\bar{u}^1, \bar{u}^2),$$

provided the jacobian of the φ's with respect to the $\bar{u}$'s is not identically zero. If then we change the coordinates $\bar{u}^\alpha$ on $\bar{S}$ into coordinates u^α by means of the above equations, corresponding points on the two surfaces have the same coordinates. In terms of these coordinates we denote by $g_{\alpha\beta}$ and $\bar{g}_{\alpha\beta}$ the components of the first fundamental tensors of S and $\bar{S}$ respectively.

The orthogonal trajectories of the curves $u^1 = $ const. on the two surfaces are the integral curves of the respective equations (see [26.2])

(36.1) $$g_{12}\,du^1 + g_{22}\,du^2 = 0, \qquad \bar{g}_{12}\,du^1 + \bar{g}_{22}\,du^2 = 0.$$

Since these trajectories correspond, we must have

$$\frac{\bar{g}_{12}}{g_{12}} = \frac{\bar{g}_{22}}{g_{22}}.$$

In like manner from the equations

(36.2) $g_{11} \, du^1 + g_{12} \, du^2 = 0,$ $\bar{g}_{11} \, du^1 + \bar{g}_{12} \, du^2 = 0$

of the orthogonal trajectories of the curves $u^2 =$ const. we must have

$$\frac{\bar{g}_{11}}{g_{11}} = \frac{\bar{g}_{12}}{g_{12}}.$$

Hence we have the three equations

(36.3) $$\frac{\bar{g}_{\alpha\beta}}{g_{\alpha\beta}} = r^2$$ $(\alpha, \beta = 1, 2),$

where r is a function of u^1 and u^2. Consequently

(36.4) $$d\bar{s}^2 = r^2 \, ds^2.$$

Conversely when (36.3) is satisfied it follows from (25.2) and (25.3) that any two corresponding angles on the two surfaces are equal, and we have

[36.1] *A necessary and sufficient condition that two surfaces having two corresponding nets of curves as coordinate curves be in conformal correspondence with points having the same coordinates corresponding is that the first fundamental forms of the two surfaces be proportional.*

We say that the correspondence is *direct* or *inverse* according as corresponding angles have the same or opposite sense.

In §30 it was shown that the first fundamental form of a surface is expressible in the form

(36.5) $$t^2(du^{1^2} + du^{2^2})$$

and that in terms of real coordinates u'^α defined by (see (30.19))

(36.6) $u'^1 + iu'^2 = f(u^1 \pm iu^2),$ $u'^1 - iu'^2 = f_0(u^1 \mp iu^2),$

where f and f_0 are conjugate functions, the first fundamental form is

(36.7) $$\frac{t^2}{f' f_0'} \, ((du'^1)^2 + (du'^2)^2).$$

Thus the equations $u'^\alpha = u^\alpha$ determine a conformal correspondence of the surface upon itself, and from theorem [30.4] it follows that the most general correspondence is defined by (36.6). Hence we have

[36.2] *When a pair of isometric coordinates u^α of a surface are known, the most general conformal correspondence of the surface with itself is obtained by making the point (u^1, u^2) correspond to the point (u'^1, u'^2), where u'^α are defined by equations (36.6).*

Also we have

[36.3] *When for each of two surfaces a pair of isometric coordinates u^α and u'^α are known, the equations $u'^\alpha = u^\alpha$ define a conformal correspondence of the surfaces, and by means of equations of the form (36.6) applied to one of the surfaces all the conformal correspondences of the two surfaces are obtained.*

Thus a conformal correspondence can be established between any two surfaces and in a great variety of ways, but it is not possible to establish an isometric correspondence of any two surfaces, that is, the particular conformal correspondence for which $r = 1$ (see §31).

If we denote by θ_0 and θ_0' the angles made at a point by a curve with the curves $u^2 = $ const. and $u'^2 = $ const. on a surface with the fundamental forms (36.5) and (36.7) respectively, it follows from (25.6) that

$$\cos\theta_0,\ \sin\theta_0 = \frac{du^1,\ du^2}{\sqrt{(du^1)^2 + (du^2)^2}},\qquad \cos\theta_0',\ \sin\theta_0' = \frac{du'^1,\ du'^2}{\sqrt{(du'^1)^2 + (du'^2)^2}},$$

which in each case is an abbreviated way of writing two equations. From these expressions we have, since $e^{i\theta_0} = \cos\theta_0 + i\sin\theta_0$,

$$e^{2i\theta_0} = \frac{e^{i\theta_0}}{e^{-i\theta_0}} = \frac{du^1 + i\,du^2}{du^1 - i\,du^2},\qquad e^{2i\theta_0'} = \frac{du'^1 + i\,du'^2}{du'^1 - i\,du'^2}.$$

From these equations according as the first or second signs are used in (36.6) we have

$$(36.8)\qquad e^{2i(\theta_0' - \theta_0)} = \frac{f'(u^1 + iu^2)}{f_0'(u^1 - iu^2)},\qquad e^{2i(\theta_0' + \theta_0)} = \frac{f'(u^1 - iu^2)}{f_0'(u^1 + iu^2)}.$$

If $\bar\theta_0$ and $\bar\theta_0'$ are the angles made by a second curve with the curves $u^2 = $ const. and $u'^2 = $ const. respectively, it follows from the first of (36.8) that $\bar\theta_0' - \theta_0' = \bar\theta_0 - \theta_0$, and from the second that $\bar\theta_0' - \theta_0' = -(\bar\theta_0 - \theta_0)$. Consequently we have

[36.4] *In the conformal correspondence of a surface with itself defined by (36.6), angles have the same or opposite sense according as the first or second signs are used in (36.6).*

From the foregoing discussion it follows that conformal correspondence is intimately related to functions of a complex variable. In treatments of the latter there is extensive study of conformal correspondence in the plane (see Exs. 1, 2).

EXERCISES

1. For the plane referred to cartesian coordinates the equation

$$\bar{x}^1 + i\bar{x}^2 = \frac{c^2}{x^1 - ix^2}$$

defines a conformal correspondence of the plane upon itself such that the lines $x^\alpha = $ const. correspond to two families of circles all passing through the origin with the centers of the circles of each family on one of the coordinate axes; also corresponding points are on a line through the origin and their distances r and $\bar{r}$ from the origin are in the relation $r\bar{r} = c^2$.

2. In terms of coordinates u^α in the plane defined by

$$u^1 = x^1 + ix^2, \qquad u^2 = x^1 - ix^2,$$

where x^α are cartesian coordinates, an equation of any real circle not passing through the origin is of the form

(i) $au^1u^2 + bu^1 + cu^2 + d = 0,$

where a and d are real numbers, and b and c are conjugate imaginaries. The equations

(ii) $u^1 = \dfrac{a_1 \bar{u}^1 + a_2}{a_3 \bar{u}^1 + a_4}, \qquad u^2 = \dfrac{b_1 \bar{u}^2 + b_2}{b_3 \bar{u}^2 + b_4}$

define a conformal correspondence of the plane upon itself such that circles correspond to circles or straight lines.

3. For the unit sphere with the equations

$$x^1 = \frac{u^1 + u^2}{1 + u^1 u^2}, \qquad x^2 = \frac{i(u^2 - u^1)}{1 + u^1 u^2}, \qquad x^3 = \frac{u^1 u^2 - 1}{1 + u^1 u^2},$$

u^1 and u^2 being conjugate imaginaries, equation (i) of Ex. 2 is an equation of a circle on the sphere, and equations (ii) of Ex. 2 define a conformal correspondence of the sphere with itself in which circles correspond to circles.

4. When in Ex. 3 the second of equations (ii) in Ex. 2 is replaced by

$$u^2 = \frac{a_4 \bar{u}^2 - a_3}{-a_2 \bar{u}^2 + a_1},$$

the correspondence of the sphere with itself is isometric, and defines a rotation of the sphere into itself, when the a's are chosen so that the correspondence is real.

5. The equations of Ex. 3 and

$$\bar{x}^1 = \frac{1}{2}(u^1 + u^2), \qquad \bar{x}^2 = \frac{i}{2}(u^2 - u^1)$$

for the plane define a conformal correspondence of the sphere and the plane, such that corresponding points are on the line with the equations

$$\frac{X^1}{u^1 + u^2} = \frac{X^2}{i(u^2 - u^1)} = \frac{X^3 - 1}{-2},$$

where X^i are cartesian coordinates in space; thus corresponding points lie on a line through the point P $(0, 0, 1)$, and consequently a line in the plane corresponds to the circle on the sphere which is its intersection by the plane determined by P and the line; this correspondence is called the *stereographic projection* of the sphere on the plane.

6. When two surfaces S and $\bar{S}$ are referred to coordinates u^α and $g = \bar{g}$, corresponding elements of area (see §25) are equal, and the correspondence is said to be *equivalent*. If $g \neq \bar{g}$ and one effects the change of coordinates $u'^1 = \varphi(u^1, u^2)$, $u'^2 = u^2$ on S, where φ is defined by

$$\sqrt{g(u^1, u^2)} = \sqrt{\bar{g}(\varphi, u^2)}\,\frac{\partial\varphi}{\partial u^1},$$

then $g'(u'^1, u'^2) = \bar{g}(u'^1, u'^2)$, and consequently an equivalent correspondence between S and $\bar{S}$ is defined by the equations $u'^1 = u^1$, $u'^2 = u^2$.

7. Let S and $\bar{S}$ be two surfaces such that the points with the same coordinates u^α correspond, and let $g_{\alpha\beta}$ and $\bar{g}_{\alpha\beta}$ be the first fundamental tensors of S and $\bar{S}$ respectively; if the correspondence is not conformal, on each surface the integral curves of the equation obtained from equation (26.24) when $a_{\alpha\beta}$ are replaced by $\bar{g}_{\alpha\beta}$ form an orthogonal net, as follows from theorem [26.4]; if the correspondence is conformal, any orthogonal net on one surface corresponds to an orthogonal net on the other surface.

37. GEODESIC CORRESPONDENCE OF TWO SURFACES

In this section we consider the relation between two surfaces S and $\bar{S}$ in one-to-one correspondence such that to each geodesic on either surface there corresponds a geodesic on the other, corresponding points having the same coordinates, as shown in §36. In this case the two surfaces are said to be in *geodesic correspondence*.

Since the arc of a curve is peculiar to the curve and thus cannot be taken as a common parameter of corresponding curves on the surfaces S and $\bar{S}$, it is necessary to express the equations of the geodesics in terms of some common parameter t, that is, u^α as functions of t. In consequence of (32.27) we have as equations of the geodesics on the surface S in terms of a general parameter t

$$(37.1) \quad \frac{du^1}{dt}\left(\frac{d^2 u^2}{dt^2} + \begin{Bmatrix} 2 \\ \beta\gamma \end{Bmatrix}\frac{du^\beta}{dt}\frac{du^\gamma}{dt}\right) - \frac{du^2}{dt}\left(\frac{d^2 u^1}{dt^2} + \begin{Bmatrix} 1 \\ \beta\gamma \end{Bmatrix}\frac{du^\beta}{dt}\frac{du^\gamma}{dt}\right) = 0.$$

In like manner equations of geodesics on $\bar{S}$ are

$$(37.2) \quad \frac{du^1}{dt}\left(\frac{d^2 u^2}{dt^2} + \begin{Bmatrix} \overline{2} \\ \beta\gamma \end{Bmatrix}\frac{du^\beta}{dt}\frac{du^\gamma}{dt}\right) - \frac{du^2}{dt}\left(\frac{d^2 u^1}{dt^2} + \begin{Bmatrix} \overline{1} \\ \beta\gamma \end{Bmatrix}\frac{du^\beta}{dt}\frac{du^\gamma}{dt}\right) = 0,$$

where the Christoffel symbols $\begin{Bmatrix} \alpha \\ \beta\gamma \end{Bmatrix}$ are formed with respect to the fundamental tensor $\bar{g}_{\alpha\beta}$ of $\bar{S}$. Since it is understood that the coordinate curves on S and $\bar{S}$ are such that the points on the two surfaces with the same coordinates correspond and that the parameter t is the same for both surfaces, on subtracting equations (37.1) from (37.2), we obtain

$$(37.3) \qquad \left(\frac{du^1}{dt}\, a^2_{\beta\gamma} - \frac{du^2}{dt}\, a^1_{\beta\gamma}\right) \frac{du^\beta}{dt}\, \frac{du^\gamma}{dt} = 0,$$

where

$$(37.4) \qquad a^\alpha_{\beta\gamma} = \left\{\overline{\begin{matrix}\alpha\\\beta\gamma\end{matrix}}\right\} - \left\{\begin{matrix}\alpha\\\beta\gamma\end{matrix}\right\}.$$

If we denote by $\left\{\begin{matrix}\alpha\\\beta\gamma\end{matrix}\right\}'$ and $\left\{\overline{\begin{matrix}\alpha\\\beta\gamma\end{matrix}}\right\}'$ the Christoffel symbols for a coordinate system u'^α, it follows from equations (32.1) and analogous equations for $\bar{S}$ that

$$a^\alpha_{\beta\gamma}\, \frac{\partial u^\beta}{\partial u'^\mu}\, \frac{\partial u^\gamma}{\partial u'^\nu} = a'^\lambda_{\mu\nu}\, \frac{\partial u^\alpha}{\partial u'^\lambda},$$

which are equivalent by (24.22) to

$$a'^\lambda_{\mu\nu} = a^\alpha_{\beta\gamma}\, \frac{\partial u^\beta}{\partial u'^\mu}\, \frac{\partial u^\gamma}{\partial u'^\nu}\, \frac{\partial u'^\lambda}{\partial u^\alpha}.$$

Hence, $a^\alpha_{\beta\gamma}$ are the components of a mixed tensor, contravariant of the first order and covariant of the second order (see §20, Ex. 11), symmetric in the indices β and γ as follows from (37.4).

Equation (37.3) must be satisfied identically; otherwise we should have an equation of the first order and third degree satisfied by all the geodesics; this would mean that through each point there would pass at most three geodesics. Thus it could not be true that a geodesic passes through a given point in any given direction. In order to obtain the conditions upon the tensor $a^\alpha_{\beta\gamma}$ so that equation (37.3) shall be an identity, we observe that this equation may be written in the form

$$(\delta^\alpha_\epsilon\, a^\mu_{\beta\gamma} - \delta^\mu_\epsilon\, a^\alpha_{\beta\gamma})\, \frac{du^\beta}{dt}\, \frac{du^\gamma}{dt}\, \frac{du^\epsilon}{dt} = 0$$

for $\alpha \neq \mu$ and that the latter is an identity for $\alpha = \mu$. If these equations are to be satisfied identically, we must have (see §19, Ex. 5)

$$\delta^\alpha_\epsilon\, a^\mu_{\beta\gamma} - \delta^\mu_\epsilon\, a^\alpha_{\beta\gamma} + \delta^\alpha_\beta\, a^\mu_{\gamma\epsilon} - \delta^\mu_\beta\, a^\alpha_{\gamma\epsilon} + \delta^\alpha_\gamma\, a^\mu_{\epsilon\beta} - \delta^\mu_\gamma\, a^\alpha_{\epsilon\beta} = 0.$$

Contracting for μ and ϵ, that is, putting $\mu = \epsilon$ and summing with respect to ϵ, and noting that $\delta^\epsilon_\epsilon = 2$ and that $a^\alpha_{\beta\gamma}$ is symmetric in β

and γ, we obtain as a necessary condition

$$3a^\alpha_{\beta\gamma} = \delta^\alpha_\beta a^\epsilon_{\gamma\epsilon} + \delta^\alpha_\gamma a^\epsilon_{\beta\epsilon}.$$

Since $a^\alpha_{\beta\gamma}$ is a tensor, $a^\epsilon_{\gamma\epsilon}$ is a covariant vector, which we denote by $3a_\gamma$ (see §19). From this result and (37.4) we have

$$(37.5) \qquad \left\{ \begin{matrix} \alpha \\ \beta\gamma \end{matrix} \right\} = \left\{ \begin{matrix} \alpha \\ \beta\gamma \end{matrix} \right\} + \delta^\alpha_\beta a_\gamma + \delta^\alpha_\gamma a_\beta.$$

When these expressions are substituted in (37.2), the resulting equation reduces to (37.1), and consequently (37.5) are sufficient as well as necessary conditions that S and $\bar{S}$ be in geodesic correspondence.

Contracting equations (37.5) for α and β, we have in consequence of the third of equations (28.2)

$$\frac{\partial}{\partial u^\gamma} \log \sqrt{\frac{\bar{g}}{g}} = 3a_\gamma.$$

From (25.8) it follows that $\bar{g}/g$ is a scalar, and consequently a_γ is the gradient, that is $a_\gamma = \dfrac{\partial\varphi}{\partial u^\gamma}$, where $\varphi = \dfrac{1}{6} \log \dfrac{\bar{g}}{g}$. Hence we have

[37.1] *A necessary and sufficient condition that two surfaces with fundamental tensors $g_{\alpha\beta}$ and $\bar{g}_{\alpha\beta}$ be in geodesic correspondence is that their Christoffel symbols be in the relation*

$$(37.6) \qquad \left\{ \begin{matrix} \alpha \\ \beta\gamma \end{matrix} \right\} = \left\{ \begin{matrix} \alpha \\ \beta\gamma \end{matrix} \right\} + \delta^\alpha_\beta \varphi_{,\gamma} + \delta^\alpha_\gamma \varphi_{,\beta};$$

the function φ is equal to $\dfrac{1}{6} \log \dfrac{\bar{g}}{g}$.

If we denote by $\bar{R}^\epsilon_{\beta\gamma\delta}$ the Riemann tensor for $\bar{g}_{\alpha\beta}$, we find from (37.6) and (20.13)

$$(37.7) \qquad \bar{R}^\epsilon_{\beta\gamma\delta} = R^\epsilon_{\beta\gamma\delta} + \delta^\epsilon_\delta \varphi_{\beta\gamma} - \delta^\epsilon_\gamma \varphi_{\beta\delta},$$

where

$$(37.8) \qquad \varphi_{\beta\gamma} = \varphi_{,\beta\gamma} - \varphi_{,\beta}\varphi_{,\gamma},$$

$\varphi_{,\beta\gamma}$ being the second covariant derivative of φ based upon $g_{\alpha\beta}$.

From (28.5) and (28.3) we have

$$(37.9) \qquad R^\epsilon_{\beta\gamma\delta} = g^{\epsilon\alpha}R_{\alpha\beta\gamma\delta} = K(\delta^\epsilon_\gamma g_{\beta\delta} - \delta^\epsilon_\delta g_{\beta\gamma}).$$

Contracting for ϵ and δ, we obtain, since $\delta^\epsilon_\epsilon = 2$,

$$(37.10) \qquad R_{\beta\gamma} = -Kg_{\beta\gamma}.$$

Contracting (37.7) for ϵ and δ, we have

(37.11) $\bar{R}_{\beta\gamma} = R_{\beta\gamma} + \varphi_{\beta\gamma}$,

from which, (37.10), and a similar equation for $\bar{R}_{\beta\gamma}$ we have

(37.12) $\varphi_{\beta\gamma} = K g_{\beta\gamma} - \bar{K}\bar{g}_{\beta\gamma}$,

where $\bar{K}$ is the Gaussian curvature of $\bar{S}$.

We consider, in particular, the case when K is a constant. From (37.8) and (37.12) we have

(37.13) $\varphi_{,\alpha\beta} = \varphi_{,\alpha}\,\varphi_{,\beta} + K g_{\alpha\beta} - \bar{K}\bar{g}_{\alpha\beta}$.

Finding the covariant derivative based upon the tensor $g_{\alpha\beta}$, we have

(37.14) $\varphi_{,\alpha\beta\gamma} = \varphi_{,\alpha\gamma}\,\varphi_{,\beta} + \varphi_{,\alpha}\,\varphi_{,\beta\gamma} - \bar{K}_{,\gamma}\bar{g}_{\alpha\beta} - \bar{K}\bar{g}_{\alpha\beta,\gamma}$.

From equations similar to (28.2) we have

$$\frac{\partial \bar{g}_{\alpha\beta}}{\partial u^{\gamma}} = \bar{g}_{\delta\beta} \begin{Bmatrix} \delta \\ \alpha\gamma \end{Bmatrix} + \bar{g}_{\alpha\delta} \begin{Bmatrix} \delta \\ \beta\gamma \end{Bmatrix},$$

from which on substituting from (37.6), we obtain

$$\bar{g}_{\alpha\beta,\gamma} = 2\bar{g}_{\alpha\beta}\,\varphi_{,\gamma} + \bar{g}_{\beta\gamma}\,\varphi_{,\alpha} + \bar{g}_{\alpha\gamma}\,\varphi_{,\beta} .$$

Substituting this expression in (37.14) and replacing the $\bar{g}_{\alpha\beta}$ by their expressions from (37.13), we obtain

(37.15) $\quad \varphi_{,\alpha\beta\gamma} = 2(\varphi_{,\alpha}\,\varphi_{,\beta\gamma} + \varphi_{,\beta}\,\varphi_{,\gamma\alpha} + \varphi_{,\gamma}\,\varphi_{,\alpha\beta}) - 4\varphi_{,\alpha}\,\varphi_{,\beta}\,\varphi_{,\gamma}$
$$- K(2g_{\alpha\beta}\,\varphi_{,\gamma} + g_{\beta\gamma}\,\varphi_{,\alpha} + g_{\gamma\alpha}\,\varphi_{,\beta}) - \bar{K}_{,\gamma}\bar{g}_{\alpha\beta} ,$$

from which it follows that

$$\varphi_{,\alpha\beta\gamma} - \varphi_{,\alpha\gamma\beta} = K(g_{\alpha\gamma}\,\varphi_{,\beta} - g_{\alpha\beta}\,\varphi_{,\gamma}) - \bar{K}_{,\gamma}\bar{g}_{\alpha\beta} + \bar{K}_{,\beta}\bar{g}_{\alpha\gamma} .$$

From the appropriate Ricci identity (22.19) and (37.9) we have

$$\varphi_{,\alpha\beta\gamma} - \varphi_{,\alpha\gamma\beta} = \varphi_{,\epsilon}R^{\epsilon}_{\alpha\beta\gamma} = \varphi_{,\epsilon}K(\delta^{\epsilon}_{\beta}g_{\alpha\gamma} - \delta^{\epsilon}_{\gamma}g_{\alpha\beta}) = K(g_{\alpha\gamma}\,\varphi_{,\beta} - g_{\alpha\beta}\,\varphi_{,\gamma}),$$

and consequently

$$\bar{K}_{,\gamma}\bar{g}_{\alpha\beta} - \bar{K}_{,\beta}\bar{g}_{\alpha\gamma} = 0.$$

From these equations we have that $\bar{K}$ is a constant, since the determinant of $\bar{g}_{\alpha\beta}$ is not equal to zero. Hence we have the theorem of Beltrami:[*]

* 1869, 2, p. 232.

[37.2] *The only surfaces in geodesic correspondence with a surface of constant Gaussian curvature are surfaces of constant Gaussian curvature.*

We consider in particular the case when $K = 0$ and assume that the coordinates u^α are cartesian. In this case the covariant derivatives of φ in (37.15) are ordinary partial derivatives. If then we put $\varphi = -\frac{1}{2} \log \theta$ and thus understand that $\theta > 0$, equation (37.15) reduces in this case to

$$(37.16) \qquad \frac{\partial^3 \theta}{\partial u^\alpha \, \partial u^\beta \, \partial u^\gamma} = 0,$$

and from (37.13) we have

$$\bar{g}_{\alpha\beta} = \frac{1}{4\bar{K}\theta^2} \left(2\theta \, \theta_{,\alpha\beta} - \theta_{,\alpha} \theta_{,\beta} \right).$$

The integral of (37.16) is

$$\theta = a_{\alpha\beta} u^\alpha u^\beta + b_\alpha u^\alpha + c,$$

where $a_{\alpha\beta}$, b_α and c are constants. By suitable transformations of the cartesian coordinates this expression for θ is reducible to one or the other of the forms*

$$\theta = a_\alpha u^{\alpha^2} + c, \qquad \theta = au^{1^2} + 2bu^2 + c,$$

depending on the values of $a_{\alpha\beta}$. The expressions for $\bar{g}_{\alpha\beta}$ for these expressions for θ are respectively

$$(37.17) \quad \bar{g}_{11}, \bar{g}_{12}, \bar{g}_{22} = \frac{a_1(a_2 u^{2^2} + c), \; -a_1 a_2 u^1 u^2, \; a_2(a_1 u^{1^2} + c)}{\bar{K}(a_\alpha u^{\alpha^2} + c)^2},$$

and

$$(37.18) \qquad \bar{g}_{11}, \bar{g}_{12}, \bar{g}_{22} = \frac{a(2bu^2 + c), \; -abu^1, \; -b^2}{\bar{K}(au^{1^2} + 2bu^2 + c)^2}.$$

From these expressions we have respectively

$$\bar{g} = \frac{a_1 a_2 c}{\bar{K}^2(a_\alpha u^{\alpha^2} + c)^3}, \qquad \bar{g} = \frac{-ab^2}{\bar{K}^2(au^{1^2} + 2bu^2 + c)^3}.$$

The constants and the domains of the variables must be such that $\bar{g} > 0$. Also both $\bar{g}_{11}$ and $\bar{g}_{22}$ must be positive. This means for (37.18) that $\bar{K}$ is negative, since θ is necessarily positive.

*C. G., p. 227.

We return to the consideration of the general case and assume that the surfaces S and $\bar{S}$ are referred to the orthogonal net on S which corresponds to an orthogonal net on $\bar{S}$ (see §36, Ex. 7). In this coordinate system equations (37.6), on putting $\varphi = -\frac{1}{2} \log \theta$, reduce to (see §28, Ex. 1)

$$\frac{\partial \log \bar{g}_{\alpha\alpha}}{\partial u^\alpha} = \frac{\partial}{\partial u^\alpha} \log \frac{g_{\alpha\alpha}}{\theta^2}, \qquad \frac{\partial \log \bar{g}_{\alpha\alpha}}{\partial u^\beta} = \frac{\partial}{\partial u^\beta} \log \frac{g_{\alpha\alpha}}{\theta},$$

$$\frac{1}{\bar{g}_{\beta\beta}} \frac{\partial \bar{g}_{\alpha\alpha}}{\partial u^\beta} = \frac{1}{g_{\beta\beta}} \frac{\partial g_{\alpha\alpha}}{\partial u^\beta} \qquad\qquad (\beta \neq \alpha).$$

Expressing the condition of integrability of the first two of these equations, we obtain $\dfrac{\partial^2 \log \theta}{\partial u^1 \partial u^2} = 0$ and consequently $\theta = U_1 U_2$, where U_1 and U_2 are functions of u^1 and u^2 respectively, both having the same sign since θ must be positive. Then from the first two of the equations we have $\bar{g}_{\alpha\alpha} = \dfrac{g_{\alpha\alpha}}{U_\alpha^2 U_\beta}$ $(\beta \neq \alpha)$. When these expressions are substituted in the third of the above equations we find that

$$g_{\alpha\alpha} = (U_\alpha - U_\beta)\varphi_\alpha(u^\alpha).$$

Hence by a suitable choice of the coordinates without changing the coordinate lines the fundamental forms of S and $\bar{S}$ are reducible to the respective forms

$$(37.19) \qquad (U_1 - U_2)(du^{1^2} + du^{2^2}), \qquad \left(\frac{1}{U_2} - \frac{1}{U_1}\right)\left(\frac{du^{1^2}}{U_1} + \frac{du^{2^2}}{U_2}\right).$$

Consequently we have the theorem of Dini* (see §32):

[37.3] *A necessary and sufficient condition that a surface be in geodesic correspondence with another surface is that it be a surface of Liouville.*

EXERCISES

1. When the equation (37.3) is written

$$a_{11}^2 \left(\frac{du^1}{dt}\right)^3 + (2a_{12}^2 - a_{11}^1)\left(\frac{du^1}{dt}\right)^2 \frac{du^2}{dt} + (a_{22}^2 - 2a_{12}^1)\frac{du^1}{dt}\left(\frac{du^2}{dt}\right)^2 - a_{22}^1\left(\frac{du^2}{dt}\right)^3 = 0,$$

the condition that it be an identity is

$$a_{11}^2 = 2a_{12}^2 - a_{11}^1 = 2a_{12}^1 - a_{22}^2 = a_{22}^1 = 0,$$

which are equivalent to

$$a_{\beta\gamma}^\alpha = \tfrac{1}{3}(\delta_\beta^\alpha a_{\gamma\epsilon}^\epsilon + \delta_\gamma^\alpha a_{\beta\epsilon}^\epsilon) = \delta_\beta^\alpha a_\gamma + \delta_\gamma^\alpha a_\beta.$$

---

* 1870, 1, p. 278.

2. The results of Ex. 1. follow also from the equation of §32, Ex. 7.

3. Since the first of the forms (37.19) is not altered when U_1 and U_2 are replaced by $U_1 + a$ and $U_2 + a$, where a is a constant, the second of (37.19) may be replaced by

$$\left(\frac{1}{U_2 + a} - \frac{1}{U_1 + a}\right)\left(\frac{du^{1^2}}{U_1 + a} + \frac{du^{2^2}}{U_2 + a}\right).$$

4. For a surface with the first fundamental form

$$(u^{1^4} - u^{2^4})\left(\varphi(u^1)\,du^{1^2} + \varphi\left(\frac{1}{u^2}\right) du^{2^2}\right),$$

the correspondence defined by $u^1 = \dfrac{1}{u^2}$, $u^2 = \dfrac{1}{u^1}$ is a geodesic correspondence of the surface with itself.

5. From (37.7) and (37.11) one finds that

$$\overline{W}^\epsilon{}_{\beta\gamma\delta} = W^\epsilon{}_{\beta\gamma\delta},$$

where

$$W^\epsilon{}_{\beta\gamma\delta} = R^\epsilon{}_{\beta\gamma\delta} + \delta^\epsilon_\gamma R_{\beta\delta} - \delta^\epsilon_\delta R_{\beta\gamma},$$

and similarly for $\overline{W}^\epsilon{}_{\beta\gamma\delta}$. Thus the tensor $W^\epsilon{}_{\beta\gamma\delta}$ has the same components for two surfaces in geodesic correspondence. Determine under what conditions it is a zero tensor.

6. From (37.6) one has $\overline{\Pi}^\alpha_{\beta\gamma} = \Pi^\alpha_{\beta\gamma}$, where

$$\Pi^\alpha_{\beta\gamma} = \left\{\begin{matrix} \alpha \\ \beta\gamma \end{matrix}\right\} - \frac{1}{3}\left(\delta^\alpha_\beta\left\{\begin{matrix} \epsilon \\ \epsilon\gamma \end{matrix}\right\} + \delta^\alpha_\gamma\left\{\begin{matrix} \epsilon \\ \epsilon\beta \end{matrix}\right\}\right)$$

and similarly for $\overline{\Pi}^\alpha_{\beta\gamma}$. Thus the quantities $\Pi^\alpha_{\beta\gamma}$ are equal for two surfaces in geodesic correspondence. In terms of these quantities equations (37.1) and (37.2) become

$$\frac{du^1}{dt}\left(\frac{d^2u^2}{dt^2} + \Pi^2_{\beta\gamma}\frac{du^\beta}{dt}\frac{du^\gamma}{dt}\right) - \frac{du^2}{dt}\left(\frac{d^2u^1}{dt^2} + \Pi^1_{\beta\gamma}\frac{du^\beta}{dt}\frac{du^\gamma}{dt}\right) = 0.$$

CHAPTER IV

Surfaces in Space

38. THE SECOND FUNDAMENTAL FORM OF A SURFACE

In the preceding chapter we considered a surface in space defined by three equations

$$(38.1) \qquad x^i = f^i(u^1, u^2) \qquad (i = 1, 2, 3),$$

where x^i are cartesian coordinates. Applying the euclidean metric we found that lengths of arcs and angles between curves in the surface are expressible in terms of the first fundamental tensor $g_{\alpha\beta}$ defined by

$$(38.2) \qquad g_{\alpha\beta} = \sum_i \frac{\partial x^i}{\partial u^\alpha} \frac{\partial x^i}{\partial u^\beta} \qquad (\alpha, \beta = 1, 2).*$$

Thus the euclidean metric of space induces a metric on a surface expressible in terms of the tensor $g_{\alpha\beta}$. Throughout the preceding chapter we considered the geometry of a surface in terms of this tensor, and without any further reference to the character of the surface as viewed from the enveloping space. From this point of view any two developable surfaces are equivalent to one another, and to a plane, that is, have the same metric properties. Also any surface isometric with a surface of revolution is equivalent to the latter. However, as viewed from the enveloping space two such surfaces may be entirely different in form. It is this question of the geometry of a surface as viewed from the enveloping space which is the subject of the present chapter. This study involves the consideration of the form of a surface in the neighborhood of an ordinary point and its position relative to the tangent plane at the point.

In §10 quantities A^{ij} were defined by

$$(38.3) \qquad A^{ij} = \frac{\partial(f^i, f^j)}{\partial(u^1, u^2)},$$

and it was shown in §11 that the tangents to all the curves through an ordinary point x^i, that is, one for which the three quantities A^{ij} are not

---

* In what follows Latin indices take the values 1, 2, 3 and Greek indices the values 1, 2.

all equal to zero, lie in a plane containing this point, called the *tangent plane*, and that an equation of the plane is

$$(\bar{x}^1 - x^1)A^{23} + (\bar{x}^2 - x^2)A^{31} + (\bar{x}^3 - x^3)A^{12} = 0,$$

where $\bar{x}^i$ are current coordinates.

In consequence of the identity

$$(38.4) \quad g \equiv |g_{\alpha\beta}| = g_{11}g_{22} - g_{12}^2 = (A^{12})^2 + (A^{23})^2 + (A^{31})^2,$$

established in §24, the quantities X^i defined by

$$(38.5) \qquad X^i = \frac{1}{\sqrt{g}} A^{jk} = \frac{1}{\sqrt{g}} \begin{vmatrix} \dfrac{\partial x^j}{\partial u^1} & \dfrac{\partial x^k}{\partial u^1} \\[2mm] \dfrac{\partial x^j}{\partial u^2} & \dfrac{\partial x^k}{\partial u^2} \end{vmatrix},$$

where i, j, k take the values 1, 2, 3 cyclically, are such that

$$(38.6) \qquad \sum_i X^i X^i = 1.$$

Accordingly an equation of the tangent plane is

$$(38.7) \qquad \sum_i X^i(\bar{x}^i - x^i) = 0,$$

and X^i are direction cosines of the *normal to the surface* at the point x^i, that is, the normal to the tangent plane at the point.

From (38.5) it follows that*

$$(38.8) \qquad \sum_i X^i \frac{\partial x^i}{\partial u^\alpha} = 0 \qquad\qquad (\alpha = 1, 2),$$

and using (38.4) one may verify that

$$(38.9) \qquad \begin{vmatrix} \dfrac{\partial x^1}{\partial u^1} & \dfrac{\partial x^2}{\partial u^1} & \dfrac{\partial x^3}{\partial u^1} \\[2mm] \dfrac{\partial x^1}{\partial u^2} & \dfrac{\partial x^2}{\partial u^2} & \dfrac{\partial x^3}{\partial u^2} \\[2mm] X^1 & X^2 & X^3 \end{vmatrix} = \sqrt{g} \,\dagger.$$

We have from (24.4) and (24.8) that $\dfrac{1}{\sqrt{g_{11}}} \dfrac{\partial x^i}{\partial u^1}$ are the direction cosines

* C. G., Theorem [26.4].

† In (38.5) and (38.9) $\sqrt{g}$ means the positive square root; this convention is always used.

of the vector from a point u^α on a surface tangent to the curve $u^2 =$ const. and in the direction in which u^1 is increasing; and similarly for $\dfrac{1}{\sqrt{g_{22}}} \dfrac{\partial x^i}{\partial u^2}$ and the curve $u^1 =$ const. Hence if the coordinate net on a surface is an orthogonal net, equation (38.9) expresses the fact that the positive tangent vectors of components $\dfrac{\partial x^i}{\partial u^1}, \dfrac{\partial x^i}{\partial u^2}$ and the normal vector X^i have the same mutual orientation as the x^1-, x^2-, and x^3-axes.* If the coordinate curves do not form an orthogonal net, we consider the

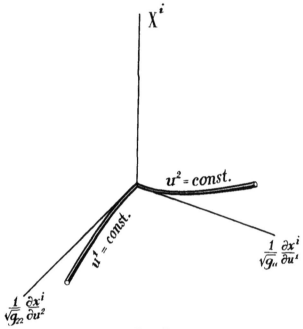

$$X^i$$

$$u^2 = const.$$

$$\frac{1}{\sqrt{g_{11}}} \frac{\partial x^i}{\partial u^1}$$

$$u^1 = const.$$

$$\frac{1}{\sqrt{g_{22}}} \frac{\partial x^i}{\partial u^2}$$

FIG. 12

result of applying a positive transformation so that the new coordinate curves shall be the given curves $u^2 =$ const. and their orthogonal trajectories, the equations of such a transformation being of the form $u'^1 = \varphi(u^1, u^2)$, $u'^2 = u^2$. Then equation (38.9) transforms into an equation of the same form in consequence of (25.13) and positive sense in the surface is unaltered as shown in §25. Hence the mutual orientation of the vectors $\dfrac{1}{\sqrt{g_{11}}} \dfrac{\partial x^i}{\partial u^1}, \dfrac{1}{\sqrt{g_{22}}} \dfrac{\partial x^i}{\partial u^2}$, X^i is as shown in Fig. 12 whatever be the coordinate curves.

* C. G., p. 162.

For a curve in the surface through two nearby points x^i and $\bar{x}^i$ we have

$$(38.10) \qquad \bar{x}^i = x^i + \left(\frac{dx^i}{ds}\right)_0 s + \frac{1}{2}\left(\frac{d^2 x^i}{ds^2}\right)_0 s^2 + \cdots,$$

where a subscript zero indicates the value of a quantity at x^i and s is the arc from the point x^i. Since x^i are functions of the coordinates u^α, we have

$$(38.11) \qquad \frac{dx^i}{ds} = \frac{\partial x^i}{\partial u^\alpha}\frac{du^\alpha}{ds}, \qquad \frac{d^2 x^i}{ds^2} = \frac{\partial^2 x^i}{\partial u^\alpha \partial u^\beta}\frac{du^\alpha}{ds}\frac{du^\beta}{ds} + \frac{\partial x^i}{\partial u^\alpha}\frac{d^2 u^\alpha}{ds^2}.$$

Consequently because of (38.8) the distance p of the point $P_1(\bar{x})$ from the tangent plane (38.7) at $P(x)$ is given by*

$$(38.12) \qquad p = \frac{1}{2}\left(d_{\alpha\beta}\frac{du^\alpha}{ds}\frac{du^\beta}{ds}\right)_0 s^2 + B,$$

where B indicates the sum of terms of the third and higher orders in s, and where by definition

$$(38.13) \qquad d_{\alpha\beta} = \sum_i X^i \frac{\partial^2 x^i}{\partial u^\alpha \partial u^\beta}.$$

In consequence of (38.8) equations (38.13) are equivalent to

$$(38.14) \qquad d_{\alpha\beta} = \sum_i X^i x^i_{,\alpha\beta},$$

where $x^i_{,\alpha\beta}$ is the second covariant derivative of x^i, that is,

$$(38.15) \qquad x^i_{,\alpha\beta} \doteq \frac{\partial^2 x^i}{\partial u^\alpha \partial u^\beta} - \frac{\partial x^i}{\partial u^\gamma}\begin{Bmatrix}\gamma\\\alpha\beta\end{Bmatrix}.$$

Since x^i being functions of u^α are scalars as regards a change in the coordinates u^α, as are also X^i, it follows from (38.14) that $d_{\alpha\beta}$ are the components of a symmetric covariant tensor of the second order, being the sum of three covariant tensors each multiplied by a scalar. It is called the *second fundamental quadratic tensor* of the surface, and $d_{\alpha\beta}\,du^\alpha\,du^\beta$ is called the *second fundamental quadratic form* of the surface. It is shown in §39 that a surface is completely characterized by its first and second fundamental forms and thus that the distinction between two isometric surfaces as viewed from the enveloping space is based upon their respective second fundamental forms. Two surfaces which have the same first fundamental form but different second fundamental forms are said to be *applicable*.

* C. G., p. 95.

Gauss,* who introduced the second fundamental form, used the expression $D\,dp^2 + 2D'\,dp\,dq + D''\,dq^2$, as have many subsequent writers. We have adopted the expression $d_{\alpha\beta}\,du^\alpha\,du^\beta$, since it enables one to write equations in simpler form with the use of the summation convention.

From (38.13) and (38.8) we have

$$(38.16) \qquad d_{\alpha\beta} = -\sum_i \frac{\partial X^i}{\partial u^\alpha}\frac{\partial x^i}{\partial u^\beta} = -\sum_i \frac{\partial X^i}{\partial u^\beta}\frac{\partial x^i}{\partial u^\alpha}.$$

Since X^i are scalars as regards a change of coordinates u^α, it follows again from (38.16) that $d_{\alpha\beta}$ are the components of a symmetric covariant tensor of the second order.

If we differentiate equations (38.2) with respect to u^γ we obtain

$$\sum_i \frac{\partial^2 x^i}{\partial u^\alpha\,\partial u^\gamma}\frac{\partial x^i}{\partial u^\beta} + \sum_i \frac{\partial^2 x^i}{\partial u^\beta\,\partial u^\gamma}\frac{\partial x^i}{\partial u^\alpha} = \frac{\partial g_{\alpha\beta}}{\partial u^\gamma}.$$

If we permute the indices α, β, γ cyclically twice, we obtain the two equations

$$\sum_i \frac{\partial^2 x^i}{\partial u^\beta\,\partial u^\alpha}\frac{\partial x^i}{\partial u^\gamma} + \sum_i \frac{\partial^2 x^i}{\partial u^\gamma\,\partial u^\alpha}\frac{\partial x^i}{\partial u^\beta} = \frac{\partial g_{\beta\gamma}}{\partial u^\alpha},$$

$$\sum_i \frac{\partial^2 x^i}{\partial u^\gamma\,\partial u^\beta}\frac{\partial x^i}{\partial u^\alpha} + \sum_i \frac{\partial^2 x^i}{\partial u^\alpha\,\partial u^\beta}\frac{\partial x^i}{\partial u^\gamma} = \frac{\partial g_{\gamma\alpha}}{\partial u^\beta}.$$

Subtracting from the sum of these two equations the preceding one, and making use of the definition (20.1) of the Christoffel symbols of the first kind, we obtain

$$(38.17) \qquad \sum_i \frac{\partial^2 x^i}{\partial u^\alpha\,\partial u^\beta}\frac{\partial x^i}{\partial u^\gamma} = [\alpha\beta, \gamma].$$

From this result and (38.15) we have (see (20.4))

$$\sum_i x^i_{,\alpha\beta}\frac{\partial x^i}{\partial u^\gamma} = [\alpha\beta, \gamma] - \sum_i \frac{\partial x^i}{\partial u^\gamma}\frac{\partial x^i}{\partial u^\delta}\begin{Bmatrix}\delta\\ \alpha\beta\end{Bmatrix} = [\alpha\beta, \gamma] - g_{\gamma\delta}\begin{Bmatrix}\delta\\ \alpha\beta\end{Bmatrix} = 0.$$

Comparing these equations with (38.8), we have that $x^i_{,\alpha\beta}$ is some multiple of X^i, say $x^i_{,\alpha\beta} = h_{\alpha\beta}X^i$. Multiplying by X^i and summing with respect to i, we have from (38.6) and (38.14) $h_{\alpha\beta} = d_{\alpha\beta}$, and consequently the following *equations of Gauss*:†

$$(38.18) \qquad x^i_{,\alpha\beta} = d_{\alpha\beta}X^i.$$

---

* 1827, 1, p. 234.
† 1827, 1, pp. 233, 234.

We wish now to find expressions for $\dfrac{\partial X^i}{\partial u^\alpha}$. From (38.6) it follows that

$$\sum_i X^i \frac{\partial X^i}{\partial u^\alpha} = 0,$$

from which it is seen that $\dfrac{\partial X^i}{\partial u^\alpha}$ for each value of α are direction numbers of a vector orthogonal to the normal. Hence each such vector is a linear homogeneous function of the two tangential vectors $\dfrac{\partial x^i}{\partial u^\alpha}$. Thus we have

$$\frac{\partial X^i}{\partial u^\alpha} = t^\beta_\alpha \frac{\partial x^i}{\partial u^\beta},$$

where the expressions for t^β_α are to be determined. In order to find these expressions, we multiply both sides of the above equation by $\dfrac{\partial x^i}{\partial u^\gamma}$ and sum with respect to i, and in consequence of (38.16) and (38.2) obtain

$$-d_{\alpha\gamma} = t^\beta_\alpha g_{\beta\gamma}.$$

Multiplying both sides of this equation by $g^{\gamma\delta}$ and summing with respect to γ, we obtain because of (24.16)

$$t^\delta_\alpha = -d_{\alpha\gamma} g^{\gamma\delta}.$$

Consequently the desired equations are

(38.19) $$\frac{\partial X^i}{\partial u^\alpha} = -d_{\alpha\gamma} g^{\gamma\beta} \frac{\partial x^i}{\partial u^\beta}.$$

EXERCISES

1. When $d_{\alpha\beta} = 0$, it follows from (38.19) that the surface is a plane.
2. When

$$\frac{d_{11}}{g_{11}} = \frac{d_{12}}{g_{12}} = \frac{d_{22}}{g_{22}} = t(\neq 0),$$

it follows from (38.19) that $\dfrac{\partial X^i}{\partial u^\alpha} = -t \dfrac{\partial x^i}{\partial u^\alpha}$. From the conditions of integrability of these equations it follows that t is a constant, since $g \neq 0$. Hence $X^i = -tx^i + a^i$, and the surface is a sphere.

3. If a straight line lies entirely on a surface, $\dfrac{d^2 x^i}{ds^2} = 0$ along the line. From

(38.11) and (38.13) it follows that the line is one of the integral curves of the equation

$$d_{\alpha\beta} \, du^\alpha \, du^\beta = 0.$$

4. For a surface of revolution with the equations

$$x^1 = u^1 \cos u^2, \qquad x^2 = u^1 \sin u^2, \qquad x^3 = \varphi(u^1),$$

one finds that

$$X^1, X^2, X^3 = \frac{-\varphi' \cos u^2, \; -\varphi' \sin u^2, \; 1}{\sqrt{1 + \varphi'^2}},$$

$$d_{11}, d_{12}, d_{22} = \frac{\varphi'', \; 0, \; u^1 \varphi'}{\sqrt{1 + \varphi'^2}}.$$

5. For a helicoid with the equations (see §24, Ex. 6)

$$x^1 = u^1 \cos u^2, \qquad x^2 = u^1 \sin u^2, \qquad x^3 = \varphi(u^1) + au^2,$$

one finds that

$$X^1, X^2, X^3 = \frac{a \sin u^2 - u^1 \varphi' \cos u^2, \; -(a \cos u^2 + u^1 \varphi' \sin u^2), \; u^1}{\sqrt{u^{1^2}(1 + \varphi'^2) + a^2}},$$

$$d_{11}, d_{12}, d_{22} = \frac{u^1 \varphi'', \; -a, \; u^{1^2} \varphi'}{\sqrt{u^{1^2}(1 + \varphi'^2) + a^2}}.$$

6. Two points x^i and $\bar{x}^i$ are symmetric with respect to the point x_0^i, if and only if $x^i + \bar{x}^i = 2x_0^i$. Two surfaces symmetric with respect to a point are isometric, and their second fundamental forms differ only in sign.

7. For any surface

$$\Delta_1 x^i = 1 - X^{i^2}, \qquad \Delta_1(x^i, x^j) = -X^i X^j \qquad (i \neq j),$$

where the differential parameters are formed with respect to $g_{\alpha\beta}$ (see §29).

39. THE EQUATION OF GAUSS AND THE EQUATIONS OF CODAZZI

In order that tensors $g_{\alpha\beta}$ and $d_{\alpha\beta}$ be the fundamental tensors of a surface it is necessary that the conditions of integrability of the equations (38.18) be satisfied, that is,

(39.1) $$x^i_{,\alpha\beta\gamma} - x^i_{,\alpha\gamma\beta} = x^i_{,\sigma} R^\sigma_{\alpha\beta\gamma},$$

as follows from equations of the form (22.19). By means of (38.19) these equations of condition are reducible to

$$(d_{\alpha\gamma} d_{\beta\epsilon} - d_{\alpha\beta} d_{\gamma\epsilon} - R_{\epsilon\alpha\beta\gamma}) g^{\sigma\epsilon} \frac{\partial x^i}{\partial u^\sigma} + (d_{\alpha\beta,\gamma} - d_{\alpha\gamma,\beta}) X^i = 0,$$

since $R^{\sigma}{}_{\alpha\beta\gamma} = g^{\sigma\epsilon}R_{\epsilon\alpha\beta\gamma}$ (see (20.15)). If these equations are multiplied by $\dfrac{\partial x^i}{\partial u^{\delta}}$ and summed with respect to i, we get, by means of (38.2), (38.8), and (24.16), the first of the following equations, and the second, when the above equations are multiplied by X^i and summed with respect to i:

$$d_{\alpha\gamma}d_{\beta\delta} - d_{\alpha\beta}d_{\gamma\delta} - R_{\delta\alpha\beta\gamma} = 0,$$
(39.2)
$$d_{\alpha\beta,\gamma} - d_{\alpha\gamma,\beta} = 0.$$

Because of (28.3) the first set of equations (39.2) which are not satisfied identically are equivalent to

$$(39.3) \qquad d_{11}d_{22} - d_{12}^2 = R_{1212}.$$

From this result and (28.5) we have *the equation of Gauss*[*]

$$(39.4) \qquad K = \frac{R_{1212}}{g} = \frac{d_{11}d_{22} - d_{12}^2}{g_{11}g_{22} - g_{12}^2}.$$

Since $d_{\alpha\beta}$ is a symmetric tensor, the second set of equations (39.2) consists of the two equations

$$(39.5) \qquad d_{\alpha\alpha,\beta} - d_{\alpha\beta,\alpha} = 0 \qquad (\alpha \neq \beta);$$

they are known as the *equations of Codazzi* because they are equivalent to equations derived by him.[†]   Equivalent equations had been derived earlier by Mainardi.[‡]   When the expressions for the covariant derivatives in (39.5) are written out, these equations become

$$(39.6) \qquad \frac{\partial d_{\alpha\alpha}}{\partial u^{\beta}} - \frac{\partial d_{\alpha\beta}}{\partial u^{\alpha}} - d_{\alpha\delta}\left\{\begin{matrix} \delta \\ \alpha\beta \end{matrix}\right\} + d_{\delta\beta}\left\{\begin{matrix} \delta \\ \alpha\alpha \end{matrix}\right\} = 0 \qquad (\alpha \neq \beta).$$

When equations (38.19) are differentiated covariantly with respect to u^{δ}, on noting that the covariant derivative of $g_{\alpha\beta}$ is zero by theorem [28.3] and making use of (38.18), one obtains

$$X^i{}_{,\alpha\delta} = -g^{\gamma\delta}\left(d_{\alpha\gamma,\delta}\frac{\partial x^i}{\partial u^{\beta}} + d_{\alpha\gamma}d_{\beta\delta}X^i\right).$$

Since X^i are scalars, the condition of integrability of (38.19) is $X^i{}_{,\alpha\delta} = X^i{}_{,\delta\alpha}$. This condition is satisfied because of the second set of (39.2), that is, the Codazzi equations.

Having shown that the first and second fundamental tensors of a

* 1827, 1, p. 234.
† 1869, 3, p. 275.
‡ 1856, 1, p. 395.

surface satisfy the equations (39.3) and (39.5), we inquire whether two tensors satisfying these equations are the fundamental tensors of a surface. This means that there must exist solutions x^i and X^i of equations (38.18) and (38.19) which satisfy the equations (38.2), (38.6) and (38.8). In order to answer this question we put

$$(39.7) \qquad \frac{\partial x^i}{\partial u^\alpha} = p^i_\alpha$$

in terms of which equations (38.18) and (38.19) are

$$(39.8) \qquad \frac{\partial p^i_\alpha}{\partial u^\beta} = p^i_\gamma \begin{Bmatrix} \gamma \\ \alpha\beta \end{Bmatrix} + d_{\alpha\beta} X^i, \qquad \frac{\partial X^i}{\partial u^\alpha} = -d_{\alpha\gamma} g^{\gamma\beta} p^i_\beta,$$

and equations (38.2), (38.6) and (38.8) are

$$(39.9) \qquad \sum_i p^i_\alpha p^i_\beta = g_{\alpha\beta}, \qquad \sum_i X^i X^i = 1, \qquad \sum_i X^i p^i_\alpha = 0.$$

Equations (39.8) and (39.9) constitute a mixed system of the type considered in §23. The conditions of integrability of equations (39.8) are satisfied because of (39.3) and (39.5), this being in fact the manner in which equations (39.3) and (39.5) were obtained. Equations (39.9) are the set E_0 in the terminology of §23. When they are differentiated it is found that the resulting equations are satisfied in consequence of (39.8), and consequently there are no sets E_1, E_2, $\cdots$, in the terminology of §23, to be satisfied. Equations (39.9) constitute six conditions upon the nine functions p^i_α and X^i, and thus by theorem [23.2] for a mixed system the solution of the mixed system involves three arbitrary constants. When such a solution is given the equations of the surface are given by the quadratures

$$(39.10) \qquad x^i = \int p^i_\alpha \, du^\alpha + b^i,$$

as follows from (39.7), where b^i are three additional constants. From the second and third of (39.9) it follows that X^i are the direction cosines of the normal to this surface.

We give an interpretation of the six arbitrary constants involved in the present problem. We observe that if x^i and X^i constitute a solution of equations (38.2), (38.6), (38.8), (38.18), and (38.19), so also do the quantities $\bar{x}^i$ and $\bar{X}^i$ defined by

$$(39.11) \qquad \bar{x}^i = a^i_j x^j + b^i, \qquad \bar{X}^i = a^i_j X^j,$$

where the a's and b's are constants and the a's are subject to the six conditions

$$(39.12) \qquad\qquad \sum_i a^i_j a^i_k = \delta_{jk}. \qquad\qquad (j, k = 1, 2, 3).$$

From the form of the first set of equations (39.11) and from (39.12) it follows that two surfaces defined by x^i and $\bar{x}^i$ may be obtained from one another by a rotation and a translation, that is, by a motion. Since these equations of motion involve the same number of arbitrary constants as the general solution of the present problem, it follows from the above discussion that any two surfaces with the same tensors $g_{\alpha\beta}$ and $d_{\alpha\beta}$ are transformable into one another by a motion in space. Hence we have

[39.1] *Two sets of functions $g_{\alpha\beta}$ and $d_{\alpha\beta}$ which satisfy the equations of Gauss and Codazzi and such that $g > 0$ are the components of the first and second fundamental tensors of a surface which is determined to within a motion in space.*

EXERCISES

1. When a surface is defined by an equation $x^3 = f(x^1, x^2)$ (see §25, Ex. 2), one finds that

$$X^1, X^2, X^3 = \frac{-p_1, -p_2, 1}{\sqrt{1 + p_1^2 + p_2^2}}, \qquad d_{11}, d_{12}, d_{22} = \frac{r_{11}, r_{12}, r_{13}}{\sqrt{1 + p_1^2 + p_2^2}},$$

where

$$p_\alpha = \frac{\partial x^3}{\partial x^\alpha}, \qquad r_{\alpha\beta} = \frac{\partial^2 x^3}{\partial x^\alpha \partial x^\beta} \qquad\qquad (\alpha, \beta = 1, 2).$$

2. For a right conoid with the equations (see §10, Ex. 4)

$$x^1 = u^1 \cos u^2, \qquad x^2 = u^1 \sin u^2, \qquad x^3 = \varphi(u^2),$$

one finds that

$$X^1, X^2, X^3 = \frac{\sin u^2 \varphi', -\cos u^2 \varphi', u^1}{\sqrt{\varphi'^2 + u^{1^2}}},$$

$$d_{11}, d_{12}, d_{22} = \frac{0, -\varphi', u^1 \varphi''}{\sqrt{\varphi'^2 + u^{1^2}}};$$

the locus of the normals to the surface along a generator is a hyperbolic paraboloid.

3. For a central quadric with the equations in §10 Ex. 3 one finds that

$$X^i = \sqrt{\frac{a_j a_k (a_i - u^1)(a_i - u^2)}{u^1 u^2 (a_i - a_j)(a_i - a_k)}},$$

where i, j, k take the values 1, 2, 3 cyclically, and

$$d_{\alpha\alpha} = -\frac{1}{4}\sqrt{\frac{a_1 a_2 a_3}{u^1 u^2}}\ \frac{u^\alpha - u^\beta}{(a_1 - u^\alpha)(a_2 - u^\alpha)(a_3 - u^\alpha)} \quad (\beta \neq \alpha), \quad d_{12} = 0.$$

4. For a paraboloid with the equations of §30, Ex. 7 one finds that

X_1, X_2, X_3

$$= \frac{a_1^{3/2}\sqrt{a_1 - a_2}\ \sqrt{u^1 u^2},\ a^2\sqrt{a_2 - a_1}\ \sqrt{(1 + a_1 u^1)(1 + a_2 u^2)},\ -\sqrt{a_1 a_2}}{A},$$

$$d_{\alpha\alpha} = \frac{1}{4}\sqrt{\frac{a_1^3}{a_2}\frac{(a_1 - a_2)(u^\alpha - u^\beta)}{Au^\alpha(1 + a_1 u^\alpha)}} \quad (\beta \neq \alpha), \quad d_{12} = 0,$$

where

$$A = \sqrt{[a_1(a_1 - a_2)u^1 - a_2][a_1(a_1 - a_2)u^2 - a_2]}.$$

40. NORMAL CURVATURE OF A SURFACE. PRINCIPAL RADII OF NORMAL CURVATURE

Consider any curve C upon a surface. Its tangent vector $\dfrac{dx^i}{ds}$ at any point P is perpendicular to the principal normal and binormal to the curve and the normal vector X^i to the surface. These three vectors are shown in Fig. 13 lying in the plane of the paper, and it is understood that the tangent vector to the curve at P is normal to the plane of the paper and is directed toward the reader; the line PR represents the line of intersection of the plane of the paper and the tangent plane to the surface at P. Denote by $\bar{\omega}$ (see Fig. 13) the angle made by the normal vector X^i with the principal normal, whose components are denoted by β^i. Hence on making use of (4.7) we have

(40.1) $$\cos\bar{\omega} = \sum_i X^i \beta^i = \rho \sum_i X^i \frac{d^2 x^i}{ds^2},$$

where ρ is the radius of curvature of C at P. In consequence of (38.11), (38.8), (38.13), and (24.6) we obtain from this equation

(40.2) $$\frac{\cos\bar{\omega}}{\rho} = \frac{d_{\alpha\beta}\,du^\alpha\,du^\beta}{g_{\alpha\beta}\,du^\alpha\,du^\beta}.$$

Thus $\dfrac{\cos\bar{\omega}}{\rho}$ is equal to the ratio of the values of the second and first fundamental quadratic forms of the surface for the direction of the curve at P, that is, for differentials du^α giving the direction of the tangent to the curve at P.

Since the right-hand member of (40.2) is completely determined by the values of the differentials du^α, it follows that the quantity $\dfrac{\cos \bar{\omega}}{\rho}$ is the same for all curves through P having a common tangent at P, denoted by PT. Since by hypothesis ρ is a positive quantity, it follows that according as $\cos \bar{\omega}$ is positive or negative for one of these curves it is the same for all of the curves. For a curve whose direction at P satisfies the equation

$$(40.3) \qquad d_{\alpha\beta}\, du^\alpha\, du^\beta = 0,$$

either $1/\rho = 0$ at P or the osculating plane at P is the tangent plane to the surface at P as follows from (40.2).

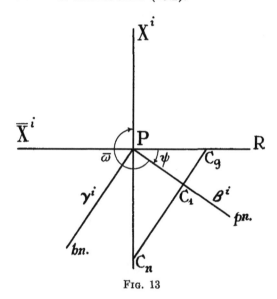

FIG. 13

Consider, in particular the plane curve in which the surface is cut by the plane determined by the normal to the surface at P and the tangent PT called the *normal section* of the surface for the direction PT. In this case $\bar{\omega}$ is $0°$ or $180°$ according as the principal normal to C and the normal X^i have the same or opposite directions. If we denote by ρ_n the radius of curvature of this plane curve at P, assumed to be positive, we have

$$(40.4) \qquad \frac{\cos \bar{\omega}}{\rho} = \frac{e}{\rho_n},$$

where e is $+1$ or -1, according as the principal normal to C and the normal X^i have the same or opposite directions.

Hence we have the *theorem of Meusnier:*[*]

[40.1] *The center of curvature at a point P for a curve on a surface is the projection upon its osculating plane of the center of curvature of that normal section of the surface which is tangent to the curve at P.*

If now we define a quantity R for a direction du^α at a point u^α by

$$(40.5) \qquad \frac{1}{R} = \frac{\cos \bar{\omega}}{\rho} = \frac{d_{\alpha\beta} \, du^\alpha \, du^\beta}{g_{\alpha\beta} \, du^\alpha \, du^\beta},$$

it follows from (40.4) that the absolute value of R is the radius of curvature of the normal section of the surface for the given direction, and that R is positive or negative according as the normal of the plane section at u^α and the normal X^i to the surface have the same or opposite directions. The quantity R defined by (40.5) is called the *radius of normal curvature* of the surface at a point u^α for the direction du^α.

A normal section of a plane being a straight line, $1/R = 0$ in this case. A normal section of a sphere being a great circle R is equal to the radius of the sphere. In both these cases the value of R does not vary with the direction du^α. Conversely, if R is not to vary with du^α, it is necessary and sufficient that $d_{\alpha\beta} = 0$, or that $d_{\alpha\beta}$ be proportional to $g_{\alpha\beta}$; in the former case the surface is a plane, and in the latter a sphere (see §38, Exs. 1 and 2).

In order to find the directions at a point on a surface for which $1/R$ is a maximum or minimum, we note that equation (40.5) is of the form (26.26). Consequently these directions are given by the values of du^α satisfying the equations

$$(40.6) \qquad \left(d_{\alpha\beta} - \frac{1}{R} \, g_{\alpha\beta} \right) du^\beta = 0,$$

as follows from (26.17) and the discussion following (26.26), and the corresponding values of $1/R$ are the roots of the determinant equation

$$(40.7) \qquad \left| d_{\alpha\beta} - \frac{1}{R} \, g_{\alpha\beta} \right| = 0.$$

Thus the directions sought are the principal directions for the tensor $d_{\alpha\beta}$; these directions are called the *principal directions of normal curvature*. They are given by the equation

$$(40.8) \qquad \begin{vmatrix} d_{1\alpha} \, du^\alpha & d_{2\alpha} \, du^\alpha \\ g_{1\alpha} \, du^\alpha & g_{2\alpha} \, du^\alpha \end{vmatrix} = 0,$$

[*] 1776, 1, p. 489.

as follows from (26.23), and they are real by theorem [26.4]. When theorem [26.5] is applied to this case, we have

[40.2] *At each point of a surface, other than a plane or a sphere, equation (40.8) determines the two directions for which the normal curvature is a maximum or minimum, these directions being perpendicular.*

The values of R for these directions, called the *principal radii of normal curvature*, are the roots of the equation (40.7), that is

(40.9) $\quad (d_{11}d_{22} - d_{12}^2)R^2 - (d_{11}g_{22} + d_{22}g_{11} - 2d_{12}g_{12})R + g = 0.$

If these roots are denoted by ρ_1 and ρ_2, it follows from (40.9) and (39.4) that

(40.10) $$\frac{1}{\rho_1 \rho_2} = \frac{d_{11}d_{22} - d_{12}^2}{g} = K,$$

(40.11) $\quad K_m = \dfrac{1}{\rho_1} + \dfrac{1}{\rho_2} = \dfrac{g_{11}d_{22} + g_{22}d_{11} - 2g_{12}d_{12}}{g} = g^{\alpha\beta}d_{\alpha\beta}.$

From (40.10) we have the following theorem of Gauss:*

[40.3] *The curvature K at a point of a surface is equal to the reciprocal of the product of the principal radii of normal curvature at the point.*

The quantity $\dfrac{1}{\rho_1} + \dfrac{1}{\rho_2}$, denoted by K_m, is called the *mean curvature of the surface* at a point. Equation (40.11) gives its expression in terms of the two fundamental tensors of the surface.

Ordinarily the principal radii at a point are unequal, but there may be points at which they are equal. Such a point is called an *umbilical point*. From (40.10) and (40.11) it follows that

[40.4] *A necessary and sufficient condition that a point u^α be an umbilical point is that at the point*

(40.12) $$(g^{\alpha\beta}d_{\alpha\beta})^2 = \frac{4}{g}(d_{11}d_{22} - d_{12}^2).$$

From geometric considerations it follows that every point on a sphere is an umbilical point. For this reason an umbilical point is sometimes called a *spherical point*.

The centers of curvature of the *principal normal sections* of a surface

* 1827, 1, p. 231.

at a point, that is, the normal sections in the principal directions, are given by

(40.13) $$x_1^i = x^i + \rho_1 X^i, \qquad x_2^i = x^i + \rho_2 X^i.$$

These centers are called the *principal centers of curvature* for the point.

When the Gaussian curvature K is positive at a point P of a surface, the principal radii have the same sign. Since in this case $d_{11}d_{22} - d_{12}^2 > 0$, as follows from (40.10), the second quadratic form $d_{\alpha\beta} \, du^\alpha \, du^\beta$ has the same sign at a point for all values of du^α, there being no real values of du^α for which the quadratic form is equal to zero. Consequently it follows from (38.12) that points in the neighborhood of P lie entirely on one side of the tangent plane. A surface at all points of which K is positive is called a *surface of positive curvature*.

When $K < 0$ at a point P, the principal radii differ in sign. In this case equation (40.3) defines two real directions at P for each of which R is infinite. From theorem [26.4] we have that the bisectors of the angles between these directions are the principal directions at P. Since the principal radii differ in sign and since the second fundamental form for a point can change sign only at the directions given by (40.3), it follows that in the neighborhood of P part of the surface lies on one side of the tangent plane and part on the other, these parts being separated by the two branches of the curve in which the surface is cut by the tangent plane at P, the tangents at P to these branches having the directions given by (40.3). A surface at all points of which $K < 0$ is called a *surface of negative curvature*. From the considerations of this and the preceding paragraph it follows from geometric considerations that ellipsoids, hyperboloids of two sheets and elliptic paraboloids are surfaces of positive curvature, and that hyperboloids of one sheet and hyperbolic paraboloids are surfaces of negative curvature.

When $K = 0$ at a point, the directions given by (40.3) coincide, since in this case $d_{11}d_{22} - d_{12}^2 = 0$, and for this direction R is infinite. In this case the coefficient of $\frac{1}{2}s^2$ in (38.12) is a perfect square and the question of whether the distance p has always the same sign depends on the terms of higher order. Consider, for example, a cylinder whose cross-section has a simple inflection at a point P. $K = 0$ for all points of the surface, and the surface lies on both sides of the tangent plane along the generator through P.

EXERCISES

1. If a segment equal to twice the radius of normal curvature for a direction PT at a point P on a surface is laid off from P on the normal to the surface in the

appropriate direction, and a sphere is described with the segment as diameter, the circle in which the sphere is met by the osculating plane of any curve through P and with the direction PT is the circle of curvature of the curve.

2. When the coordinate curves of a surface are such that $g_{12} = d_{12} = 0$, $1/R$ is a maximum or minimum in the direction of the coordinate curves, and

$$\frac{1}{\rho_1} = \frac{d_{11}}{g_{11}}, \qquad \frac{1}{\rho_2} = \frac{d_{22}}{g_{22}};$$

in this coordinate system the points, if any, for which $d_{11}g_{22} - d_{22}g_{11} = 0$ are umbilical points.

3. Show analytically that each point of a sphere is an umbilical point.

4. The tangent to the meridians and parallels of a surface of revolution are the principal directions at each point of the surface; the principal radii of a surface with the equations of §10, Ex. 2 are given by

$$\frac{1}{\rho_1} = \frac{\varphi''}{(1 + \varphi'^2)^{3/2}}, \qquad \frac{1}{\rho_2} = \frac{\varphi'}{u^1(1 + \varphi'^2)^{1/2}},$$

(see §24, Ex. 3 and §38, Ex. 4); ρ_2 is the segment of the normal between the point of the surface and the axis of the surface.

5. Let P_0 denote the center of normal curvature in the direction bisecting the principal directions at a point P of a surface, and P_1 and P_2 the centers of normal curvature in directions equally inclined to this bisector; then P, P_1, P_0, P_2 form a harmonic range.

6. Let $R_1, R_2, \cdots, R_m$ denote the radii of normal curvature for $m \ (> 2)$ directions such that the angle of two adjoining directions is $2\pi/m$; then

$$\frac{1}{m}\left(\frac{1}{R_1} + \frac{1}{R_2} + \cdots + \frac{1}{R_m}\right) = \frac{1}{2}\left(\frac{1}{\rho_1} + \frac{1}{\rho_2}\right).$$

7. From §30, Ex. 6 and §39, Ex. 3 it can be shown that the points

$$x^1 = \pm\sqrt{\frac{a_1(a_1 - a_2)}{a_1 - a_3}}, \qquad x^2 = 0, \qquad x^3 = \pm\sqrt{\frac{a_3(a_2 - a_3)}{a_1 - a_3}}$$

are umbilical points of an ellipsoid, and

$$x^1 = \pm\sqrt{\frac{a_1(a_1 - a_3)}{a_1 - a_2}}, \qquad x^2 = \pm\sqrt{\frac{a_2(a_2 - a_3)}{a_2 - a_1}}, \qquad x^3 = 0$$

of a hyperboloid of two sheets; and that there are no umbilical points on a hyperboloid of one sheet.

8. The principal radii of normal curvature of the surface of revolution of a parabola about its directrix are in constant ratio.

9. The surface of revolution of a circle of radius a about a line in the plane of the circle at the distance $b(> a)$ from the center of the circle is called a *torus*; the Gaussian curvature of the surface at a point P is positive, zero, or negative according as the distance of P from the axis is greater than, equal to, or less than b.

41. LINES OF CURVATURE OF A SURFACE

The integral curves of the differential equation (40.8), that is, the equation

(41.1)
$$(d_{11}g_{12} - d_{12}g_{11})\, du^{1^2} + (d_{11}g_{22} - d_{22}g_{11})\, du^1\, du^2$$
$$+ (d_{12}g_{22} - d_{22}g_{12})\, du^{2^2} = 0,$$

are called the *lines of curvature* of a surface. We have seen that they form a real orthogonal net, and that the tangents to the two curves of the net at a point are the principal directions of normal curvature at the point. We now give another characteristic property of the lines of curvature.

The normals to a surface at points x^i of a curve C form a ruled surface, whose equations in parametric form are

(41.2)
$$\bar{x}^i = x^i + tX^i,$$

the quantities x^i, X^i being expressed in terms of the arc s of C, and t being a second parameter, namely the distance of the point $\bar{x}^i$ from the point x^i. In order that the ruled surface be the tangent surface of a curve Γ, and thus a developable surface, the normals to the given surface being the tangents to Γ, the equation $t = f(s)$ of the curve Γ must be such that $\dfrac{d\bar{x}^i}{ds}$ are proportional to X^i. From (41.2) it follows that this condition is

(41.3)
$$\frac{dx^i}{ds} + t\frac{dX^i}{ds} + \frac{dt}{ds}X^i = hX^i.$$

Multiplying these equations by X^i and summing with respect to i, we have in consequence of (38.6) and (38.8) that

(41.4)
$$\frac{dt}{ds} = h,$$

and consequently (41.3) reduce to

(41.5)
$$\frac{dx^i}{ds} + t\frac{dX^i}{ds} = 0.$$

In order that the normals to the surface along a curve shall form a cone, it is necessary that the equation $t = f(s)$ be such that the point $\bar{x}^i$ defined by (41.2) be the vertex of the cone, in which case $\dfrac{d\bar{x}^i}{ds} = 0$. In this case we have (41.3) with $h = 0$, and from (41.4) $t = $ const., and again we have equations (41.5).

When equations (41.5) written in the form

(41.5)′ $$\left(\frac{\partial x^i}{\partial u^\alpha} + t \frac{\partial X^i}{\partial u^\alpha}\right) \frac{du^\alpha}{ds} = 0$$

are multiplied by $\dfrac{\partial x^i}{\partial u^\beta}$ and summed with respect to i, we obtain in consequence of (38.16)

(41.6) $$(g_{\alpha\beta} - t\, d_{\alpha\beta}) \frac{du^\alpha}{ds} = 0.$$

Since these equations are equivalent to (40.6) with $t = R$, and the equations (40.8) are a consequence of (40.6), we have

[41.1] *A necessary and sufficient condition that the normals to a surface along a curve form a developable surface is that the curve be a line of curvature; when the developable surface is the tangent surface of a curve, equations of this curve are (41.2) where t is the principal radius of normal curvature of the line of curvature concerned; the developable surface is a cone, if and only if the corresponding principal radius is a constant along the curve.*

The proof of this theorem is not complete until one considers the possibility of the normals forming a cylinder, that is, $\dfrac{dX^i}{ds} = 0$. In this case we have, using (38.16),

$$\sum_i \frac{\partial x^i}{\partial u^\alpha} \frac{dX^i}{ds} = \sum_i \frac{\partial x^i}{\partial u^\alpha} \frac{\partial X^i}{\partial u^\beta} \frac{du^\beta}{ds} = -d_{\alpha\beta} \frac{du^\beta}{ds} = 0 \quad (\alpha = 1, 2).$$

A curve for which these two equations hold is evidently an integral curve of equation (40.8), and consequently is a line of curvature. In order that the above equations for $\alpha = 1, 2$ be consistent we must have $d_{11}d_{22} - d_{12}^2 = 0$ along the curve. Hence from (39.4) and theorem [28.1] either the surface is a developable surface and the curve is a generator, or $K = 0$ at least at all points of the curve.

Since the condition $\dfrac{dX^i}{ds} = 0$ holds along each generator of a developable surface, we have

[41.2] *The lines of curvature of a developable surface are its generators and their orthogonal trajectories.*

In order that the coordinate curves be lines of curvature, it is necessary and sufficient that

(41.7) $d_{11}\, g_{12} - d_{12}\, g_{11} = 0,$ $d_{12}\, g_{22} - d_{22}\, g_{12} = 0,$

as follows from equation (41.1). Consequently $g_{12} = d_{12} = 0$, unless the two fundamental forms are proportional, which is the case only when the surface is a plane or sphere (see §38, Exs. 1, 2). Hence we have

[41.3] *A necessary and sufficient condition that the lines of curvature be coordinate, that is, the coordinate curves, on a surface other than a plane or sphere is that*

$$(41.8) \qquad d_{12} = g_{12} = 0;$$

any orthogonal net on a plane or sphere satisfies these conditions.

When the lines of curvature are coordinate, equations (38.19) reduce to

$$(41.9) \qquad \frac{\partial X'}{\partial u^\alpha} = -\frac{1}{\rho_\alpha} \frac{\partial x^i}{\partial u^\alpha} \quad (\alpha = 1, 2; \alpha \text{ not summed}),$$

in consequence of (41.8) and (24.18), the principal radii being given by

$$(41.10) \qquad \frac{1}{\rho_1} = \frac{d_{11}}{g_{11}}, \qquad \frac{1}{\rho_2} = \frac{d_{22}}{g_{22}}.$$

When the lines of curvature are coordinate, the Codazzi equations (39.6) reduce to (see §28, Ex. 1)

$$(41.11) \qquad \begin{aligned} \frac{\partial d_{11}}{\partial u^2} - \frac{1}{2}\left(\frac{d_{11}}{g_{11}} + \frac{d_{22}}{g_{22}}\right)\frac{\partial g_{11}}{\partial u^2} &= 0, \\ \frac{\partial d_{22}}{\partial u^1} - \frac{1}{2}\left(\frac{d_{11}}{g_{11}} + \frac{d_{22}}{g_{22}}\right)\frac{\partial g_{22}}{\partial u^1} &= 0. \end{aligned}$$

In consequence of (41.10) these equations are expressible in the form

$$(41.12) \qquad \begin{aligned} \frac{\partial}{\partial u^2}\left(\frac{1}{\rho_1}\right) + \frac{1}{2}\left(\frac{1}{\rho_1} - \frac{1}{\rho_2}\right)\frac{\partial \log g_{11}}{\partial u^2} &= 0, \\ \frac{\partial}{\partial u^1}\left(\frac{1}{\rho_2}\right) + \frac{1}{2}\left(\frac{1}{\rho_2} - \frac{1}{\rho_1}\right)\frac{\partial \log g_{22}}{\partial u^1} &= 0. \end{aligned}$$

EXERCISES

1. When a line of curvature is a plane curve and the normals to the surface along the curve lie in the plane of the curve, the developable surface of the normals consists of the tangent lines to the plane evolute of the curve.

2. The lines of curvature of the tangent surface of a curve are the tangents to the curve and its involutes.

3. The meridians and parallels of a surface of revolution are the lines of curvature of the surface (see §24, Ex. 3 and §38, Ex. 4).

4. The coordinate curves on a central quadric with the parametric equations in §10, Ex. 3 are the lines of curvature of the quadric (see §30, Ex. 6 and §39, Ex. 3).

5. The coordinate curves on a paraboloid with the parametric equations in §30, Ex. 7 are the lines of curvature of the paraboloid (see §39, Ex. 4).

6. The lines of curvature of a spiral surface as defined in §26, Ex. 8 can be found by quadratures.

42. CONJUGATE DIRECTIONS AND CONJUGATE NETS. ISOMETRIC-CONJUGATE NETS

Consider a curve C on a surface and the tangent planes to the surface at points of the curve. The envelope of this one parameter family of tangent planes is a developable surface (§12). Through each point of the curve there passes a generator of the developable surface. The generator of the surface and the tangent to the curve at the point are said to have *conjugate directions*. In order to find an analytic expression for conjugate directions we consider the equation of the tangent plane

$$(42.1) \qquad \sum_i X^i(\bar{x}^i - x^i) = 0,$$

where $\bar{x}^i$ are current coordinates, and X^i and x^i are functions of the arc along the curve. On differentiating (42.1) with respect to s we have in consequence of (38.8)

$$(42.2) \qquad \sum_i \frac{\partial X^i}{\partial u^\alpha} (\bar{x}^i - x^i) \frac{du^\alpha}{ds} = 0.$$

This equation and (42.1) are equations of the generator of the developable through the point x^i. The quantities $\bar{x}^i - x^i$ are direction numbers of the generator, and they are proportional to $\frac{\partial x^i}{\partial u^\alpha} \delta u^\alpha$, where δu^α are differentials in the direction conjugate to the tangent to the curve at the point (see (24.4)). In consequence of this observation and equations (38.16) we have from (42.2)

$$(42.3) \qquad d_{\alpha\beta} du^\alpha \delta u^\beta = 0,$$

which gives the relation between the differentials du^α for C and δu^α for its conjugate direction. From the form of (42.3) it is seen that conjugacy of directions is reciprocal, that is, if a direction is conjugate to a given direction, the latter is conjugate to the former.

Comparing equations (42.3) and (26.7), we observe that the second fundamental tensor bears to conjugacy the same relation which the first fundamental tensor bears to orthogonality.

The directions conjugate to each of the curves

(42.4) $\varphi(u^1, u^2) = \text{const.}$

at each point of the curve are given by (see (26.2))

(42.5) $\left(d_{11} \dfrac{\partial \varphi}{\partial u^2} - d_{12} \dfrac{\partial \varphi}{\partial u^1} \right) \delta u^1 + \left(d_{12} \dfrac{\partial \varphi}{\partial u^2} - d_{22} \dfrac{\partial \varphi}{\partial u^1} \right) \delta u^2 = 0.$

Looking upon this equation as a differential equation in δu^α, we have that its integral curves and the curves $\varphi = \text{const.}$ form a net, such that a curve of each family passes through a point of the surface and at the point their directions are conjugate. Such a net is called a *conjugate net*.

From (42.3) it follows that the curves conjugate to the curves $u^2 = \text{const.}$ are the integral curves of the equation

(42.6) $d_{11}\delta u^1 + d_{12}\delta u^2 = 0.$

In order that they be the curves $u^1 = \text{const.}$ we must have $d_{12} = 0$, and conversely. Hence we have

[42.1] *A necessary and sufficient condition that the coordinate curves form a conjugate net is that d_{12} be zero.*

From this theorem and theorem [41.3] we have

[42.2] *The lines of curvature of a surface form a conjugate net, and the only conjugate net which is an orthogonal net.*

From equations (38.18) it follows that when the coordinate curves form a conjugate net each of the coordinates x^i is a solution of an equation of the form

(42.7) $\dfrac{\partial^2 \theta}{\partial u^1 \partial u^2} + a^1 \dfrac{\partial \theta}{\partial u^1} + a^2 \dfrac{\partial \theta}{\partial u^2} = 0,$

where in general a^1 and a^2 are functions of u^α.

Conversely, if x^i for $i = 1, 2, 3$ are linearly independent solutions of an equation of this type, then

$$\begin{vmatrix} \dfrac{\partial^2 x^1}{\partial u^1 \partial u^2} & \dfrac{\partial^2 x^2}{\partial u^1 \partial u^2} & \dfrac{\partial^2 x^3}{\partial u^1 \partial u^2} \\[2ex] \dfrac{\partial x^1}{\partial u^1} & \dfrac{\partial x^2}{\partial u^1} & \dfrac{\partial x^3}{\partial u^1} \\[2ex] \dfrac{\partial x^1}{\partial u^2} & \dfrac{\partial x^2}{\partial u^2} & \dfrac{\partial x^3}{\partial u^2} \end{vmatrix} = 0,$$

from which and from (38.5) and (38.13) it follows that $d_{12} = 0$.   Hence we have

[42.3] *If $f^i(u^1, u^2)$ for $i = 1, 2, 3$ are three linearly independent solutions of an equation of the form (42.7), the coordinate curves form a conjugate net on the surface with the equations $x^i = f^i(u^1, u^2)$.*

When the lines of curvature are coordinate, we have, since $g_{12} = 0$,

$$\sum_i \frac{\partial x^i}{\partial u^1} \frac{\partial x^i}{\partial u^2} = 0.$$

In this case not only are x^i solutions of an equation of the form (42.7) but also the sum of their squares, that is $\sum_i x^i x^i$, is a solution as one verifies by substitution, and conversely.   Hence we have the theorem of Darboux:*

[42.4] *A necessary and sufficient condition that the coordinate curves be the lines of curvature on the surface defined by three linearly independent solutions of an equation of the form (42.7) is that the sum of the squares of these solutions is also a solution.*

Darboux† has applied this result to the proof of the following theorem:

[42.5] *When a surface S is transformed into a surface S_1 by an inversion, the lines of curvature of S are transformed into the lines of curvature of S_1 .*

An *inversion*, or a *transformation by reciprocal radii*, is defined by

$$(42.8) \qquad x_1^i - x_0^i = \frac{c^2(x^i - x_0^i)}{\sum_j (x^j - x_0^j)(x^j - x_0^j)},$$

where c is a constant, and the point x_0^i is the *center* of the transformation.   From (42.8) it follows that

$$\sum_i [(x_1^i - x_0^i)(x_1^i - x_0^i)] \sum_j (x^j - x_0^j)(x^j - x_0^j) = c^4,$$

$$(42.9)$$

$$x^i - x_0^i = \frac{c^2(x_1^i - x_0^i)}{\sum_j (x_1^j - x_0^j)(x_1^j - x_0^j)},$$

which shows the reciprocal character of the transformation.

Suppose now that x^i are solutions of an equation (42.7), then x_1^i given by (42.8) are functions of u^α which are solutions of the equation in θ_1 resulting from the substitution

$$(42.10) \qquad \theta = \frac{\theta_1}{\sum_j (x_1^j - x_0^j)(x_1^j - x_0^j)}$$

* 1887, 1, p. 136.
† 1887, 1, p. 208.

in equation (42.7), as follows from the second of equations (42.9), x_0^i being constants. If $\sum_i x^i x^i$ is a solution of (42.7), then as follows from the first of (42.9) and (42.10) c^4 is a solution of the equation in θ_1, which consequently is of the form (42.7). Since (42.7) admits unity for a solution, it follows from (42.10) that the equation in θ_1 admits the solution $\sum_i x_1^i x_1^i$, and theorem [42.5] follows in consequence of theorem [42.4].

For a surface of positive or negative curvature, $d_{11}d_{22} - d_{12}^2$, denoted by d, is not equal to zero. Consequently quantities $d^{\alpha\beta}$ are uniquely defined by

$$(42.11) \qquad d^{\alpha\beta}d_{\beta\gamma} = \delta_\gamma^\alpha ;$$

in fact

$$(42.12) \qquad d^{11} = \frac{d_{22}}{d}, \qquad d^{12} = -\frac{d_{12}}{d}, \qquad d^{22} = \frac{d_{11}}{d}.$$

If then we define quantities $\bar{\Delta}_1\varphi$ and $\bar{\Delta}_1(\varphi, \psi)$ by

$$(42.13) \qquad \bar{\Delta}_1\varphi = d^{\alpha\beta}\varphi_{,\alpha}\varphi_{,\beta}, \qquad \bar{\Delta}_1(\varphi, \psi) = d^{\alpha\beta}\varphi_{,\alpha}\psi_{,\beta},$$

these are scalars for a transformation of coordinates u^α, since it follows from (42.11) as in §14 that $d^{\alpha\beta}$ is a symmetric contravariant tensor. From (42.13) we have

$$(42.14) \qquad \begin{aligned} \bar{\Delta}_1 u^1 &= d^{11} = \frac{d_{22}}{d}, \qquad \bar{\Delta}_1 u^2 = d^{22} = \frac{d_{11}}{d}, \\ \bar{\Delta}_1(u^1, u^2) &= d^{12} = -\frac{d_{12}}{d}, \end{aligned}$$

and consequently in terms of any net $\varphi = $ const., $\psi = $ const., the second fundamental form may be written

$$(42.15) \qquad \frac{\bar{\Delta}_1\psi \, d\varphi^2 - 2\bar{\Delta}_1(\varphi, \psi) \, d\varphi \, d\psi + \bar{\Delta}_1\varphi \, d\psi^2}{\bar{\Delta}_1\varphi\bar{\Delta}_1\psi - \bar{\Delta}_1^2(\varphi, \psi)}.$$

We define also the quantity $\bar{\Delta}_2\varphi$ by an equation analogous to (29.16), namely

$$\bar{\Delta}_2\varphi = \frac{1}{\sqrt{ed}} \frac{\partial}{\partial u^\beta} (\sqrt{ed} \, d^{\alpha\beta}\varphi_{,\alpha}),$$

where e is $+1$ or -1 according as K is positive or negative for the portion of the surface under consideration. If then φ is any real solution of the equation $\bar{\Delta}_2\varphi = 0$, a real function ψ is obtained from (see (30.5))

(42.16) $$\sqrt{ed}\, d^{\alpha 1}\varphi_{,\alpha} = \psi_{,2}, \qquad \sqrt{ed}\, d^{\alpha 2}\varphi_{,\alpha} = -\psi_{,1}$$

by a quadrature. Proceeding as in §30, we find from these equations that

$$\varphi_{,1} = e\sqrt{ed}\, d^{\beta 2}\psi_{,\beta}, \qquad \varphi_{,2} = -e\sqrt{ed}\, d^{\beta 1}\psi_{,\beta},$$

and

$$\bar{\Delta}_1\varphi = e\bar{\Delta}_1\psi, \qquad \bar{\Delta}_1(\varphi, \psi) = 0,$$

so that the second fundamental form (42.15) becomes

$$\frac{1}{\bar{\Delta}_1\varphi}\,(d\varphi^2 + e\,d\psi^2).$$

In case the coordinates are such that the second fundamental form is

(42.17) $$\lambda(du^{1^2} + e\,du^{2^2}),$$

the coordinate curves are said to form an *isometric-conjugate net*.

When for a conjugate net on a surface

(42.18) $$\frac{d_{11}}{d_{22}} = \frac{f_1(u^1)}{f_2(u^2)},$$

new coordinates can be chosen in terms of which the second fundamental form assumes the form (42.17). The coordinates are called *isometric-conjugate* when the second fundamental form is of the particular form (42.17).

Proceeding as in §30, we have analogously to theorem [30.3]

[42.6] *A necessary and sufficient condition that a family of curves* $\varphi =$ const. *and their conjugate trajectories form an isometric-conjugate net is that* $\bar{\Delta}_2\varphi = 0$ *or that the ratio of* $\bar{\Delta}_2\varphi$ *and* $\bar{\Delta}_1\varphi$ *is a function of* φ.

EXERCISES

1. The coordinate curves on a surface defined by

(i) $$x^i = f_1^i(u^1) + f_2^i(u^2)$$

form a conjugate net. Each of the curves $u^2 =$ const. may be obtained from the curve $x^i = f_1^i(u^1)$ by a suitable translation, and the curves $u^1 =$ const. by a suitable translation of the curve $x^i = f_2^i(u^2)$; and the tangents to one coordinate family at points of a curve of the other family are parallel. A surface with equations of the form (i) is called a *surface of translation*, and the curves $x^i = f_1^i(u^1)$ and $x^i = f_2^i(u^2)$ are called the *generating curves*.

2. The coordinate curves form a conjugate net on the surface

$$x^i = \frac{f_1^i(u^1) + f_2^i(u^2)}{f(u^1) + \varphi(u^2)}.$$

3. The meridians and parallels on a surface of revolution form an isometric-conjugate net (see §38, Ex. 4.)

4. The lines of curvature on a central quadric and on a paraboloid form an isometric-conjugate net (see §39, Exs. 3 and 4).

5. The coordinate curves form an isometric-conjugate net on the surface

$$x^1 = f_1(u^1)\varphi_1(u^2), \qquad x^2 = f_2(u^1)\varphi_1(u^2), \qquad x^3 = \varphi_2(u^2).$$

6. When the coordinate curves of a surface form an isometric-conjugate net and the coordinates are isometric-conjugate, then $K = \dfrac{e\lambda^2}{g}$. If one puts $\dfrac{1}{\rho} = \dfrac{\lambda}{\sqrt{g}}$,

then $d_{11} = \dfrac{\sqrt{g}}{\rho}$, $d_{22} = \dfrac{e\sqrt{g}}{\rho}$, and the equations of Codazzi (39.6) are

$$\frac{\partial}{\partial u^\alpha} \log \frac{\sqrt{g}}{\rho} = \left\{ \begin{matrix} \beta \\ \alpha\beta \end{matrix} \right\} - e \left\{ \begin{matrix} \alpha \\ \beta\beta \end{matrix} \right\} \qquad (\alpha, \beta = 1, 2; \beta \neq \alpha).$$

7. When the coordinate curves on a surface are isometric-conjugate and the coordinates are isometric-conjugate, equations (42.16) reduce to

$$\varphi_{,1} = \psi_{,2}, \qquad \varphi_{,2} = -e\psi_{,1}.$$

When $e = 1$, a solution of these equations is $\varphi + i\psi = f(u^1 + iu^2)$, and when $e = -1$,

$$\varphi = f_1(u^1 + u^2) + f_2(u^1 - u^2), \qquad \psi = f_1(u^1 + u^2) - f_2(u^1 - u^2).$$

8. When a plane is subjected to an inversion it is tranformed into a plane or a sphere according as the given plane passes through the center of the inversion or not; when a line is subjected to an inversion it is transformed into a line or a circle according as the given line passes through the center of the inversion or not.

9. From (42.8) one obtains

$$\sum_i dx_1^i dx_1^i = \frac{c^4 \sum_i dx^i dx^i}{\left[\sum_k (x^k - x_0^k)(x^k - x_0^k) \right]^2},$$

and consequently a transformation by reciprocal radii is conformal.

10. When a cone of revolution is subjected to a transformation by reciprocal radii whose center is not on the cone, the lines of curvature on the new surface are circles, and this surface is the envelope of a family of spheres having a point in common, the spheres being the transforms of the tangent planes to the cone.

11. A necessary and sufficient condition that a conjugate net on a surface be a Tchebychef net (see §29, Ex. 12) is that the surface be a surface of translation.

43. ASYMPTOTIC DIRECTIONS AND ASYMPTOTIC LINES. MEAN CONJUGATE DIRECTIONS. THE DUPIN INDICATRIX

A direction at a point in a surface which is self-conjugate is called *asymptotic*. From equation (42.3) it follows that the asymptotic directions at each point of a surface are given by

$$(43.1) \qquad\qquad d_{\alpha\beta}\, du^{\alpha}\, du^{\beta} = 0.$$

The integral curves of this differential equation of the first order and second degree are called the *asymptotic lines* of the surface. Accordingly an asymptotic line is a curve whose tangent at any point is an asymptotic direction at the point. From (43.1) it follows that the asymptotic directions upon a surface are conjugate imaginary, real and distinct, or real and coincident according as $d_{11}d_{22} - d_{12}^2$ is positive, negative, or equal to zero. From this result and (39.4) we have

[43.1] *The asymptotic lines are conjugate imaginary, real and distinct, or real and coincident on a surface, or any portion of a surface, for which the Gaussian curvature is positive, negative, or equal to zero.*

From theorem [26.4] applied to equation (43.1) and equation (40.8) we have

[43.2] *On a surface, or portion of a surface, for which the Gaussian curvature is not equal to zero, the lines of curvature bisect the angles formed by the asymptotic lines.*

When a straight line lies entirely in a surface, the tangent plane to the surface at each point of the line contains the line. From the definition of asymptotic lines it follows that (see §38, Ex. 3):

[43.3] *When a straight line lies entirely in a surface, it is an asymptotic line of the surface.*

In §41 we saw that the generators of a developable surface are also lines of curvature. From (40.1) and (40.2) we have

[43.4] *The osculating planes of a curved asymptotic line are tangent to the surface along the curve.*

From equation (43.1) we have

[43.5] *A necessary and sufficient condition that the asymptotic lines be coordinate is that*

$$(43.2) \qquad\qquad d_{11} = d_{22} = 0.$$

From this result and (40.11) it follows that the asymptotic lines form an orthogonal net, if and only if $K_m = \dfrac{1}{\rho_1} + \dfrac{1}{\rho_2} = 0$. A surface for which the mean curvature K_m is equal to zero is called a *minimal surface*; the property of such a surface which justifies the term minimal is established in §50. From (40.11) it follows also that when the minimal curves (§2) on such a surface are coordinate, that is, $g_{11} = g_{22} = 0$, then $d_{12} = 0$. Hence we have

[43.6] *On a minimal surface the asymptotic lines form an orthogonal net and the minimal lines a conjugate net; each of these properties characterizes a minimal surface.*

From theorem [43.5] and equations (38.18) it follows that, when the asymptotic lines are coordinate, the quantities x^i as functions of u^1 and u^2 are solutions of two equations of the form

$$(43.3) \qquad \frac{\partial^2 \theta}{\partial u^{1^2}} = a_{11} \frac{\partial \theta}{\partial u^1} + a_{12} \frac{\partial \theta}{\partial u^2}, \qquad \frac{\partial^2 \theta}{\partial u^{2^2}} = a_{21} \frac{\partial \theta}{\partial u^1} + a_{22} \frac{\partial \theta}{\partial u^2},$$

where in general the a's are functions of u^α. By an argument similar to that which led to theorem [42.3] we have

[43.7] *When two equations of the form (43.3) admit three linearly independent solutions $f^i(u^1, u^2)$ for $i = 1, 2, 3$, the coordinate curves are the asymptotic lines on the surface with the equations $x^i = f^i(u^1, u^2)$.*

When equations (43.2) are satisfied, and we put $1/\rho = d_{12}/\sqrt{g}$, then the Gaussian curvature is given by

$$(43.4) \qquad\qquad K = -\frac{1}{\rho^2},$$

and the Codazzi equations (39.6) reduce to

$$(43.5) \qquad \frac{\partial \log \sqrt{\rho}}{\partial u^1} = \begin{Bmatrix} 2 \\ 12 \end{Bmatrix}, \qquad \frac{\partial \log \sqrt{\rho}}{\partial u^2} = \begin{Bmatrix} 1 \\ 12 \end{Bmatrix},$$

by means of equations (28.2), namely

$$(43.6) \qquad\qquad \frac{\partial \log \sqrt{g}}{\partial u^\alpha} = \begin{Bmatrix} \beta \\ \beta\alpha \end{Bmatrix}.$$

The conditions of integrability of (43.5) are

$$(43.7) \qquad\qquad \frac{\partial}{\partial u^1} \begin{Bmatrix} 1 \\ 12 \end{Bmatrix} = \frac{\partial}{\partial u^2} \begin{Bmatrix} 2 \\ 12 \end{Bmatrix},$$

from which and (43.6) we have also

(43.8)
$$\frac{\partial}{\partial u^1}\left\{\begin{matrix}2\\22\end{matrix}\right\} = \frac{\partial}{\partial u^2}\left\{\begin{matrix}1\\11\end{matrix}\right\}.$$

Conversely, if three real functions $g_{\alpha\beta}$ such that $g > 0$ satisfy the condition (43.7) and these functions and the function ρ determined to within sign by the quadrature (43.5) satisfy the equation

(43.9)
$$\frac{R_{1212}}{g} = -\frac{1}{\rho^2},$$

then $g_{\alpha\beta}$, $d_{11} = d_{22} = 0$, $d_{12} = \sqrt{g}/\rho$ satisfy the equations of Gauss and Codazzi, and in accordance with theorem [39.1] we have

[43.8] *If a real tensor $g_{\alpha\beta}$ for which $g > 0$ satisfies the equation (43.7), and $g_{\alpha\beta}$ and a function ρ determined to within sign by (43.5) satisfy the equation (43.9), then $g_{\alpha\beta}$ is the first fundamental tensor of a surface of negative curvature upon which the asymptotic lines are coordinate; this surface is determined in space to within a motion and a symmetry with respect to a point.*

The possibility of symmetry with respect to a point (see §38, Ex. 6) is due to the fact that the sign of ρ is not determined by (43.5).

When the lines of curvature of a surface are coordinate, and the angles which a pair of conjugate directions at a point make with the tangent to the curve $u^2 = $ const. at the point are denoted by θ and θ', that is (see (25.7)),

(43.10)
$$\tan \theta = \sqrt{\frac{g_{22}}{g_{11}}\frac{du^2}{du^1}}, \qquad \tan \theta' = \sqrt{\frac{g_{22}}{g_{11}}\frac{\delta u^2}{\delta u^1}},$$

the equation (42.3) with $d_{12} = 0$ may be written in the form

(43.11)
$$\tan \theta \tan \theta' = -\frac{\rho_2}{\rho_1},$$

where

(43.12)
$$\frac{1}{\rho_1} = \frac{d_{11}}{g_{11}}, \qquad \frac{1}{\rho_2} = \frac{d_{22}}{g_{22}}.$$

From this equation we obtain

(43.13)
$$\tan (\theta - \theta') = \frac{\rho_2 \cot \theta + \rho_1 \tan \theta}{\rho_1 - \rho_2}.$$

If we equate to zero the derivative with respect to $\tan \theta$ of the right-hand member of this equation, we find that the values of $\tan \theta$ for which this derivative is zero are $\pm\sqrt{\rho_2/\rho_1}$, and that for $\tan \theta = \sqrt{\rho_2/\rho_1}$, when $|\rho_1| > |\rho_2|$ then $\tan (\theta - \theta^1)$ is a minimum. This direction is real only at a point for which K is positive, and from (43.11) we have that in this case $\tan \theta' = -\tan \theta$, that is, the two conjugate directions are equally inclined to the lines of curvature. Conversely if the latter condition holds, it follows from (43.11) that $\tan^2 \theta = \rho_2/\rho_1$. Hence we have

[43.9] *A necessary and sufficient condition that there exist a pair of real conjugate directions at a point P in a surface which make equal angles with the tangents to the lines of curvature at P is that the Gaussian curvature be positive at P; the angle θ of one of these directions is given by $\tan \theta = \sqrt{\rho_2/\rho_1}$; an angle between these directions is the minimum of the angles between conjugate directions at P.*

From (43.10) it follows that these conjugate directions are given by

$$\frac{du^2}{du^1} = \sqrt{\frac{g_{11}\,\rho_2}{g_{22}\,\rho_1}}, \qquad \frac{\delta u^2}{\delta u^1} = -\sqrt{\frac{g_{11}\,\rho_2}{g_{22}\,\rho_1}}.$$

For both of these directions the radius of normal curvature is given by

$$(43.14) \qquad R = \frac{g_{11}\,du^{1^2} + g_{22}\,du^{2^2}}{d_{11}\,du^{1^2} + d_{22}\,du^{2^2}} = \frac{\rho_1 + \rho_2}{2},$$

in consequence of (43.12). Thus the radius of normal curvature is the mean of the principal radii of curvature. Accordingly we call these directions the *mean-conjugate directions*.

When the lines of curvature are coordinate and θ denotes the angle which a direction at a point P makes with the curve $u^2 = $ const. at P, the equation (40.5) may be written in the form

$$(43.15) \qquad \frac{1}{R} = \frac{\cos^2 \theta}{\rho_1} + \frac{\sin^2 \theta}{\rho_2},$$

in consequence of (25.6) and (43.12). This equation is called *the equation of Euler*. From it we have that the angle θ which an asymptotic direction makes with the curve $u^2 = $ const. at the point is given by

$$(43.16) \qquad \tan^2 \theta = -\frac{\rho_2}{\rho_1}.$$

This result follows also from (43.11) and the definition of asymptotic directions as self-conjugate.

We have remarked in §40 that at a point of a surface at which $K > 0$ the principal radii have the same sign, and that R for any other normal section has the same sign. If in the tangent plane at such a point P, we take P for the origin of a cartesian coordinate system, and the tangents to the directions of the principal radii ρ_1 and ρ_2 for the x- and y-axes respectively, and lay off on each line through P and in both directions from P line segments of length $\sqrt{|R|}$, where R is the corresponding radius of normal curvature, an equation of the locus of the end points of these segments is found from (43.15) to be

(43.17)
$$\frac{x^2}{|\rho_1|} + \frac{y^2}{|\rho_2|} = 1.$$

Thus the locus is an ellipse, whose principal axes are the principal directions at P, and from (43.11) it follows that conjugate diameters of the ellipse* are conjugate directions on the surface.

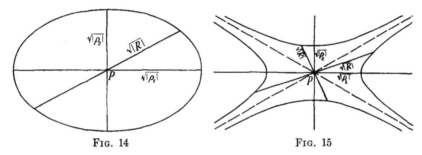

FIG. 14   FIG. 15

When $K < 0$, the principal radii ρ_1 and ρ_2 differ in sign, and certain values of R are positive and others are negative. In this case the locus of the end points of the segments $\sqrt{|R|}$, as shown in Fig. 15, consists of the two conjugate hyperbolas with the equations

(43.18)
$$\frac{x^2}{\rho_1} + \frac{y^2}{\rho_2} = 1, \qquad \frac{x^2}{\rho_1} + \frac{y^2}{\rho_2} = -1.$$

The asymptotes of these hyperbolas are given by $\dfrac{x^2}{\rho_1} + \dfrac{y^2}{\rho_2} = 0$, which equation is equivalent to equation (43.16), that is, *the asymptotes are the asymptotic directions at the point*, which accounts for the name of these directions. In this case also equation (43.11) gives conjugate diameters of the hyperbolas† as well as conjugate directions on the surface.

According as K is positive or negative at a point the conic defined by equation (43.17) or (43.18) is called the *Dupin indicatrix at the point*.

* C. G., p. 192.
† C. G., p. 192.

Hence a point of a surface at which K is positive or negative is sometimes called *elliptic* or *hyperbolic*.

EXERCISES

1. The asymptotic lines on a surface of revolution can be found by quadratures (see §38, Ex. 4).

2. When the coordinate curves on a surface form an isometric-conjugate net, the asymptotic lines can be found by quadratures.

3. The coordinate curves on a surface of translation with the equations

$$x^1 = f(u^1), \qquad x^2 = \varphi(u^2), \qquad x^3 = f_1(u^1) + \varphi_1(u^2)$$

are plane curves, the planes of one family being perpendicular to those of the other family, and the curves form an isometric-conjugate net.

4. The catenoid (see §24, Ex. 4) and the skew helicoid (see §24, Ex. 7) are applicable minimal surfaces.

5. From (43.8) and theorem [43.7] it follows that a necessary condition that two equations of the form (43.3) admit three linearly independent solutions is that
$$\frac{\partial a_{11}}{\partial u^2} = \frac{\partial a_{22}}{\partial u^1}.$$

6. For a surface of constant negative curvature $-1/a^2$ referred to its asymptotic lines the coordinates can be chosen so that

$$g_{11} = g_{22} = a^2, \qquad g_{12} = a^2 \cos \omega,$$

where ω is a solution of the equation

$$\frac{\partial^2 \omega}{\partial u^1 \partial u^2} = \sin \omega.$$

7. The integral curves of an equation $a_{\alpha\beta} du^\alpha du^\beta = 0$ form a conjugate net, if and only if

$$a_{11}d_{22} + a_{22}d_{11} - 2a_{12}d_{12} = 0.$$

8. The surface with the equations

$$x^1 = ac(u^1 + u^2), \qquad x^2 = bc(u^1 - u^2), \qquad x^3 = 2cu^1u^2,$$

where a, b, c are constants, is a hyperbolic paraboloid referred to its rulings, that is, its asymptotic lines; find the equations (43.3) in this case.

9. The surface with the equations

$$\frac{x^1}{a}, \frac{x^2}{b}, \frac{x^3}{c} = \frac{u^1 + u^2, \; 1 - u^1u^2, \; u^1 - u^2}{1 + u^1u^2}$$

where a, b, c are constants, is a hyperboloid of one sheet referred to its rulings, that is, its asymptotic lines; find the equations (43.3) in this case.

10. A necessary and sufficient condition that the asymptotic lines on a surface form a Tchebychef net (see §29, Ex. 12) is that the surface have constant Gaussian curvature.

11. For a minimal surface referred to its lines of curvature one has from equations (41.12) that

$$g_{11} = \rho_1 U_1, \qquad g_{22} = \rho_1 U_2,$$

where U_1 and U_2 are functions of u^1 and u^2 respectively; hence the lines of curvature on a minimal surface form an isometric net (see (30.11)).

12. Given two surfaces S and $\bar{S}$ with the respective second fundamental forms $d_{\alpha\beta}\,du^\alpha\,du^\beta$ and $\bar{d}_{\alpha\beta}\,du^\alpha\,du^\beta$; a necessary and sufficient condition that the asymptotic lines on S correspond to a conjugate system on $\bar{S}$, corresponding points being those with the same values of u^α on the two surfaces, is that

$$d_{11}\,\bar{d}_{22} + d_{22}\,\bar{d}_{11} - 2d_{12}\,\bar{d}_{12} = 0;$$

when this condition is satisfied the asymptotic lines on $\bar{S}$ correspond to a conjugate system on S.

13. On the two surfaces of Ex. 12 the integral curves of the equation

$$\begin{vmatrix} \bar{d}_{1\alpha}\,du^\alpha & \bar{d}_{2\alpha}\,du^\alpha \\ d_{1\alpha}\,du^\alpha & d_{2\alpha}\,du^\alpha \end{vmatrix} = 0$$

form a conjugate net on each surface (see theorem [26.4]).

14. From (43.14), (40.10) and (40.11) it follows that the radius of normal curvature in either mean-conjugate direction is given by

$$R = \frac{1}{2}\frac{gg^{\alpha\beta}\,d_{\alpha\beta}}{d_{11}\,d_{22} - d_{12}^2};$$

when this expression for R is substituted in (40.5) the integral curves of the resulting equation are the *mean-conjugate curves of the surface*.

15. When the mean-conjugate curves of a surface of positive curvature are coordinate

$$\frac{d_{11}}{g_{11}} = \frac{d_{22}}{g_{22}}, \qquad d_{12} = 0.$$

44. GEODESIC CURVATURE AND GEODESIC TORSION OF A CURVE

Consider any curve C upon a surface and the tangent plane to the surface at a point P of C. Project orthogonally upon this tangent plane the portion of the curve in the neighborhood of P and let C' denote the resulting plane curve. The curve C' is a normal section of the projecting cylinder and C is a curve upon this cylinder, and consequently theorem [40.1] may be applied to the cylinder to determine the relation between the radius of curvature ρ of C and r of C'. This relation is

(44.1)
$$\frac{1}{r} = \frac{\cos\psi}{\rho},$$

where ψ is the angle which the principal normal to C makes with the tangent plane to the surface at P, that is, with the normal to C' in this plane. From this result it follows that the normal to the osculating plane of C at its center of curvature for the point P meets the tangent plane to the surface at P in the center of curvature of C' for the point P, and by theorem [40.1] this same normal meets the normal plane section of the surface at P and tangential to C in the center of curvature of the normal section. In Fig. 13 C_n is the latter point, and C_1 and C_g the centers of curvature of C and C' respectively.

In order to find the coordinates of the center of curvature of C', we make use of the components $\bar{X}^i$ of the unit normal vector to the projecting cylinder. Since this normal is perpendicular to the tangent to C and lies in the normal plane, we have

$$\sum_i \bar{X}_i \frac{dx^i}{ds} = 0, \qquad \sum_i \bar{X}^i X^i = 0.$$

These conditions are satisfied by

(44.2)
$$\bar{X}^i = X^j \frac{dx^k}{ds} - X^k \frac{dx^j}{ds},$$

where i, j, k take the values 1, 2, 3 cyclically, from which we have $\sum_i \bar{X}^i \bar{X}^i = 1$. By the choice (44.2) the positive sense given to the vector $\bar{X}^i$ is such that the tangent to C, that is, the vector $\frac{dx^i}{ds}$, the vector $\bar{X}^i$, and the vector X^i have the same mutual orientation as the x^1-, x^2-, and x^3-axes*, since

(44.3)
$$\begin{vmatrix} \dfrac{dx^1}{ds} & \dfrac{dx^2}{ds} & \dfrac{dx^3}{ds} \\ \bar{X}^1 & \bar{X}^2 & \bar{X}^3 \\ X^1 & X^2 & X^3 \end{vmatrix} = 1.$$

If we put

(44.4)
$$\bar{X}^i = \mu^\alpha \frac{\partial x^i}{\partial u^\alpha}, \qquad \frac{dx^i}{ds} = \frac{du^\alpha}{ds} \frac{\partial x^i}{\partial u^\alpha},$$

then μ^α and $\frac{du^\alpha}{ds}$ are the components in the surface of the tangent vectors $\bar{X}^i$ and $\frac{dx^i}{ds}$ respectively (see (24.23)). From the discussion following

* C. G., p. 162.

(38.9) we have from (44.3) that the vector μ^α makes a right-angle with the vector $\dfrac{du^\alpha}{ds}$. Hence from (25.19) and the first of (44.4) we have

(44.5)
$$\bar{X}^i = \epsilon^{\gamma\alpha} g_{\gamma\beta} \frac{du^\beta}{ds} \frac{\partial x^i}{\partial u^\alpha}$$

$$= \frac{1}{\sqrt{g}} \left(g_{1\beta} \frac{\partial x^i}{\partial u^2} - g_{2\beta} \frac{\partial x^i}{\partial u^1} \right) \frac{du^\beta}{ds}.$$

The angle ψ in (44.1) is equal or supplementary to the angle between the principal normal to C and the vector $\bar{X}^i$ according as the latter is in the direction PC_g in Fig. 13 or in the opposite direction. Hence from (44.1) we have

(44.6)
$$\frac{e}{r} = \frac{1}{\rho} \sum_i \bar{X}^i \beta^i,$$

where β^i are direction cosines of the principal normal to C at P, and where e is $+1$ or -1 according as the angle between the principal normal and the vector $\bar{X}^i$ is acute or obtuse. In consequence of (4.7), (44.5), and (38.17) we have

$$\frac{e}{r} = \epsilon^{\gamma\alpha} g_{\gamma\beta} \frac{du^\beta}{ds} \sum_i \frac{\partial x^i}{\partial u^\alpha} \frac{d^2 x^i}{ds^2}$$

$$= \epsilon^{\gamma\alpha} g_{\gamma\beta} \frac{du^\beta}{ds} \sum_i \frac{\partial x^i}{\partial u^\alpha} \left(\frac{\partial^2 x^i}{\partial u^\delta \partial u^\epsilon} \frac{du^\delta}{ds} \frac{du^\epsilon}{ds} + \frac{\partial x^i}{\partial u^\delta} \frac{d^2 u^\delta}{ds^2} \right)$$

$$= \epsilon^{\gamma\alpha} g_{\gamma\beta} \frac{du^\beta}{ds} \left([\delta\epsilon, \alpha] \frac{du^\delta}{ds} \frac{du^\epsilon}{ds} + g_{\alpha\delta} \frac{d^2 u^\delta}{ds^2} \right),$$

where $[\delta\epsilon, \alpha]$ are Christoffel symbols of the first kind formed with respect to the tensor $g_{\alpha\beta}$. Since $g^{\alpha\beta}[\gamma\delta, \beta] = \left\{ \begin{matrix} \alpha \\ \gamma\delta \end{matrix} \right\}$ by (20.2), by means of the identities of §25, Ex. 8, the above equation may be written with a change of indices in the form

(44.7)
$$\frac{e}{r} = \epsilon_{\alpha\beta} \frac{du^\alpha}{ds} \left(\frac{d^2 u^\beta}{ds^2} + \left\{ \begin{matrix} \beta \\ \gamma\delta \end{matrix} \right\} \frac{du^\gamma}{ds} \frac{du^\delta}{ds} \right).$$

Comparing this result with equation (34.3) we have that e/r is equal to the geodesic curvature κ_g of the curve C, defined intrinsically in §34, and that κ_g is positive or negative according as the vector $\bar{X}^i$ is in the direction PC_g in Fig. 13, or in the opposite direction. Accordingly the center of curvature of the projected curve C' for its point of contact with the given curve C is called the *center of geodesic curvature* of C for this point. In view of the above results we have

[44.1] *The normal to the osculating plane of a curve C on a surface at the center of curvature for a point P of C meets the tangent plane to the surface at P in the center of geodesic curvature of C for the point P, and the length of the line segment with end points P and the center of geodesic curvature is equal to the reciprocal of the absolute value of the geodesic curvature of C for the point P.*

From the above discussion, equation (44.1), and theorem [34.1] we have

[44.2] *A necessary and sufficient condition that a curve be a geodesic on a surface is that its principal normal at each point of the curve be normal to the surface at the point.*

When a curve C is defined by an equation $\varphi(u^1, u^2) = 0$, or as an integral curve of a differential equation $M_\alpha \, du^\alpha = 0$, its geodesic curvature κ_g can be found directly (see (34.7) and §34, Ex. 7). From (40.2) we have that $\dfrac{\cos \bar{\omega}}{\rho}$ may be found directly. Also from (44.1) we have

$$(44.8) \qquad\qquad \kappa_g = \frac{\sin \bar{\omega}}{\rho},$$

since $\psi = \bar{\omega} - \dfrac{\pi}{2}$ or $\dfrac{3\pi}{2} - \bar{\omega}$ according as e is $+1$ or -1. Hence ρ and $\bar{\omega}$ can be found directly. We shall show that the torsion τ of the curve also may be obtained in terms of $\bar{\omega}$ and the fundamental tensors of the surface. Consequently the intrinsic equations in space of the curve C can be found directly.

From the definition of $\bar{\omega}$ in §40 it follows that

$$(44.9) \qquad\qquad \sin \bar{\omega} = \sum_i X^i \gamma^i,$$

where γ^i are the direction cosines of the binormal of C. Differentiating equation (44.9) with respect to s, and making use of the Frenet formulas (6.1) and equation (40.1), we obtain

$$(44.10) \qquad \cos \bar{\omega} \left(\frac{d\bar{\omega}}{ds} - \tau \right) = \sum_i \gamma^i \frac{dX^i}{ds} = -d_{\alpha\delta} g^{\gamma\delta} \frac{du^\alpha}{ds} \sum_i \gamma^i \frac{\partial x^i}{\partial u^\gamma},$$

the last expression following from (38.19). Since the binormal is in the plane of the vectors X^i and $\bar{X}^i$, we have

$$\gamma^i = aX^i + b\bar{X}^i,$$

where a and b are to be determined. The vector γ^i makes with the vectors X^i and $\bar{X}^i$ the angles $\bar{\omega} - \dfrac{\pi}{2}$ and $\bar{\omega} - \pi$ respectively (see Fig. 13). On multiplying the above equations by X^i and summing with respect to i, and also by $\bar{X}^i$ and summing with respect to i, we find that $a = \sin \bar{\omega}$, $b = -\cos \bar{\omega}$. Consequently

$$\gamma^i = \sin \bar{\omega}\, X^i - \cos \bar{\omega}\, \bar{X}^i.$$

When these expressions are substituted in the right-hand member of (44.10), the result is reducible, in consequence of (38.8) and (44.5), to

$$\cos \bar{\omega}\, d_{\alpha\delta}\, g^{\gamma\delta}\, \epsilon^{\sigma\tau}\, g_{\sigma\beta}\, \frac{du^\alpha}{ds} \frac{du^\beta}{ds} \sum_i \frac{\partial x^i}{\partial u^\tau} \frac{\partial x^i}{\partial u^\gamma} = -\cos \bar{\omega}\, \tau_g,$$

where

$$\tau_g = \epsilon^{\delta\gamma}\, d_{\alpha\delta}\, g_{\beta\gamma}\, \frac{du^\alpha}{ds} \frac{du^\beta}{ds}$$

$$(44.11) \qquad = \frac{1}{\sqrt{g}\ g_{\alpha\beta}\, du^\alpha\, du^\beta}\, [(d_{11}g_{12} - d_{12}g_{11})\, du^{1^2}$$

$$+ (d_{11}g_{22} - d_{22}g_{11})\, du^1\, du^2$$

$$+ (d_{12}g_{22} - d_{22}g_{12})\, du^{2^2}].$$

Consequently (44.10) reduces to

$$\cos \bar{\omega} \left(\frac{d\bar{\omega}}{ds} - \tau + \tau_g \right) = 0.$$

When $\cos \bar{\omega} \neq 0$, that is, when the curve is not an asymptotic line, we have

$$(44.12) \qquad \tau = \frac{d\bar{\omega}}{ds} + \tau_g.$$

We consider now the exceptional case when the curve is an asymptotic line, and assume that the asymptotic lines are coordinate, that is, $d_{11} = d_{22} = 0$. With this choice the direction cosines of the tangent α^i and of the binormal γ^i of a curve $u^2 = \text{const.}$ are given by

$$\alpha^i = \frac{1}{\sqrt{g_{11}}} \frac{\partial x^i}{\partial u^1}, \qquad \gamma^i = eX^i,$$

where e is $+1$ or -1 as the case may be. From (5.7) we have for the direction cosines of the principal normal

$$\beta^i = \frac{e}{\sqrt{g_{11}}} \left(X^j \frac{\partial x^k}{\partial u^1} - X^k \frac{\partial x^j}{\partial u^1} \right)$$

as i, j, k take the values 1, 2, 3 cyclically.

From the Frenet formulas (6.1) we have

$$\tau = \sum_i \beta^i \frac{d\gamma^i}{ds} = \frac{e}{\sqrt{g_{11}}} \sum_i \beta^i \frac{\partial X^i}{\partial u^1} = -\frac{e}{\sqrt{g_{11}}} d_{12} g^{2\beta} \sum_i \beta^i \frac{\partial x^i}{\partial u^\beta},$$

the last expression following from (38.19). Substituting the above expressions for β^i, we obtain in consequence of (38.9)

$$\tau = -\frac{d_{12}}{\sqrt{g}} = -\sqrt{-K}.$$

In like manner the torsion of an asymptotic line $u^1 =$ const. can be shown to be equal to $\sqrt{-K}$. Hence we have the theorem of Enneper:

[44.3] *The square of the torsion of an asymptotic line at a point of a surface is equal to the absolute value of the Gaussian curvature at the point; the torsions of the two asymptotic lines through a point differ in sign.*

In both of the above cases $\tau = \tau_g$ as follows from (44.11). Consequently equation (44.12) holds also in this case since $\bar{\omega}$ is constant for an asymptotic line. Since τ_g is a function only of a point and a direction at a point as follows from (44.11), the quantity $\tau - \dfrac{d\bar{\omega}}{ds}$ is the same for all curves at a point having a common tangent. It is equal to τ for any curve for which $\bar{\omega}$ is a constant, and in particular for the geodesic through the point in this direction. Accordingly τ_g is called the *geodesic torsion* of a curve. From (44.11) and (41.1) we have

[44.4] *A necessary and sufficient condition that the geodesic torsion of a curve be zero at a point is that the curve be tangent to a line of curvature at the point.*

EXERCISES

1. A plane curve on a surface is a geodesic, if and only if the tangent planes to the surface along the curve are perpendicular to the plane of the curve.

2. On the rectifying developable of a twisted curve (see §12, Ex. 2) the given curve is a geodesic, and thus becomes a straight line when the developable surface is rolled out upon a plane.

3. Straight lines on a surface are the only geodesic asymptotic lines.

4. At each point of an orthogonal net on a surface the geodesic torsions of the two curves of the net differ only in sign.

5. A geodesic line of curvature is a plane curve, and a plane geodesic line is a line of curvature.

6. When two surfaces meet under a constant angle, the geodesic torsions of the curve of intersection with respect to the two surfaces are equal; if the curve of intersection is a line of curvature of one of the surfaces, it is for the other also;

if two surfaces intersect along a curve which is a line of curvature of each, they intersect under constant angle.

7. When a surface is met by a plane or a sphere under constant angle, the curve of intersection is a line of curvature of the surface.

8. The geodesic torsion of a curve is given by

$$\tau_g = \frac{1}{2}\left(\frac{1}{\rho_1} - \frac{1}{\rho_2}\right)\sin 2\theta,$$

where θ is the angle made by the curve and the line of curvature at the point for which ρ_1 is the radius of normal curvature.

9. A necessary and sufficient condition that the curves $u^2 = $ const. on a surface be straight lines, and thus that the surface be a ruled surface, is that

$$d_{11} = 0, \qquad \begin{Bmatrix} 2 \\ 11 \end{Bmatrix} = 0.$$

10. Upon a surface straight lines are the only plane asymptotic lines.

11. Show that

$$\Delta_2 x^i = \left(\frac{1}{\rho_1} + \frac{1}{\rho_2}\right) X^i;$$

from this it follows that any family of parallel planes intersect a minimal surface in curves which together with their orthogonal trajectories form an isometric net.

45. PARALLEL VECTORS IN A SURFACE

In §35 we gave an intrinsic definition of parallelism in a surface. We now interpret parallelism from the view-point of the space in which the surface may be considered as imbedded. Denoting by ξ^i the components in space referred to cartesian coordinates x^i of the unit vectors λ^α at points of a curve C, we have

$$(45.1) \qquad \qquad \xi^i = \lambda^\alpha \frac{\partial x^i}{\partial u^\alpha}.$$

Differentiating with respect to s, we obtain

$$(45.2) \qquad \begin{aligned} \frac{d\xi^i}{ds} &= \frac{d\lambda^\alpha}{ds}\frac{\partial x^i}{\partial u^\alpha} + \lambda^\gamma \frac{\partial^2 x^i}{\partial u^\gamma \partial u^\beta}\frac{du^\beta}{ds} \\ &= \left(\frac{d\lambda^\alpha}{ds} + \lambda^\gamma \begin{Bmatrix} \alpha \\ \gamma\beta \end{Bmatrix}\frac{du^\beta}{ds}\right)\frac{\partial x^i}{\partial u^\alpha} + \lambda^\gamma d_{\gamma\delta}\frac{du^\delta}{ds}\,X^i, \end{aligned}$$

the last expression being a consequence of (38.18). Since $\sum_i \xi^i \xi^i = 1$, we have analogously to (35.5) that the vector η^i associate to the vector ξ^i in space is given by

$$(45.3) \qquad \qquad \frac{d\xi^i}{ds} = \eta^i.$$

The vector η^i is in general different from the associate vector of λ^α in the surface (see theorem [45.1] and Ex. 3). If then equation (45.2) is multiplied by $\dfrac{\partial x^i}{\partial u^\beta}$ and summed with respect to i, one obtains in consequence of (38.2), (38.8), and (45.3)

$$(45.4) \qquad \sum_i \eta^i \frac{\partial x^i}{\partial u^\beta} = g_{\alpha\beta}\left(\frac{d\lambda^\alpha}{ds} + \lambda^\gamma \begin{Bmatrix} \alpha \\ \gamma\beta \end{Bmatrix} \frac{du^\beta}{ds}\right).$$

If the vectors at points of a curve are parallel as viewed from the enveloping space, their components in this space are constant, that is $\dfrac{d\xi^i}{ds} = 0$. Hence the right-hand member of (45.4) is equal to zero, and, since $g \neq 0$, the quantities λ^α satisfy (35.13). The same result follows if $\sum_i \eta^i \dfrac{\partial x^i}{\partial u^\beta} = 0$, that is, if the vectors associate to the vectors ξ^i are normal to the surface. Conversely, if equations (35.13) are satisfied, either the associate vector η^i is a zero vector, or it is normal to the surface. Hence we have

[45.1] *A necessary and sufficient condition that vectors along a curve in a surface be parallel with respect to the curve is that the vectors be parallel as viewed from the enveloping space or that the associate vectors in this space be normal to the surface.*

We have remarked that the tangents to a curve are parallel with respect to the curve, if and only if the latter is a geodesic. In this case the vector η^i in (45.3) is the principal normal (see (4.7)), and consequently theorem [44.2] is a particular case of theorem [45.1].

As a consequence of theorem [45.1] we have

[45.2] *If two surfaces are tangent to one another along a curve, vectors parallel with respect to this curve in either surface are parallel also in the other surface.*

From this theorem we have that if the vectors λ^α at points of a curve C in a surface S are parallel with respect to C, they are parallel with respect to C in the developable surface which is the envelope of the tangent planes to S at the points of C, and conversely. In consequence of theorem [35.4], when this developable surface is rolled out upon the plane the set of parallel vectors go into vectors which are parallel in the euclidean sense. If then one has such a developable surface and C' is the curve into which C goes, in order to find geometrically a set of vectors parallel with respect to C, one has only to take a set of vectors parallel in the euclidean sense at points on C', and when the developable is rolled back over the surface the resulting vectors are parallel with respect to C.

The notion of parallelism of vectors in a Riemannian space of any number of dimensions is due to Levi-Civita.* He introduced the notion of infinitesimal parallelism which in terms of a surface imbedded in a euclidean space is as follows: Given two nearby points P_0 and P_1 of a curve in a surface corresponding to values s and $s + ds$ of the arc of the curve, and vectors $\lambda^\alpha(s)$ and

$$\lambda^\alpha(s + ds) = \lambda^\alpha(s) + \frac{d\lambda^\alpha}{ds} ds,$$

terms of higher degree in ds being neglected; by definition the vector at P_1 is parallel to the vector λ^α at P_0, if as viewed from the enveloping space it makes the same angle with an arbitrary vector tangential to the surface at P_0 as the vector λ^α at P_0 does. Such a tangential vector is given by $a^\beta \frac{\partial x^i}{\partial u^\beta}$. The components of the vector at P_1 as a vector of the enveloping space are given by

$$\xi^i + \frac{d\xi^i}{ds} ds = \xi^i + \left(\nu^\alpha \frac{\partial x^i}{\partial u^\alpha} + \lambda^\gamma d_{\gamma\delta} \frac{du^\delta}{ds} X^i \right)_0 ds,$$

which follows from (45.2) and (35.5). The condition that the two angles mentioned be equal is

$$\sum_i a^\beta \frac{\partial x^i}{\partial u^\beta} \left(\nu^\alpha \frac{\partial x^i}{\partial u^\alpha} + \lambda^\gamma d_{\gamma\delta} \frac{du^\delta}{ds} X^i \right)_0 ds = 0.$$

In consequence of (38.2) and (38.8) this equation reduces to

$$a^\beta (g_{\alpha\beta} \nu^\alpha)_0 ds = 0.$$

Since the a's are arbitrary, and the determinant $g > 0$, we have $\nu_0^\alpha = 0$, that is, the vector λ^α must satisfy equations (35.13) at P_0, and consequently depends upon the direction P_0P_1 on the surface. Having arrived at this result, Levi-Civita† used equations (35.13) to define a family of vectors λ^α parallel with respect to a curve.

EXERCISES

1. When the coordinates x^i of the enveloping space are any whatever, the equations (45.3) are

$$\frac{d\xi^i}{ds} + \xi^h \begin{Bmatrix} i \\ hj \end{Bmatrix}_a \frac{dx^j}{ds} = \eta^i,$$

---

* 1917, 1.
† 1917, 1, p. 179.

where the Christoffel symbols $\left\{\begin{matrix} i \\ hj \end{matrix}\right\}_a$ are formed with respect to the metric tensor a_{ij} of space.

2. A necessary and sufficient condition that vectors in a surface parallel with respect to a curve C be parallel as viewed from the enveloping space is that the direction of the vector at each point of C be conjugate to C.

3. A necessary and sufficient condition that the associate vector with respect to a curve C of a tangent vector to a surface coincide with the associate vector with respect to C of the tangent vector as looked upon as a vector in the surface is that the latter be conjugate to the curve C.

4. In order that the normals to a surface along a curve C be parallel, it is necessary that $d_{11}d_{22} - d_{12}^2 = 0$ along C; in this case C is an asymptotic line.

5. Show by means of equations (38.19) that a necessary and sufficient condition that the vector associate to the normal vector to a surface with respect to a curve C on the surface be tangent to C is that C be a line of curvature.

6. Show by means of equations (38.19) that a necessary and sufficient condition that the vector associate to the normal vector to a surface with respect to a curve C on the surface be perpendicular to C is that the curve be an asymptotic line.

46. SPHERICAL REPRESENTATION OF A SURFACE. THE GAUSSIAN CURVATURE OF A SURFACE

The Gaussian curvature of a surface has been derived as an intrinsic property of a surface and it has been shown that it is equal to the reciprocal of the product of the principal radii of normal curvature, when the surface is considered as imbedded in space. These results are due to Gauss, who has also given an interpretation of curvature as a generalization of the concept of curvature of a curve. In arriving at this result Gauss[*] introduced the concept of the *spherical representation* of a surface. In accordance with this concept one takes the unit sphere, that is, the sphere of unit radius with center at the origin, and makes correspond to each point P of the surface the point $\overline{P}$ in which the sphere is met by the radius parallel to and with the same sense as the normal X^i to the surface at P. Thus to a point of coordinates x^i corresponds the point of coordinates $\bar{x}^i$, where

$$(46.1) \qquad\qquad \bar{x}^i = X^i.$$

We denote by $d\bar{s}$ the linear element of the sphere, written

$$(46.2) \quad d\bar{s}^2 = \sum_i d\bar{x}^i \, d\bar{x}^i = \sum_i \frac{\partial X^i}{\partial u^\alpha} \frac{\partial X^i}{\partial u^\beta} \, du^\alpha \, du^\beta = h_{\alpha\beta} \, du^\alpha \, du^\beta,$$

[*] 1827, 1, p. 226.

$h_{\alpha\beta}$ being thus defined. In consequence of (38.19) the quantities $h_{\alpha\beta}$ are expressible in the form

$$(46.3) \qquad h_{\alpha\beta} = \sum_i \frac{\partial X^i}{\partial u^\alpha} \frac{\partial X^i}{\partial u^\beta} = d_{\alpha\gamma} g^{\gamma\delta} d_{\beta\mu} g^{\mu\nu} \sum_i \frac{\partial x^i}{\partial u^\delta} \frac{\partial x^i}{\partial u^\nu},$$

which in consequence of (24.7) and (24.18) reduces to

$$(46.4) \qquad h_{\alpha\beta} = d_{\alpha\gamma} d_{\beta\delta} \, g^{\gamma\delta}.$$

In order to give these expressions another form, we have

$$h_{11} = d_{1_1} d_{1\delta} \, g^{1\delta} + d_{12} d_{1\delta} \, g^{2\delta} = d_{11} d_{\gamma\delta} \, g^{\gamma\delta} + g^{2\delta}(d_{12} d_{1\delta} - d_{11} d_{2\delta})$$

$$= d_{11} d_{\gamma\delta} \, g^{\gamma\delta} + g^{22}(d_{12}^2 - d_{11} d_{22}) = d_{11} K_m - g_{11} K,$$

in consequence of (40.10) and (40.11). Proceeding in like manner we have

$$(46.5) \qquad h_{\alpha\beta} = d_{\alpha\beta} K_m - g_{\alpha\beta} K.$$

Accordingly the linear element of the spherical representation may be written in the form

$$(46.6) \qquad d\bar{s}^2 = K_m \, d_{\alpha\beta} \, du^\alpha \, du^\beta - K g_{\alpha\beta} \, du^\alpha \, du^\beta,$$

and in consequence of (40.5) in the form

$$(46.7) \qquad d\bar{s}^2 = \left(\frac{K_m}{R} - K\right) ds^2.$$

If h denotes the determinant of the quantities $h_{\alpha\beta}$, we have from (46.4) and the expression (40.10) for K

$$(46.8) \qquad \sqrt{h} = eK\sqrt{g},$$

where e is $+1$ or -1 according as K is positive or negative at the point under consideration.

If we denote by $\bar{X}^i$ the vector normal to the sphere at the point $\bar{x}^i$, we have analogously to (38.5)

$$(46.9) \qquad \bar{X}^i = \frac{1}{\sqrt{h}} \begin{vmatrix} \dfrac{\partial X^j}{\partial u^1} & \dfrac{\partial X^k}{\partial u^1} \\ \dfrac{\partial X^j}{\partial u^2} & \dfrac{\partial X^k}{\partial u^2} \end{vmatrix},$$

as i, j, k take the values 1, 2, 3 cyclically. Substituting from (38.19) and making use of (38.5) and (46.8), we have

$$\bar{X}^i = \frac{1}{\sqrt{h}} \begin{vmatrix} d_{1\gamma} g^{\gamma 1} & d_{1\gamma} g^{\gamma 2} \\ d_{2\gamma} g^{\gamma 1} & d_{2\gamma} g^{\gamma 2} \end{vmatrix} \sqrt{g} X^i = \frac{e}{K} \begin{vmatrix} d_{11} d_{12} \\ d_{12} d_{22} \end{vmatrix} \cdot \begin{vmatrix} g^{11} g^{12} \\ g^{12} g^{22} \end{vmatrix} X^i,$$

which reduces in consequence of (15.5) and (39.4) to

(46.10) $$\bar{X}^i = eX^i,$$

where e is $+1$ or -1 according as $K > 0$ or $K < 0$.

From this result it follows that according as a point of a surface is *elliptic* or *hyperbolic*, that is, $K > 0$ or $K < 0$, the vectors X^i and $\bar{X}^i$ have the same or opposite sense, that is, the positive sides of the tangent planes to the surface and to the unit sphere are the same or opposite. This is seen also from equations (41.9) when the lines of curvature are coordinate, since for an elliptic point the vectors $\dfrac{\partial X^i}{\partial u^\alpha}$ have the same sense as the corresponding vectors $\dfrac{\partial x^i}{\partial u^\alpha}$ or both have the opposite sense, whereas for a hyperbolic point one has the same sense and the other opposite sense.

If then we have a bounded portion of a surface such that K has the same sign at all points and on the boundary C, and we denote by $\bar{C}$ the contour of the corresponding bounded portion of the unit sphere, it follows that as a point P describes the curve C the corresponding point $\bar{P}$ on the sphere describes the curve $\bar{C}$ in the same or opposite sense according as K is positive or negative. The areas of two such portions of the sphere and surface are given by

$$\int \sqrt{h}\, du^1\, du^2 = \int\int eK\sqrt{g}\, du^1\, du^2, \qquad \int \sqrt{g}\, du^1\, du^2$$

respectively, the limits of integration being the same for the two integrals. From this result it follows that the limit of the ratio of these two integrals as the portion of the surface shrinks to a point (and consequently also the portion of the sphere) is equal to the value of eK at the point. When $K = 0$ at all points of a bounded portion of a surface and on the boundary, this portion of the surface is developable. Since the normals to a developable surface at all points of a generator are parallel, it follows that the spherical representation of $\bar{C}$ for a developable surface is the segment of a curve, and consequently of zero area. Hence the above result applies to the case $K = 0$ also, and we have the following theorem of Gauss*:

[46.1] *The limit of the ratio of the area of the spherical representation of a bounded portion of a surface and the area of this bounded portion as the latter shrinks to a point is equal to the absolute value of the Gaussian curvature at the point.*

* 1827, 1, p. 226.

We consider now further consequences of the preceding results. Since the radius of normal curvature of a surface at a point depends upon the direction of the normal section unless the surface is a plane or a sphere (see §40), it follows from (46.7) that the correspondence between a surface other than a plane or a sphere and its spherical representation is conformal (see §36), if and only if $K_m = 0$, that is, when the surface is a minimal surface. For a sphere $\dfrac{1}{R} = \dfrac{1}{\rho_1} = \dfrac{1}{\rho_2} = \pm\dfrac{1}{a}$, where a is the radius of the sphere; in this case $d\bar{s}^2 = \dfrac{1}{a^2}\, ds^2$. Hence we have

[46.2] *A surface and its spherical representation are in conformal correspondence, if and only if the surface is a sphere or a minimal surface.*

In both of these cases to an orthogonal net on the surface corresponds an orthogonal net on the unit sphere. From (46.5) it follows that the coordinate curves on a surface and the unit sphere are orthogonal nets only in case $K_m = 0$ or $d_{12} = 0$. Hence we have

[46.3] *The lines of curvature of a surface are represented on the unit sphere by an orthogonal net; this is a characteristic property of the lines of curvature of a surface which is not a minimal surface nor a sphere.*

This theorem is also a corollary of the following theorem:

[46.4] *The tangents to a curve in a surface and to its spherical representation at corresponding points are parallel, if and only if the curve is a line of curvature.*

In order to establish this theorem we assume that the surface is referred to a coordinate system such that the given curve is a curve $u^2 = $ const. and that the coordinate curves form an orthogonal net. For the given curve the quantities $\dfrac{\partial X^i}{\partial u^1}$ and $\dfrac{\partial x^i}{\partial u^1}$ must be proportional, being direction numbers of parallel lines by hypothesis. Hence from (38.19) we have $-d_{1\gamma}g^{\gamma 2} = 0$. Since $g^{12} = 0$ for an orthogonal coordinate net and $g^{22} \neq 0$, we have $d_{12} = 0$ and the theorem is proved.

In what follows we derive for a surface various results which are expressible in terms of the two tensors $d_{\alpha\beta}$ and $h_{\alpha\beta}$. Equations (41.5)′ express the condition that the normals to a surface along the curve whose unit tangent vector has components $\dfrac{du^\alpha}{ds}$ form a developable surface. If these equations are multiplied by $\dfrac{\partial X^i}{\partial u^\beta}$ and summed with respect to i, the resulting equation in consequence of (38.16) and (46.3) is

$$(46.11) \qquad (d_{\alpha\beta} - t\,h_{\alpha\beta})\frac{du^\alpha}{ds} = 0 \qquad (\beta = 1, 2).$$

Eliminating t from these two equations, we obtain

$$\begin{vmatrix} d_{1\alpha}\,du^\alpha & d_{2\alpha}\,du^\alpha \\ h_{1\alpha}\,du^\alpha & h_{2\alpha}\,du^\alpha \end{vmatrix} = 0,$$

which in consequence of (46.5) is equivalent to the equation (40.8) of the lines of curvature, as was to be expected.

In order that quantities $\dfrac{du^\alpha}{ds}$ be determined by (46.11), t must be a solution of the determinant equation

$$| d_{\alpha\beta} - t\,h_{\alpha\beta} | = 0.$$

In §41 preceeding theorem [41.1] it was shown that t is one of the principal radii of normal curvature (see Ex. 7). Hence the principal radii ρ_1 and ρ_2 are roots of the equation

$$(46.12) \qquad (h_{11}h_{22} - h_{12}^2)R^2 - (d_{11}h_{22} + d_{22}h_{11} - 2d_{12}h_{12})R \\ + (d_{11}d_{22} - d_{12}^2) = 0.$$

Hence we have

$$(46.13) \qquad \begin{aligned} \rho_1 + \rho_2 &= \frac{d_{11}\,h_{22} + d_{22}\,h_{11} - 2d_{12}\,h_{12}}{h} = d_{\alpha\beta}h^{\alpha\beta}, \\ \rho_1\rho_2 &= \frac{d_{11}\,d_{22} - d_{12}^2}{h}, \end{aligned}$$

where the quantities $h^{\alpha\beta}$ are defined uniquely by

$$h^{\alpha\beta}h_{\beta\gamma} = \delta_\gamma^\alpha,$$

that is (see (24.20)),

$$(46.14) \qquad h^{11} = \frac{h_{22}}{h}, \qquad h^{12} = -\frac{h_{12}}{h}, \qquad h^{22} = \frac{h_{11}}{h}.$$

From (38.16), (46.10), and (46.3) it follows that the coefficients of the second fundamental form of the spherical representation are given by

$$-\sum_i \frac{\partial \bar{X}^i}{\partial u^\alpha}\frac{\partial X^i}{\partial u^\beta} = -\sum_i e\frac{\partial X^i}{\partial u^\alpha}\frac{\partial X^i}{\partial u^\beta} = -e h_{\alpha\beta}.$$

From this result, (46.10), and equations of the form (38.18) we have for the spherical representation

$$(46.15) \qquad \frac{\partial^2 X^i}{\partial u^\alpha \partial u^\beta} - \frac{\partial X^i}{\partial u^\gamma} \left\{ \begin{matrix} \gamma \\ \alpha\beta \end{matrix} \right\} = - h_{\alpha\beta} X^i,$$

where the Christoffel symbols are formed with respect to the form (46.2). By means of equations (38.19), that is,

$$(46.16) \qquad \frac{\partial X^i}{\partial u^\alpha} = - d_{\alpha\epsilon} g^{\epsilon\delta} \frac{\partial x^i}{\partial u^\delta},$$

equations (46.15) may be given, the form

$$(46.17) \qquad \frac{\partial^2 X^i}{\partial u^\alpha \partial u^\beta} + d_{\gamma\epsilon} g^{\epsilon\delta} \frac{\partial x^i}{\partial u^\delta} \left\{ \begin{matrix} \gamma \\ \alpha\beta \end{matrix} \right\} = - h_{\alpha\beta} X^i.$$

If we differentiate (46.16) covariantly based upon $g_{\alpha\beta}$ and make use of (38.18), we obtain

$$\frac{\partial^2 X^i}{\partial u^\alpha \partial u^\beta} - \frac{\partial X^i}{\partial u^\gamma} \left\{ \begin{matrix} \gamma \\ \alpha\beta \end{matrix} \right\} = - d_{\alpha\epsilon,\beta} g^{\epsilon\delta} \frac{\partial x^i}{\partial u^\delta} - g^{\epsilon\delta} d_{\alpha\epsilon} d_{\delta\beta} X^i,$$

on noting that the covariant derivative of $g^{\gamma\delta}$ is equal to zero. In consequence of (46.16) and (46.4) these equations may be written

$$\frac{\partial^2 X^i}{\partial u^\alpha \partial u^\beta} + \left(d_{\gamma\epsilon} \left\{ \begin{matrix} \gamma \\ \alpha\beta \end{matrix} \right\} + d_{\alpha\epsilon,\beta} \right) g^{\epsilon\delta} \frac{\partial x^i}{\partial u^\delta} = - h_{\alpha\beta} X^i.$$

From these equations and (46.17) we have

$$\left(d_{\alpha\epsilon,\beta} + d_{\gamma\epsilon} \left\{ \begin{matrix} \gamma \\ \alpha\beta \end{matrix} \right\} - d_{\gamma\epsilon} \left\{ \begin{matrix} \overline{\gamma} \\ \alpha\beta \end{matrix} \right\} \right) g^{\epsilon\delta} \frac{\partial x^i}{\partial u^\delta} = 0.$$

Multiplying by $\dfrac{\partial x^i}{\partial u^\nu}$ and summing with respect to i, we obtain

$$d_{\alpha\nu,\beta} + d_{\gamma\nu} \left\{ \begin{matrix} \gamma \\ \alpha\beta \end{matrix} \right\} - d_{\gamma\nu} \left\{ \begin{matrix} \overline{\gamma} \\ \alpha\beta \end{matrix} \right\} = 0,$$

that is,

$$(46.18) \qquad \frac{\partial d_{\alpha\nu}}{\partial u^\beta} = d_{\alpha\gamma} \left\{ \begin{matrix} \gamma \\ \nu\beta \end{matrix} \right\} + d_{\gamma\nu} \left\{ \begin{matrix} \overline{\gamma} \\ \alpha\beta \end{matrix} \right\}.$$

If we multiply these equations by $d^{\nu\delta}$, as defined by (42.11), and sum with respect to ν, we obtain

$$(46.19) \qquad \left\{ \begin{matrix} \overline{\delta} \\ \alpha\beta \end{matrix} \right\} = d^{\nu\delta} \frac{\partial d_{\alpha\nu}}{\partial u^\beta} - d^{\nu\delta} d_{\alpha\gamma} \left\{ \begin{matrix} \gamma \\ \nu\beta \end{matrix} \right\}.$$

The left-hand member of equations (46.15) is the second covariant derivative of the scalars X^i based on $h_{\alpha\beta}$. We indicate such covariant derivatives by placing a bar over the index of covariant differentiation; thus we write (46.15) in the form

$$(46.20) \qquad X^i_{,\bar\alpha\bar\beta} = - h_{\alpha\beta} X^i.$$

Differentiating covariantly with respect to u^γ and noting that $h_{\alpha\beta,\bar\gamma} = 0$, we obtain

$$X^i_{,\bar\alpha\bar\beta\bar\gamma} = - h_{\alpha\beta} \frac{\partial X^i}{\partial u^\gamma}.$$

Applying the appropriate Ricci identities (22.19) to this covariant differentiation, namely

$$X^i_{,\bar\alpha\bar\beta\bar\gamma} - X^i_{,\bar\alpha\bar\gamma\bar\beta} = \frac{\partial X^i}{\partial u^\delta} \bar R^\delta_{\alpha\beta\gamma}$$

where $\bar R^\delta_{\alpha\beta\gamma}$ are of the form (20.13) in terms of the symbols $\left\{ \begin{matrix} \bar\gamma \\ \alpha\beta \end{matrix} \right\}$, we obtain

$$- h_{\alpha\beta} \frac{\partial X^i}{\partial u^\gamma} + h_{\alpha\gamma} \frac{\partial X^i}{\partial u^\beta} = \frac{\partial X^i}{\partial u^\delta} \bar R^\delta_{\alpha\beta\gamma}.$$

Multiplying these equations by $\dfrac{\partial X^i}{\partial u^\epsilon}$ and summing for i, we obtain in consequence of (46.3)

$$(46.21) \qquad h_{\alpha\gamma} h_{\beta\epsilon} - h_{\alpha\beta} h_{\gamma\epsilon} = h_{\delta\epsilon} \bar R^\delta_{\alpha\beta\gamma} = \bar R_{\epsilon\alpha\beta\gamma},$$

which expresses the fact that the Gaussian curvature of the sphere is equal to $+1$.

In terms of this covariant differentiation of $d_{\alpha\nu}$ based upon $h_{\alpha\beta}$ equations (46.18) are expressible in the form

$$(46.22) \qquad d_{\alpha\nu,\bar\beta} + d_{\alpha\gamma} \left(\left\{ \begin{matrix} \bar\gamma \\ \nu\beta \end{matrix} \right\} - \left\{ \begin{matrix} \gamma \\ \nu\beta \end{matrix} \right\} \right) = 0,$$

from which it follows that

$$(46.23) \qquad d_{\alpha\nu,\bar\beta} - d_{\alpha\beta,\bar\nu} = 0.$$

We now derive an important set of formulas. Since

$$\sum_i X^i \frac{\partial X^i}{\partial u^\alpha} = 0,$$

and the equations

$$\sum X^i \left(X^j \frac{\partial X^k}{\partial u^\alpha} - X^k \frac{\partial X^j}{\partial u^\alpha} \right) = 0,$$

where $\sum$ indicates the sum as i, j, k that the values 1, 2, 3 cyclically, are identities, it follows that

(46.24) $$X^j \frac{\partial X^k}{\partial u^\alpha} - X^k \frac{\partial X^j}{\partial u^\alpha} = a_{\alpha 1} \frac{\partial X^i}{\partial u^1} + a_{\alpha 2} \frac{\partial X^i}{\partial u^2},$$

where i, j, k take the values 1, 2, 3 cyclically, and the a's are to be determined. Multiplying these equations by $\dfrac{\partial X^i}{\partial u^\alpha}$ and summing with respect to i, the resulting left-hand member is equal to zero identically, and we have

$$0 = a_{\alpha 1} h_{1\alpha} + a_{\alpha 2} h_{2\alpha}.$$

Again multiplying the equations (46.24) by $\dfrac{\partial X^i}{\partial u^\beta}$ $(\beta \neq \alpha)$ and summing with respect to i, we have in consequence of (46.9) and (46.10)

$$\pm \sqrt{h} e = a_{\alpha 1} h_{1\beta} + a_{\alpha 2} h_{2\beta},$$

where the sign on the left is $+$ or $-$ according as $\alpha = 1$, $\beta = 2$ or $\alpha = 2$, $\beta = 1$. From these two sets of equations we obtain

$$a_{11} = -\frac{h_{12} e}{\sqrt{h}}, \qquad a_{12} = \frac{h_{11} e}{\sqrt{h}}, \qquad a_{21} = -\frac{h_{22} e}{\sqrt{h}}, \qquad a_{22} = \frac{h_{12} e}{\sqrt{h}},$$

and consequently equations (46.24) are

$$\begin{aligned}
&X^j \frac{\partial X^k}{\partial u^1} - X^k \frac{\partial X^j}{\partial u^1} = \frac{e}{\sqrt{h}} \left(-h_{12} \frac{\partial X^i}{\partial u^1} + h_{11} \frac{\partial X^i}{\partial u^2} \right) = e\sqrt{h}\, h^{2\alpha} \frac{\partial X^i}{\partial u^\alpha},\\
&X^j \frac{\partial X^k}{\partial u^2} - X^k \frac{\partial X^j}{\partial u^2} = \frac{e}{\sqrt{h}} \left(-h_{22} \frac{\partial X^i}{\partial u^1} + h_{12} \frac{\partial X^i}{\partial u^2} \right) = -e\sqrt{h}\, h^{1\alpha} \frac{\partial X^i}{\partial u^\alpha},
\end{aligned}$$

(46.25)

as i, j, k take the values 1, 2, 3 cyclically, where e is $+1$ or -1 according as the Gaussian curvature is positive or negative.

EXERCISES

1. The spherical representation of the lines of curvature of a surface of revolution is an isometric orthogonal net.

2. The osculating planes of a line of curvature and its spherical representation are parallel (see §6, Ex. 9).

3. If κ and $\bar{\kappa}$ denote the radii of curvature of a line of curvature and its spherical

representation, and κ_g and $\bar{\kappa}_g$ the radii of geodesic curvature of these respective curves, then

$$\kappa \, ds = \bar{\kappa} \, d\bar{s}, \qquad \kappa_g \, ds = \bar{\kappa}_g \, d\bar{s}.$$

4. The spherical representation of a plane line of curvature is a circle.

5. The tangents to a curve on a surface and to its spherical representation at corresponding points are orthogonal, if and only if the curve is an asymptotic line.

6. The angles between the asymptotic lines at a point on a surface and their spherical representation are equal or supplementary according as the Gaussian curvature of the surface is positive or negative at the point.

7. The principal directions for the tensor $h_{\alpha\beta}$ (see §26) are the principal directions of normal curvature; find the relation between r in (26.17) and R for this case.

8. When the asymptotic lines are coordinate, it follows from (46.18) that

$$\overline{\left\{\begin{matrix} \alpha \\ \alpha\alpha \end{matrix}\right\}} = \left\{\begin{matrix} \alpha \\ \alpha\alpha \end{matrix}\right\} - 2\left\{\begin{matrix} \beta \\ \alpha\beta \end{matrix}\right\}, \qquad \overline{\left\{\begin{matrix} \alpha \\ \alpha\beta \end{matrix}\right\}} = -\left\{\begin{matrix} \alpha \\ \alpha\beta \end{matrix}\right\}, \qquad \overline{\left\{\begin{matrix} \alpha \\ \beta\beta \end{matrix}\right\}} = -\left\{\begin{matrix} \alpha \\ \beta\beta \end{matrix}\right\} \qquad (\beta \neq \alpha).$$

9. When the coordinate curves form a conjugate net, it follows from (46.18) that

$$\overline{\left\{\begin{matrix} \alpha \\ \alpha\alpha \end{matrix}\right\}} = \frac{\partial \log d_{\alpha\alpha}}{\partial u^\alpha} - \left\{\begin{matrix} \alpha \\ \alpha\alpha \end{matrix}\right\}, \qquad \overline{\left\{\begin{matrix} \alpha \\ \alpha\beta \end{matrix}\right\}} = -\frac{d_{\beta\beta}}{d_{\alpha\alpha}}\left\{\begin{matrix} \beta \\ \alpha\alpha \end{matrix}\right\}, \qquad \overline{\left\{\begin{matrix} \alpha \\ \beta\beta \end{matrix}\right\}} = -\frac{d_{\beta\beta}}{d_{\alpha\alpha}}\left\{\begin{matrix} \beta \\ \alpha\beta \end{matrix}\right\} \qquad (\beta \neq \alpha).$$

10. The asymptotic lines on a minimal surface form an isometric orthogonal net, as do also their spherical representation.

11. From (46.4) one obtains

$$g_{\alpha\beta} = d_{\alpha\gamma}d_{\beta\delta}h^{\gamma\delta}$$

and from (46.5)

$$g_{\alpha\beta} = (\rho_1 + \rho_2)d_{\alpha\beta} - \rho_1\rho_2 h_{\alpha\beta}.$$

12. A necessary and sufficient condition that the linear element of a surface referred to a conjugate net be expressible in the form

$$ds^2 = \rho^2(h_{11}du^{1^2} - 2h_{12}du^1du^2 + h_{22}du^{2^2})$$

is that the Gaussian curvature of the surface be positive and that the coordinate conjugate net be the mean-conjugate net (see §43).

13. Theorem [46.4] and Ex. 5 are equivalent respectively to Exs. 5 and 6 of §45.

47. TANGENTIAL COORDINATES OF A SURFACE

From the equation of the tangent plane to a surface at the point x^i, namely

$$(47.1) \qquad \sum_i X^i(\bar{x}^i - x^i) = 0,$$

where $\bar{x}^i$ are current coordinates, we have that the algebraic distance W of the origin from the tangent plane is given by

$$(47.2) \qquad W = \sum_i X^i x^i.$$

Looking upon this equation as an identity in u^α, we have in consequence of (38.8)

$$(47.3) \qquad \frac{\partial W}{\partial u^\alpha} = \sum_i \frac{\partial X^i}{\partial u^\alpha} x^i.$$

For each point P on a surface the quantities X^i, $\dfrac{\partial X^i}{\partial u^1}$ and $\dfrac{\partial X^i}{\partial u^2}$ are the components of independent vectors, as follows from (38.6) and

$$(47.4) \qquad \sum_i X^i \frac{\partial X^i}{\partial u^\alpha} = 0.$$

Accordingly the direction numbers x^i of the line segment OP, where O is the origin, are expressible in the form

$$x^i = aX^i + b^\alpha \frac{\partial X^i}{\partial u^\alpha}.$$

In order to determine a and b^α we multiply these equations by X^i and sum with respect to i, and again by $\dfrac{\partial X^i}{\partial u^\beta}$ and sum with respect to i, with the result, in consequence of (47.2), (47.3), and (46.3),

$$a = W, \qquad b^\alpha h_{\alpha\beta} = \frac{\partial W}{\partial u^\beta}.$$

If the second set of these equations is multiplied by $h^{\beta\gamma}$, where the latter are defined by (46.14), and summed with respect to β, we obtain

$$h^{\beta\gamma} \frac{\partial W}{\partial u^\beta} = b^\alpha h_{\alpha\beta} h^{\beta\gamma} = b^\gamma.$$

Hence we have

$$(47.5) \qquad x^i = WX^i + h^{\alpha\beta} \frac{\partial X^i}{\partial u^\alpha} \frac{\partial W}{\partial u^\beta}.$$

If these equations are differentiated covariantly based on $h_{\alpha\beta}$, we have in consequence of (46.20) and the fact that the covariant derivative of $h^{\alpha\beta}$ is equal to zero

$$(47.6) \qquad \begin{aligned} \frac{\partial x^i}{\partial u^\gamma} &= W \frac{\partial X^i}{\partial u^\gamma} + X^i \frac{\partial W}{\partial u^\gamma} + h^{\alpha\beta} \frac{\partial X^i}{\partial u^\alpha} W_{,\bar\beta\bar\gamma} - h^{\alpha\beta} h_{\alpha\gamma} X^i \frac{\partial W}{\partial u^\beta} \\ &= W \frac{\partial X^i}{\partial u^\gamma} + h^{\alpha\beta} \frac{\partial X^i}{\partial u^\alpha} W_{,\bar\beta\bar\gamma}. \end{aligned}$$

From these equations we have (see (38.2) and (38.16))

$$(47.7) \quad g_{\alpha\beta} = \sum_i \left(W \frac{\partial X^i}{\partial u^\alpha} + h^{\gamma\delta} \frac{\partial X^i}{\partial u^\gamma} W_{,\delta\alpha} \right) \left(W \frac{\partial X^i}{\partial u^\beta} + h^{\mu\nu} \frac{\partial X^i}{\partial u^\nu} W_{,\mu\beta} \right)$$

$$= W^2 h_{\alpha\beta} + 2WW_{,\alpha\beta} + h^{\delta\mu} W_{,\delta\alpha} W_{,\mu\beta},$$

and

$$(47.8) \quad d_{\alpha\beta} = -\sum_i \frac{\partial X^i}{\partial u^\alpha} \left(W \frac{\partial X^i}{\partial u^\beta} + h^{\gamma\delta} \frac{\partial X^i}{\partial u^\gamma} W_{,\delta\beta} \right)$$

$$= -(W_{,\alpha\beta} + h_{\alpha\beta} W).$$

Conversely, given any four functions X^i and W of coordinates $\overset{\alpha}{u}$ such that $\sum_i X^i X^i = 1$, we have (47.4), and equations (46.20) follow from the definition (46.3) of $h_{\alpha\beta}$ and the discussion in §46. From (47.6) we have

$$\sum_i X^i \frac{\partial x^i}{\partial u^\alpha} = 0.$$

Consequently X^i are direction cosines of the normal to the surface for which x^i as functions of u^α are given by (47.5). Since (47.2) follows from (47.5), W is the algebraic distance of the origin from the tangent plane at the corresponding point. Four such functions X^i and W are called *tangential coordinates* of the surface. There can be no relation between the X's and W of the form $a_i X^i + bW = 0$, where the a's and b are constants. For, if there were, it would follow from (47.8) and (46.20) that $d_{\alpha\beta} = 0$, and consequently the surface for which x^i are given by (47.5) would be a plane (see §38, Ex. 1), in which case the X's would be constants. Hence we say that X^i and W are *linearly independent (constant coefficients)* and we have

[47.1] *Four linearly independent (constant coefficients) functions X^i and W of coordinates u^α for which $\sum_i X^i X^i = 1$ are tangential coordinates of a surface whose equations $x^i = f^i(u^1, u^2)$ are given by (47.5), and whose two fundamental tensors are given by (47.7) and (47.8).*

If one has any four functions θ^1, θ^2, θ^3, θ^4 linearly independent (constant coefficients) and one puts

$$(47.9) \qquad X^i = \frac{\theta^i}{\varphi}, \qquad W = \frac{\theta^4}{\varphi} \qquad (i = 1, 2, 3),$$

where $\varphi^2 = \theta^{1^2} + \theta^{2^2} + \theta^{3^2}$, then X^i and W satisfy the conditions of theorem [47.1]. Consequently from four linearly independent (constant coefficients) functions of u^1 and u^2 one obtains the tangential coordinates of four different surfaces according as one chooses three of the functions to define X^i by (47.9), that is, as one chooses the three functions which are direction numbers of the normal to the surface.

When the coordinate lines on a surface form a conjugate net, the tangential coordinates X^i and W are solutions of an equation of the form

$$(47.10) \qquad \frac{\partial^2 \theta}{\partial u^1 \partial u^2} = a^\alpha \frac{\partial \theta}{\partial u^\alpha} + b\theta,$$

as follows from (46.20) and (47.8). Conversely, if X^i and W are four linearly independent (constant coefficients) solutions of an equation (47.10) such that $\sum_i X^i X^i = 1$, it follows from equation (47.10) and (46.20) that

$$\left(a^\alpha - \begin{Bmatrix} \alpha \\ 12 \end{Bmatrix}\right) \frac{\partial X^i}{\partial u^\alpha} + (b + h_{12})X^i = 0.$$

From these equations we have

$$a^\alpha = \begin{Bmatrix} \alpha \\ 12 \end{Bmatrix}, \qquad b = -h_{12},$$

and consequently for the surface with equations (47.5) we have $d_{12} = 0$, as follows from (47.8). Hence we have

[47.2] *Four linearly independent (constant coefficients) solutions X^i, W of an equation of the form* (47.10) *such that* $\sum_i X^i X^i = 1$ *are the tangential coordinates of a surface upon which the coordinate curves form a conjugate net.*

If one has any four linearly independent (constant coefficients) solutions $\theta^1, \cdots, \theta^4$ of an equation of the form (47.10), the functions X^i, W defined by (47.9) are solutions of an equation of the form (47.10) and satisfy the conditions of theorem [47.2]. Hence we have

[47.3] *Four linearly independent (constant coefficients) solutions of an equation of the form* (47.10) *determine four surfaces upon each of which the coordinate curves form a conjugate net; each surface is determined by which three of the four solutions are direction numbers of the normals to the surface.*

When the asymptotic lines upon a surface of negative Gaussian curvature are coordinate, that is, $d_{11} = d_{22} = 0$, the tangential coordinates X^i and W are solutions of two equations of the form

$$(47.11) \quad \frac{\partial^2 \theta}{\partial u^{1^2}} = a_{11} \frac{\partial \theta}{\partial u^1} + a_{12} \frac{\partial \theta}{\partial u^2} + b_1\theta, \quad \frac{\partial^2 \theta}{\partial u^{2^2}} = a_{21} \frac{\partial \theta}{\partial u^1} + a_{22} \frac{\partial \theta}{\partial u^2} + b_2\theta,$$

as follows from (46.20) and (47.8). By an argument similar to that which led to theorem [47.2] we obtain

[47.4] *Four linearly independent (constant coefficients) solutions X^i, W of two equations of the form (47.11) such that $\sum_i X^i X^i = 1$ are the tangential coordinates of a surface upon which the coordinate curves are the asymptotic lines.*

Also by an argument similar to that which led to theorem [47.3] we obtain

[47.5] *When two equations of the form (47.11) admit four linearly independent (constant coefficients) solutions, these solutions determine four surfaces upon each of which the coordinate curves are the asymptotic lines; each surface is determined by which three of the four solutions are direction numbers of the normals to the surface.*

From (47.6) and (47.8) we have

$$(47.12) \qquad \frac{\partial x^i}{\partial u^\alpha} = - d_{\alpha\beta} h^{\beta\gamma} \frac{\partial X^i}{\partial u^\gamma}.$$

For a surface of negative Gaussian curvature referred to its asymptotic lines we have

$$(47.13) \qquad d_{11} = d_{22} = 0, \qquad d_{12} = \sqrt{h}\rho,$$

where from (46.13) $\rho^2 = - \rho_1 \rho_2$. In consequence of (46.14) equations (47.12) are in this case expressible in the form

$$(47.14) \qquad \begin{aligned} \frac{\partial x^i}{\partial u^1} &= \frac{\rho}{\sqrt{h}}\left(h_{12} \frac{\partial X^i}{\partial u^1} - h_{11} \frac{\partial X^i}{\partial u^2} \right), \\ \frac{\partial x^i}{\partial u^2} &= \frac{\rho}{\sqrt{h}}\left(h_{12} \frac{\partial X^i}{\partial u^2} - h_{22} \frac{\partial X^i}{\partial u^1} \right), \end{aligned}$$

which in consequence of (46.25) are equivalent to

$$(47.15) \qquad \begin{aligned} \frac{\partial x^i}{\partial u^1} &= \rho\left(X^j \frac{\partial X^k}{\partial u^1} - X^k \frac{\partial X^j}{\partial u^1} \right), \\ \frac{\partial x^i}{\partial u^2} &= -\rho\left(X^j \frac{\partial X^k}{\partial u^2} - X^k \frac{\partial X^j}{\partial u^2} \right), \end{aligned}$$

where i, j, k take the values 1, 2, 3 cyclically. If then we define functions v^i by

$$(47.16) \qquad v^i = \sqrt{\rho}\, X^i,$$

these equations become

$$(47.17) \qquad \frac{\partial x^i}{\partial u^1} = \left(\nu^j \frac{\partial \nu^k}{\partial u^1} - \nu^k \frac{\partial \nu^j}{\partial u^1} \right), \qquad \frac{\partial x^i}{\partial u^2} = -\left(\nu^j \frac{\partial \nu^k}{\partial u^2} - \nu^k \frac{\partial \nu^j}{\partial u^2} \right),$$

where i, j, k take the values 1, 2, 3 cyclically.

For the values (47.13) equations (46.23) reduce to

$$(47.18) \qquad \frac{\partial}{\partial u^1} \log \sqrt{h}\rho = \left\{ \begin{matrix} 1 \\ 11 \end{matrix} \right\} - \left\{ \begin{matrix} 2 \\ 12 \end{matrix} \right\}, \qquad \frac{\partial}{\partial u^2} \log \sqrt{h}\rho = \left\{ \begin{matrix} 2 \\ 22 \end{matrix} \right\} - \left\{ \begin{matrix} 1 \\ 12 \end{matrix} \right\}.$$

Analogously to (28.2) we have

$$(47.19) \qquad \frac{\partial \log \sqrt{h}}{\partial u^\alpha} = \left\{ \begin{matrix} \beta \\ \alpha\beta \end{matrix} \right\},$$

and consequently (47.18) are equivalent to

$$(47.20) \qquad \frac{\partial \log \sqrt{\rho}}{\partial u^1} = -\left\{ \begin{matrix} 2 \\ 12 \end{matrix} \right\}, \qquad \frac{\partial \log \sqrt{\rho}}{\partial u^2} = -\left\{ \begin{matrix} 1 \\ 12 \end{matrix} \right\}.$$

In consequence of this result and equations (46.15) we have from (47.16)

$$(47.21) \qquad \frac{\partial^2 \nu^i}{\partial u^1 \partial u^2} = \left(\frac{1}{\sqrt{\rho}} \frac{\partial^2 \sqrt{\rho}}{\partial u^1 \partial u^2} - h_{12} \right) \nu^i,$$

from which it follows that the conditions of integrability of (47.17) are satisfied.

Conversely, we have

[47.6] *If* ν^1, ν^2, ν^3 *are any three linearly independent (constant coefficients) solutions of an equation of the form*

$$(47.22) \qquad \frac{\partial^2 \theta}{\partial u^1 \partial u^2} = \lambda\theta,$$

where λ *is a function of* u^α, *the coordinate curves on the surface whose coordinates are given by the corresponding quadratures (47.17) are the asymptotic lines, and the Gaussian curvature of the surface is equal to* $-1/(\sum_i \nu^i \nu^i)^2$.

For, the conditions of integrability of (47.17) are satisfied by solutions of (47.22). If then we define X^i by (47.16) with $\rho = \sum_i \nu^i \nu^i$, equations (47.17) are reducible to (47.15), from which we obtain

$$\sum_i \frac{\partial x^i}{\partial u^1} \frac{\partial X^i}{\partial u^1} = \sum_i \frac{\partial x^i}{\partial u^2} \frac{\partial X^i}{\partial u^2} = 0, \qquad d_{12} = -\sum \frac{\partial x^i}{\partial u^1} \frac{\partial X^i}{\partial u^2} = \sqrt{h}\rho,$$

the last being a consequence of (46.9) and (46.10). From this result and (46.13) it follows that $K = -1/(\sum_i \nu^i \nu^i)^2$.

Equations (47.17) are called the *formulas of Lelieuvre.*[*]

EXERCISES

1. On the envelope of the family of planes

$$\sum_i (U_1^i + U_2^i)x^i + U_1 + U_2 = 0,$$

where U_1^i, U_1 are functions of u^1, and U_2^i, U_2 of u^2, the coordinate curves form a conjugate net of plane curves.

2. On the envelope of the family of planes

$$\cos u^1 x^1 + \sin u^1 x^2 + \cot u^2 x^3 + U_1 + U_2 = 0,$$

where U_1 and U_2 are functions of u^1 and u^2 respectively, the coordinate curves are plane lines of curvature.

3. The conditions of integrability of equations (47.12) are equivalent to equations (46.23).

4. In consequence of (47.19) equations (46.23) are equivalent to

$$\frac{\partial}{\partial u^\beta}\frac{d_{\alpha\gamma}}{\sqrt{h}} - \frac{\partial}{\partial u^\gamma}\frac{d_{\alpha\beta}}{\sqrt{h}} + \frac{d_{\alpha\gamma}}{\sqrt{h}}\left\{\begin{matrix}\overline{\delta}\\ \beta\delta\end{matrix}\right\} + \frac{d_{\beta\delta}}{\sqrt{h}}\left\{\begin{matrix}\overline{\delta}\\ \alpha\gamma\end{matrix}\right\} - \frac{d_{\alpha\beta}}{\sqrt{h}}\left\{\begin{matrix}\overline{\delta}\\ \gamma\delta\end{matrix}\right\} - \frac{d_{\gamma\delta}}{\sqrt{h}}\left\{\begin{matrix}\overline{\delta}\\ \alpha\beta\end{matrix}\right\} = 0.$$

5. When a surface is referred to its asymptotic lines, the equations of Ex. 4 reduce to

$$\frac{\partial}{\partial u^\alpha} \log \frac{d_{12}}{\sqrt{h}} = -2\left\{\begin{matrix}\overline{\beta}\\ \alpha\beta\end{matrix}\right\} \qquad (\beta \neq \alpha).$$

6. A necessary condition that the coordinate curves on the unit sphere are the representation of the asymptotic lines on a surface is that

$$\frac{\partial}{\partial u^1}\left\{\begin{matrix}\overline{1}\\ 12\end{matrix}\right\} = \frac{\partial}{\partial u^2}\left\{\begin{matrix}\overline{2}\\ 12\end{matrix}\right\};$$

when this condition is satisfied, the coordinate curves on the sphere are the representation of the asymptotic lines on a family of surfaces, which are homothetic transforms of one another, and the equations $x^i = f^i(u^1, u^2)$ of the surfaces can be found by quadratures.

7. In order that equations of the form (47.11) admit four linearly independent (constant coefficients) solutions it is necessary that

$$\frac{\partial}{\partial u^2} a_{11} = \frac{\partial}{\partial u^1} a_{22}.$$

8. The most general right conoid is defined by equations (47.17) for the values

$$\nu^1 = u^1, \qquad \nu^2 = u^2, \qquad \nu^3 = \varphi(u^2).$$

---

[*] 1888, 1, p. 126.

9. When a surface is referred to its asymptotic lines, a necessary and sufficient condition that the lines u^2 = const. be straight, and thus that the surface be ruled, is that the normals to the surface along each curve u^2 = const. be parallel to a plane, as can be shown by means of equations (47.17).

10. When the asymptotic lines in one system on a surface are represented on the sphere by great circles, the surface is a ruled surface.

11. When the equations of the unit sphere are of the form of §28, Ex. 7 with $a = 1$, the coordinate curves are the asymptotic lines of the sphere and equation (47.21) is

$$(1 + u^1 u^2)^2 \frac{\partial^2 \theta}{\partial u^1 \partial u^2} = -2\theta,$$

of which the general integral is

$$\theta = 2 \frac{u^2 \varphi(u^1) + u^1 \psi(u^2)}{1 + u^1 u^2} - \varphi' - \psi',$$

where φ and ψ are arbitrary functions of u^1 and u^2 respectively.

12. Find the equation (47.21) when the surface is a hyperboloid of one sheet; when a hyperbolic paraboloid. (See §43, Exs. 8, 9).

13. When the coordinate curves on the unit sphere satisfy the condition

$$\frac{\partial}{\partial u^1} \left\{ \frac{1}{12} \right\} = \frac{\partial}{\partial u^2} \left\{ \frac{2}{12} \right\} = 2 \left\{ \frac{1}{12} \right\} \left\{ \frac{2}{12} \right\},$$

they represent the asymptotic lines on a surface whose total curvature is of the form

$$K = \frac{-1}{[\varphi(u^1) + \psi(u^2)]^2}.$$

14. When the coordinate curves on the unit sphere form an orthogonal net a necessary and sufficient condition that the curves u^2 = const. be circles, that is, curves of constant geodesic curvature, is that

$$\frac{1}{\sqrt{h_{22}}} \frac{\partial}{\partial u^2} \log \sqrt{h_{11}} = \varphi(u^2).$$

15. When upon the unit sphere the curves of one family of an orthogonal isometric net are circles, so also are the curves of the other family (see §34, Ex. 6).

16. When the equations of the unit sphere are

$$X^1, X^2, X^3 = \frac{2u^1, 2u^2, u^{1^2} + u^{2^2} - 1}{u^{1^2} + u^{2^2} + 1},$$

the coordinate curves are circles whose planes are

$$x^1 + u^1(x^3 - 1) = 0, \qquad x^2 + u^2(x^3 - 1) = 0,$$

and they form an orthogonal net, since the linear element is

$$d\bar{s}^2 = \frac{4(du^{1^2} + du^{2^2})}{(u^{1^2} + u^{2^2} + 1)^2}.$$

These curves are the spherical representation of the plane lines of curvature of surfaces for which

$$W = 2 \frac{U_1 + U_2}{u^{1^2} + u^{2^2} + 1},$$

where U_1 and U_2 are arbitrary functions of u^1 and u^2 respectively.

17. When the equations of the unit sphere are

$$X^1, \ X^2, \ X^3 = \frac{\sqrt{1 - a^2} \sin u^1, \ -\sqrt{1 - a^2} \sinh u^2, \ \cos u^1 + a \cosh u^2}{\cosh u^2 + a \cos u^1},$$

where a is a constant ($|\,a\,| < 1$), the coordinate curves are circles whose planes are

$$\sqrt{1 - a^2}\, x^1 - \tan u^1\, (x^3 - a) = 0, \qquad \sqrt{1 - a^2}\, x^2 - a \tanh u^2 \left(x^3 - \frac{1}{a}\right) = 0,$$

and they form an orthogonal net, since the linear element is

$$d\bar{s}^2 = \frac{(1 - a^2)(du^{1^2} + du^{2^2})}{(\cosh u^2 + a \cos u^1)^2}.$$

These curves are the spherical representation of the plane lines of curvature of surfaces for which

$$W = \frac{\sqrt{1 - a^2}\ (U_1 + U_2)}{\cosh u^2 + a \cos u^1},$$

where U_1 and U_2 are arbitrary functions of u^1 and u^2 respectively.

18. When $a = 0$ in Ex. 17, the curves $u^2 = $ const. on a surface with this spherical representation lie in parallel planes and the planes of the curves $u^1 = $ const. envelope a cylinder.

48. SURFACES OF CENTER OF A SURFACE. PARALLEL SURFACES

In §41 it was shown that the normals to a surface S along the lines of curvature C_1 and C_2 through a point P on a surface form two developable surfaces D_1 and D_2, and that the coordinates of the points P_1 and P_2 on the edges of regression of these surfaces corresponding to P, that is, on the normal to S at P, are given by

$$(48.1) \qquad x_1^i = x^i + \rho_1 X^i, \qquad x_2^i = x^i + \rho_2 X^i,$$

where ρ_1 and ρ_2 are the principal radii of normal curvature at P. The surfaces S_1 and S_2, which are the loci of the points P_1 and P_2 respectively, are called the *surfaces of center* of the surface S. When S is a sphere, in place of two surfaces of center, there is only the center of the sphere. Also it is evident geometrically that the normals to a surface of revolution at points of a parallel form a right circular cone with vertex on the axis of the surface, and consequently, the axis is one of the surfaces of center (see Exs. 1, 2, 3). In what follows we consider only the case where neither of the surfaces S_1 and S_2 is degenerate.

The edges of regression Γ_1 and Γ_2 of the developable surfaces D_1 and D_2 lie on the respective surfaces S_1 and S_2. Since the normal at P is tangent to Γ_1 at P_1 and to Γ_2 at P_2, this normal is a common tangent to S_1 and S_2, and consequently the normals to S are common tangents of the surfaces S_1 and S_2. Since the generators of D_1 are tangent to S_2, it follows that D_1 is the envelope of the tangent planes to S_2 along a curve Γ_2', and that at P_2 the directions of Γ_2 and Γ_2' are conjugate as follows from the definition of conjugate directions in §42. In like manner the developable D_2 envelops S_1 along a curve Γ_1' and at P_1 the directions of Γ_1 and Γ_1' are conjugate.

The tangent planes to D_1 and D_2 along their common generator, namely the normal at P, are perpendicular, since they are determined by this normal and the tangents to C_1 and C_2 at P. This tangent plane to D_1 is the osculating plane of Γ_1 at P_1, and is perpendicular to the tangent plane to S_1 at P_1, since the tangent plane to D_2 is this tangent plane to S_1. Hence Γ_1 is a geodesic on S_1 by theorem [44.2], and similarly Γ_2 is a geodesic on S_2. Accordingly we have

[48.1] *The edges of regression of the developable surfaces consisting of the normals to a surface along the lines of curvature of one family are geodesics on the surface which is the locus of these edges; the developable surfaces which consist of the normals to the surface along the lines of curvature of the other family envelope this surface of center along the curves conjugate to the edges; the osculating planes of the edges on one surface of centers are tangent to the other surface of centers.*

We now obtain in an analytical manner the results which have just been deduced geometrically. We assume that the surface S is referred to its lines of curvature, and that ρ_1 and ρ_2 are the radii of principal curvature for the directions of the respective curves $u^2 = $ const. and $u^1 = $ const. at a point. Accordingly we have

$$(48.2) \qquad \frac{1}{\rho_1} = \frac{d_{11}}{g_{11}}, \qquad \frac{1}{\rho_2} = \frac{d_{22}}{g_{22}}, \qquad d_{12} = g_{12} = 0.$$

Making use of (41.9), we obtain from the first of (48.1)

$$(48.3) \qquad \frac{\partial x_1^i}{\partial u^1} = \frac{\partial \rho_1}{\partial u^1} X^i, \qquad \frac{\partial x_1^i}{\partial u^2} = \frac{\partial \rho_1}{\partial u^2} X^i + \left(1 - \frac{\rho_1}{\rho_2}\right) \frac{\partial x^i}{\partial u^2},$$

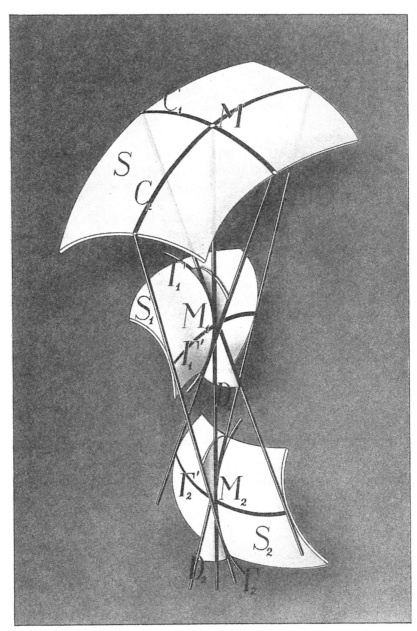

Fig. 16. A surface and its two surfaces of center

from which one finds for the components $g'_{\alpha\beta}$ of the first fundamental tensor of the surface of centers S_1

$$(48.4) \quad g'_{11} = \left(\frac{\partial\rho_1}{\partial u^1}\right)^2, \quad g'_{12} = \frac{\partial\rho_1}{\partial u^1}\frac{\partial\rho_1}{\partial u^2}, \quad g'_{22} = \left(\frac{\partial\rho_1}{\partial u^2}\right)^2 + \left(1 - \frac{\rho_1}{\rho_2}\right)^2 g_{22}.$$

Hence the linear element of S_1 may be written

$$(48.5) \qquad ds_1^2 = d\rho_1^2 + \left(1 - \frac{\rho_1}{\rho_2}\right)^2 g_{22}\,du^{2^2},$$

from which it follows that the curves $u^2 = $ const., that is, the edges of regression of the developable surfaces D_1, are geodesics (see (32.15)), and their orthogonal trajectories are the curves for which $\rho_1 = $ const.

From (48.3) we have $\sum_i \dfrac{\partial x^i}{\partial u^1}\dfrac{\partial x_1^i}{\partial u^\alpha} = 0$ $(\alpha = 1, 2)$, from which it follows that $\dfrac{\partial x^i}{\partial u^1}$ are direction numbers of the normal to S_1, a result which we obtained previously from geometrical considerations. Hence the direction cosines of the normal to S_1 are given by

$$(48.6) \qquad\qquad X_1^i = \frac{\epsilon_1}{\sqrt{g_{11}}}\frac{\partial x^i}{\partial u^1},$$

where ϵ_1 is $+1$ or -1 so that there shall hold for S_1 the results analogous to (38.9). By substitution it is found that this means that ϵ_1 is $+1$ or -1 according as $\left(\dfrac{\rho_1}{\rho_2} - 1\right)\dfrac{\partial\rho_1}{\partial u^1}$ is positive or negative.

If $d'_{\alpha\beta}$ denotes the second fundamental tensor of S_1, one finds from equations analogous to (38.13) on making use of (41.9) and (38.17)

$$(48.7) \quad\begin{aligned} & d'_{11} = -\epsilon_1\frac{\sqrt{g_{11}}}{\rho_1}\frac{\partial\rho_1}{\partial u^1}, \qquad d'_{12} = 0, \\[2mm] & d'_{22} = \frac{-\epsilon_1}{2\sqrt{g_{11}}}\left(1 - \frac{\rho_1}{\rho_2}\right)\frac{\partial g_{22}}{\partial u^1} = \frac{\epsilon_1\rho_1}{\rho_2^2}\frac{g_{22}}{\sqrt{g_{11}}}\frac{\partial\rho_2}{\partial u^1}, \end{aligned}$$

the last expression being a consequence of (41.12). Since $d'_{12} = 0$, the second part of theorem [48.1] is established anew.

In like manner from the second of equations (48.1) we obtain as the fundamental tensors of the second surface of centers S_2

$$(48.8) \quad g''_{11} = \left(\frac{\partial\rho_2}{\partial u^1}\right)^2 + \left(1 - \frac{\rho_2}{\rho_1}\right)^2 g_{11}, \quad g''_{12} = \frac{\partial\rho_2}{\partial u^1}\frac{\partial\rho_2}{\partial u^2}, \quad g''_{22} = \left(\frac{\partial\rho_2}{\partial u^2}\right)^2,$$

$$(48.9) \quad d''_{11} = \epsilon_2\frac{\rho_2}{\rho_1^2}\frac{g_{11}}{\sqrt{g_{22}}}\frac{\partial\rho_1}{\partial u^2}, \qquad d''_{12} = 0, \qquad d''_{22} = -\frac{\epsilon_2\sqrt{g_{22}}}{\rho_2}\frac{\partial\rho_2}{\partial u^2},$$

where the normal X_2^i to S_2 is given by

(48.10) $$X_2^i = \frac{\epsilon_2}{\sqrt{g_{22}}} \frac{\partial x^i}{\partial u^2}.$$

From the foregoing results we have by (40.10) that the total curvatures K_1 and K_2 of S_1 and S_2 respectively are given by

(48.11) $$K_1 = -\frac{1}{(\rho_1 - \rho_2)^2} \frac{\dfrac{\partial \rho_2}{\partial u^1}}{\dfrac{\partial \rho_1}{\partial u^1}}, \qquad K_2 = -\frac{1}{(\rho_2 - \rho_1)^2} \frac{\dfrac{\partial \rho_1}{\partial u^2}}{\dfrac{\partial \rho_2}{\partial u^2}}.$$

We inquire next under what conditions the normals to a surface S are normal to a second surface $\bar{S}$. The equations of such a surface $\bar{S}$ are given by

(48.12) $$\bar{x}^i = x^i + tX^i,$$

where t is to be determined so that

$$\sum_i X^i \frac{\partial \bar{x}^i}{\partial u^\alpha} = 0,$$

that is,

$$\sum_i X^i \left(\frac{\partial x^i}{\partial u^\alpha} + \frac{\partial t}{\partial u^\alpha} X^i + t \frac{\partial X^i}{\partial u^\alpha} \right) = 0.$$

It follows from these equations in consequence of (38.8) and (47.4) that t is a constant, and that for any constant t these equations are satisfied. Hence we have

[48.2] *If segments of constant length are laid off along the normals to a surface from points of the surface, the locus of their other end points is a surface with the same normals as the given surface.*

Two surfaces in such relation are said to be *parallel*. It is evident that there is an endless number of surfaces parallel to a given surface, that the lines of curvature correspond on all these surfaces in consequence of theorem [41.1], and that the family of parallel surfaces have the same surfaces of center.

If $\bar{g}_{\alpha\beta}$ and $\bar{d}_{\alpha\beta}$ denote the first and second fundamental tensors of the surface $\bar{S}$ with equations (48.12), we obtain in consequence of (38.16) and (46.3)

(48.13)
$$\bar{g}_{\alpha\beta} = \sum_i \frac{\partial \bar{x}^i}{\partial u^\alpha} \frac{\partial \bar{x}^i}{\partial u^\beta} = g_{\alpha\beta} - 2t\, d_{\alpha\beta} + t^2 h_{\alpha\beta},$$

$$\bar{d}_{\alpha\beta} = -\sum_i \frac{\partial \bar{x}^i}{\partial u^\alpha} \frac{\partial X^i}{\partial u^\beta} = d_{\alpha\beta} - t\, h_{\alpha\beta}.$$

Since the surfaces S and $\bar{S}$ have the same surfaces of center, it follows from (48.1) and similar equations for $\bar{S}$ that the principal radii of normal curvature of $\bar{S}$ are given by

(48.14) $\bar{\rho}_1 = \rho_1 - t, \qquad \bar{\rho}_2 = \rho_2 - t.$

If S is a surface of constant total curvature $1/a^2$, we have

$$(\bar{\rho}_1 + t)(\bar{\rho}_2 + t) = a^2.$$

When in particular $t = \pm a$, this equation reduces to

$$\frac{1}{\bar{\rho}_1} + \frac{1}{\bar{\rho}_2} = \mp\frac{1}{a},$$

and we have the following theorem of Bonnet*:

[48.3] *Among the surfaces parallel to a surface of constant total curvature $1/a^2$, there are two of constant mean curvature $\pm 1/a$ respectively at the distances $\mp a$ from the given surface.*

Also we have conversely

[48.4] *Among the surfaces parallel to a surface of constant mean curvature one has constant positive total curvature and another constant mean curvature.*

From (48.3) it follows that the normals to the surface S are tangents to the curves $u^2 = $ const. on S_1, which from (48.5) are seen to be geodesics. We shall prove the converse theorem

[48.5] *The tangents to the geodesics on any surface are normal to a family of parallel surfaces.*

We assume that the surface is referred to a family of geodesics $u^2 = $ const. and their orthogonal trajectories, and write the linear element in the form (see (32.15))

(48.15) $ds^2 = du^{1^2} + g_{22}\, du^{2^2}.$

The tangents to the geodesics have equations of the form

(48.16) $\bar{x}^i = x^i + r\dfrac{\partial x^i}{\partial u^1},$

from which we have

(48.17)
$$d\bar{x}^i = \frac{\partial x^i}{\partial u^1}\left(1 + \frac{\partial r}{\partial u^1}\right)du^1 + \left(\frac{\partial x^i}{\partial u^2} + \frac{\partial r}{\partial u^2}\frac{\partial x^i}{\partial u^1}\right)du^2$$
$$+ r\left(\frac{\partial^2 x^i}{\partial u^{1^2}}\, du^1 + \frac{\partial^2 x^i}{\partial u^1\, \partial u^2}\, du^2\right).$$

* 1853, 1, p. 437.

In order that the tangents to the geodesics be normal to a surface (48.16) r must be such that $\sum_i \frac{\partial x^i}{\partial u^1} d\bar{x}^i = 0$ for all values of du^1 and du^2. Since

(48.18) $\quad \sum_i \frac{\partial x^i}{\partial u^1} \frac{\partial x^i}{\partial u^1} = 1, \qquad \sum_i \frac{\partial x^i}{\partial u^1} \frac{\partial x^i}{\partial u^2} = 0, \qquad \sum_i \frac{\partial x^i}{\partial u^2} \frac{\partial x^i}{\partial u^2} = g_{22},$

from which we have

(48.19) $\qquad \sum_i \frac{\partial x^i}{\partial u^1} \frac{\partial^2 x^i}{\partial u^{1^2}} = 0, \qquad \sum_i \frac{\partial x^i}{\partial u^1} \frac{\partial^2 x^i}{\partial u^1 \partial u^2} = 0,$

one finds that the conditions upon r are

$$1 + \frac{\partial r}{\partial u^1} = 0, \qquad \frac{\partial r}{\partial u^2} = 0.$$

Consequently

(48.20) $\qquad\qquad r = c - u^1,$

where c is an arbitrary constant. Each value of c determines a particular one of the family of parallel surfaces normal to the tangents to the geodesics $u^2 = $ const. on S, and the theorem is established.

The given surface S is one of the surfaces of center of these parallel surfaces. In order to find the other surface of center $\bar{S}$ we note from theorem [48.1] that (48.16) are equations of $\bar{S}$ provided that r is such that $d\bar{x}^i$ are direction numbers of the tangent to the edge of regression of the envelope of the tangent planes to S along a curve conjugate to the geodesics $u^2 = $ const. Since this tangent must lie in the corresponding tangent plane to $\bar{S}$ which is also the corresponding osculating plane of the edge on S, the direction numbers of whose normal are $\frac{\partial x^i}{\partial u^2}$, r must be such that $\sum_i \frac{\partial x^i}{\partial u^2} d\bar{x}^i = 0$. In consequence of (48.18), (48.19), and

$$\sum_i \frac{\partial^2 x^i}{\partial u^{1^2}} \frac{\partial x^i}{\partial u^2} + \sum_i \frac{\partial x^i}{\partial u^1} \frac{\partial^2 x^i}{\partial u^1 \partial u^2} = 0, \qquad \sum_i \frac{\partial x^i}{\partial u^2} \frac{\partial^2 x^i}{\partial u^1 \partial u^2} = \frac{1}{2} \frac{\partial g_{22}}{\partial u^1}$$

which result from (48.18), we have

$$\sum_i \frac{\partial x^i}{\partial u^2} d\bar{x}^i = \left(g_{22} + \frac{r}{2} \frac{\partial g_{22}}{\partial u^1} \right) du^2 = 0.$$

Hence

(48.21) $\qquad\qquad \frac{1}{r} = -\frac{\partial \log \sqrt{g_{22}}}{\partial u^1},$

and equations of $\bar{S}$ are

(48.22)
$$\bar{x}^i = x^i - \frac{1}{\dfrac{\partial \log \sqrt{g_{22}}}{\partial u^1}} \frac{\partial x^i}{\partial u^1}.$$

From (34.5) it follows that r given by (48.21) is the radius of geodesic curvature of the curves $u^1 = $ const. and consequently (48.22) gives the centers of geodesic curvature of these curves. If then we say that the surface $\bar{S}$ defined by equations (48.22) is the surface *complementary* to the surface S as determined by the family of geodesics $u^2 = $ const. on S, we have

[48.6] *A complementary surface of a surface S is the locus of the centers of geodesic curvature of the orthogonal trajectories of the geodesics on S which determine the complementary surface.*

It follows from the above definition that either surface of centers of a given surface is a complementary surface of the other. Moreover, from equation (48.5) of the linear element of the surface of center S_1 it follows that the geodesics which determine the other surface of center S_2 as a complementary surface of S_1 are the orthogonal trajectories of the curves $\rho_1 = $ const. on S_1. Also from the linear element of the surface S_2, namely,

$$ds_2^2 = g_{11}\left(1 - \frac{\rho_2}{\rho_1}\right)^2 du^{1^2} + d\rho_2^2,$$

which follows from (48.8), one has that the surface S_1 is the complementary surface of S_2 which is determined by the geodesics orthogonal to the curves $\rho_2 = $ const. on S_2. Hence we have the following theorem of Beltrami*:

[48.7] *The centers of geodesic curvature of the curves $\rho_1 = $ const. on S_1 and of $\rho_2 = $ const. on S_2 are corresponding points on S_2 and S_1 respectively, corresponding points being the points of tangency of a common tangent to the two surfaces.*

EXERCISES

1. One of the surfaces of center of a surface of revolution is the axis, and the other is the surface of revolution of the evolute of the meridian curve of the surface.

2. For the envelope of a one-parameter family of spheres one of the surfaces of center is the curve of centers of the spheres.

---

* 1865, 1, p. 18.

3. In order that the surface of center S_1 be a curve, it is necessary and sufficient that $g_{11}'g_{22}' - g_{12}'^2 = 0$; it follows from (48.4) that in this case ρ_1 is a function of u^2 alone.

4. The surfaces of center of a helicoid (see §38, Ex. 5) are helicoids with the same axis and the same parameter as the given helicoid.

5. The surface

$$x^1 = \frac{a^2 - b^2}{ab} \frac{u^1 u^2}{u^1 + u^2}, \quad x^2 = \frac{\sqrt{a^2 - b^2}}{b} \frac{\sqrt{b^2 - u^{1^2}}}{u^1 + u^2}, \quad x^3 = \frac{\sqrt{a^2 - b^2}}{a} \frac{u^1 \sqrt{u^{2^2} - a^2}}{u^1 + u^2}$$

has the following properties: the coordinate curves are plane lines of curvature; $\rho_1 = u^2$, $\rho_2 = -u^1$; the surface is algebraic of the fourth order; the surfaces of center are focal conics.

6. A necessary and sufficient condition that the asymptotic lines correspond on the two surfaces of center of a surface is that there be a functional relation between the principal radii of the surface. When there is such a functional relation the surface is called a *surface of Weingarten*.

7. Equations of the lines of curvature on S_1 and S_2 are respectively

$$\frac{g_{11}}{\rho_1^2} \frac{\partial \rho_1}{\partial u^1} \frac{\partial \rho_1}{\partial u^2} du^{1^2} + \left[\frac{g_{11}}{\rho_1^2} \left(\frac{\partial \rho_1}{\partial u^2} \right)^2 + \frac{g_{22}}{\rho_2^2} \frac{\partial \rho_1}{\partial u^1} \frac{\partial \rho_2}{\partial u^1} + \frac{g_{11} g_{22}}{\rho_1^2 \rho_2^2} (\rho_1 - \rho_2)^2 \right] du^1 du^2$$

$$+ \frac{g_{22}}{\rho_2^2} \frac{\partial \rho_1}{\partial u^2} \frac{\partial \rho_2}{\partial u^1} du^{2^2} = 0,$$

$$\frac{g_{11}}{\rho_1^2} \frac{\partial \rho_1}{\partial u^2} \frac{\partial \rho_2}{\partial u^1} du^{1^2} + \left[\frac{g_{11}}{\rho_1^2} \frac{\partial \rho_1}{\partial u^2} \frac{\partial \rho_2}{\partial u^2} + \frac{g_{22}}{\rho_2^2} \left(\frac{\partial \rho_2}{\partial u^1} \right)^2 + \frac{g_{11} g_{22}}{\rho_1^2 \rho_2^2} (\rho_1 - \rho_2)^2 \right] du^1 du^2$$

$$+ \frac{g_{22}}{\rho_2^2} \frac{\partial \rho_2}{\partial u^1} \frac{\partial \rho_2}{\partial u^2} du^{2^2} = 0;$$

a necessary and sufficient condition that the lines correspond on the two surfaces is that $\rho_1 - \rho_2 = a$, where a is a constant, in which case $K_1 = K_2 = 1/a^2$, and the asymptotic lines on the two surfaces correspond.

8. Show that, in consequence of (41.12), $\Delta_1' \rho_1 = 1$, $\Delta_2' \rho_1 = \dfrac{1}{\rho_1 - \rho_2}$, where the differential parameters are formed with respect to (48.4); then from (34.6) it follows that the radius of geodesic curvature of the curves $\rho_1 = $ const. on S_1 is equal to $\rho_2 - \rho_1$, from which follows theorem [48.7].

9. The surfaces parallel to a developable surface are developable surfaces.

10. The surfaces parallel to a surface of revolution are surfaces of revolution.

11. Lines of curvature on two parallel surfaces are the only corresponding conjugate nets.

12. A necessary and sufficient condition that the asymptotic lines on a surface correspond to a conjugate system on a parallel surface is that the two surfaces be surfaces of constant mean curvature in the relation of theorem [48.4].

13. From (48.20) and (48.21) it follows that the principal radii of normal curva-

ture of the parallel surfaces normal to the tangents to the geodesics $u^2 = $ const. of a surface S with the linear element (48.15) are given by

$$\rho_1 = u^1 - c, \qquad \rho_2 = u^1 - c - \cfrac{1}{\cfrac{\partial \log \sqrt{g_{22}}}{\partial u^1}}.$$

14. If S is a surface applicable to a surface of revolution, the tangents to the curves corresponding to the meridians of the surface of revolution are normal to a family of surfaces of Weingarten as follows from Exs. 6 and 13.

15. From theorem [26.4] it follows that the left-hand member of equation (41.1) of the lines of curvature of a surface is an indefinite quadratic form (see §28). When the lines of curvature are coordinate, this form divided by $\sqrt{g}$ is reducible by (41.10) to

$$\left(\frac{1}{\rho_1} - \frac{1}{\rho_2}\right) \sqrt{g_{11} g_{22}}\, du^1\, du^2 \; ;$$

by means of (41.12) and §28, Ex. 15 the curvature of this form is reducible to

$$\frac{2\rho_1 \rho_2}{\sqrt{g_{11} g_{22}}\, (\rho_1 - \rho_2)^3} \; \frac{\partial\,(\rho_1, \rho_2)}{\partial\,(u^1, u^2)} \; ;$$

hence the curvature is equal to zero for a Weingarten surface and in consequence of theorem [28.5] the lines of curvature on a surface of Weingarten can be found by quadratures.

49. SPHERICAL AND PSEUDOSPHERICAL SURFACES

A surface whose Gaussian curvature K is a constant not zero is called a *surface of constant curvature*. According as K is positive or negative the surface is called *spherical* or *pseudospherical*.

We consider first spherical surfaces and put $K = 1/a^2$. When such a surface is referred to a family of geodesics $u^2 = $ const. and their orthogonal trajectories, the linear element is of the form

$$(49.1) \qquad\qquad ds^2 = du^{1^2} + g_{22}\, du^{2^2}.$$

From (28.5) and §28, Ex. 1 we have that g_{22} is such that

$$(49.2) \qquad\qquad \frac{\partial^2 \sqrt{g_{22}}}{\partial u^{1^2}} = -\frac{1}{a^2} \sqrt{g_{22}}.$$

The integral of this equation is

$$(49.3) \qquad\qquad \sqrt{g_{22}} = \varphi(u^2) \cos \frac{u^1}{a} + \psi(u^2) \sin \frac{u^1}{a}.$$

From (34.5) we have that the geodesic curvature κ_{g2} of the coordinate curves $u^1 = $ const. is given by

$$(49.4) \qquad\qquad \kappa_{g2} = -\frac{\partial \log \sqrt{g_{22}}}{\partial u^1}.$$

If, in particular, the coordinate curves are chosen so that the curve $u^1 = 0$ is a geodesic, the curves $u^1 = $ const. the geodesic parallels to this geodesic, and the curves $u^2 = $ const. their geodesic orthogonal trajectories, it follows from (49.4) for the curve $u^1 = 0$ that $\psi(u^2) = 0$ in (49.3). Hence by a suitable choice of the coordinate u^2 the linear element becomes

$$(49.5) \qquad\qquad ds^2 = du^{1^2} + c^2 \cos^2 \frac{u^1}{a}\, du^{2^2},$$

where c is a constant. From this it follows that $c(u_2^2 - u_1^2)$ is the arc of the geodesic $u^1 = 0$ between the curves $u^2 = u_2^2$ and $u^2 = u_1^2$.

From the foregoing discussion it follows that all spherical surfaces of the same Gaussian curvature are isometric, and consequently all these surfaces have the same intrinsic properties. However, there is a distinction between spherical surfaces of the same curvature as viewed from the enveloping space. This is seen, in particular, when we consider spherical surfaces of revolution.

A surface of revolution with the x^3-axis for axis of revolution is defined by the equations

$$(49.6) \qquad x^1 = \bar{u}^1 \cos \bar{u}^2, \qquad x^2 = \bar{u}^1 \sin \bar{u}^2, \qquad x^3 = \varphi(\bar{u}^1),$$

the function φ determining the character of the rotated curve. In terms of these coordinates the linear element is

$$(49.7) \qquad ds^2 = (1 + \varphi'(\bar{u}^1)^2)\, d\bar{u}^{1^2} + \bar{u}^{1^2}\, d\bar{u}^{2^2}.$$

When we compare this equation with (49.5), we have

$$du^{1^2} = (1 + \varphi'(\bar{u}^1)^2)\, d\bar{u}^{1^2}, \qquad \bar{u}^1 = c \cos\frac{u^1}{a}, \qquad \bar{u}^2 = u^2,$$

from which it follows that

$$\varphi(\bar{u}^1) = \int \sqrt{du^{1^2} - d\bar{u}^{1^2}} = \int \sqrt{1 - \frac{c^2}{a^2}\sin^2\frac{u^1}{a}}\, du^1.$$

Consequently equations of a surface of revolution with the linear element (49.5) are

$$x^1 = c \cos\frac{u^1}{a} \cos u^2, \qquad x^2 = c \cos\frac{u^1}{a} \sin u^2,$$

$$(49.8)$$

$$x^3 = \int \sqrt{1 - \frac{c^2}{a^2}\sin^2\frac{u^1}{a}}\, du^1.$$

There are three cases to be considered according as c is equal to, greater than, or less than a.

CASE 1°. $c = a$. Now equations (49.8) are

$$(49.9) \quad x^1 = a \cos \frac{u^1}{a} \cos u^2, \qquad x^2 = a \cos \frac{u^1}{a} \sin u^2, \qquad x^3 = a \sin \frac{u^1}{a},$$

which are seen to be equations of the sphere of radius a with center at the origin.

We observe before taking up the other two cases that the expression for x^3 in (49.8) is an elliptic integral and consequently in each case x^3 is expressible by means of appropriate elliptic functions.

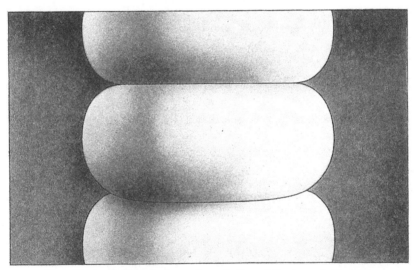

FIG. 17. A spherical surface of revolution of the hyperbolic type

CASE 2°. $c > a$. From the expression in (49.8) for x^3 it follows that $\sin^2 \frac{u^1}{a} \leqq \frac{a^2}{c^2}$ and consequently $\sqrt{c^2 - a^2} \leqq \bar{u}^1 \leqq c$. Since x^3 is periodic, the surface consists of a succession of like parts or *zones* each of which is bounded by minimum parallels of radius $\sqrt{c^2 - a^2}$, the greatest parallel of each zone being of radius c, as shown in Fig. 17. The angle which the tangent to a meridian curve makes with the plane $x^3 = 0$ is given by

$$\tan \theta = - \frac{\sqrt{1 - \frac{c^2}{a^2} \sin^2 \frac{u^1}{a}}}{\frac{c}{a} \sin \frac{u^1}{a}}.$$

These spherical surfaces are said to be of the *hyperbolic* type.

CASE 3°. $c < a$. In this case $0 \leqq \bar{u}^1 \leqq c$, $\bar{u}^1$ being equal to zero when $u^1 = ma\pi/2$ where m is any odd integer. At these points the meridian curve meets the axis under the angle $\sin^{-1} \dfrac{c}{a}$. Since x^3 is periodic, the surface consists of a succession of like zones each being spindle-shaped as shown in Fig. 18, the greatest parallel of each zone being of radius c. These spherical surfaces are said to be of the *elliptic* type.

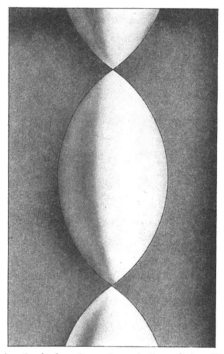

FIG. 18. A spherical surface of revolution of the elliptic type

If the linear element of the sphere (49.9) is written in the form

$$ds^2 = dv^{1^2} + a^2 \cos^2 \frac{v^1}{a} \, dv^{2^2},$$

u^α having been replaced by v^α, the equations

$$v^1 = u^1, \qquad v^2 = \frac{c}{a} u^2$$

establish an isometric correspondence between the sphere and the surfaces of cases 2° and 3° such that meridians of the latter correspond to great circles $v^2 =$ const. of the sphere. The points of either of the former surfaces whose coordinates are such that

$$u_1^1 \geqq u^1 \geqq u_2^1, \qquad 2\pi \geqq u^2 \geqq 0,$$

are the points of the zone of the surface between the parallels $u^1 = u_2^1$ and $u^1 = u_1^1$. The coordinates v^α of the corresponding points of the sphere are such that

$$u_1^1 \geqq v^1 \geqq u_2^1, \qquad \frac{2\pi c}{a} \geqq v^2 \geqq 0.$$

Hence when $c > a$ and the surface is applied upon the sphere the zone of the surface not only covers the corresponding zone of the sphere, but there is an overlapping; whereas when $c < a$ the zone of the surface fails to cover the zone of the sphere.

The form (49.5) was obtained from (49.3) by taking $\varphi(u^2) = c$, $\psi(u^2) = 0$. For any constant values of φ and ψ the expression (49.3) can be written in the form

(49.10) $$\sqrt{g_{22}} = c \cos\left(\frac{u^1}{a} + b\right),$$

where b and c are arbitrary constants. According as we take $b = 0$, $-\pi/2$, or $-\pi/4$ we get the following respective forms of the linear element

(i) $$ds^2 = du^{1^2} + c^2 \cos^2 \frac{u^1}{a} du^{2^2},$$

(49.11)   (ii) $$ds^2 = du^{1^2} + c^2 \sin^2 \frac{u^1}{a} du^{2^2},$$

(iii) $$ds^2 = du^{1^2} + c^2 \cos^2\left(\frac{u^1}{a} - \frac{\pi}{4}\right) du^{2^2}.$$

We have seen that for first of these forms the curve $u^1 = 0$ is a geodesic. For the form (ii) u^1 and cu^2/a are polar geodesic coordinates by theorem [33.1], and for the form (iii) the curve $u^1 = 0$ has geodesic curvature $1/a$.

For a pseudospherical surface of curvature $-1/a^2$ one has in place of (49.3)

(49.12) $$\sqrt{g_{22}} = \varphi(u^2) \cosh \frac{u^1}{a} + \psi(u^2) \sinh \frac{u^1}{a}.$$

When (i) the curve $u^1 = 0$ is a geodesic, (ii) the coordinates u^1, cu^2/a are polar geodesic, or (iii) the curve $u^1 = 0$ has geodesic curvature $-1/a$, the linear element has the following forms respectively:

$$\text{(i)} \qquad ds^2 = du^{1^2} + c^2 \cosh^2 \frac{u^1}{a}\, du^{2^2},$$

$$(49.13) \qquad \text{(ii)} \qquad ds^2 = du^{1^2} + c^2 \sinh^2 \frac{u^1}{a}\, du^{2^2},$$

$$\text{(iii)} \qquad ds^2 = du^{1^2} + c^2 e^{2u^1/a}\, du^{2^2}.$$

Any case other than these for which $\varphi(u^2)$ and $\psi(u^2)$ are constants may be obtained by taking for $\sqrt{g_{22}}$ either of the values $c \cosh \left(\frac{u^1}{a} + b \right)$ or $c \sinh \left(\frac{u^1}{a} + b \right)$, where b and c are constants. By change of the coordinate u^1, the corresponding linear elements are reducible to (i) or (ii). Hence the forms (49.13) are general for the case when $\varphi(u^2)$ and $\psi(u^2)$ are constants.

The forms (49.13) are linear elements of surfaces of revolution whose equations are given by (49.6) where for the respective forms we have

$$\text{(i)} \qquad \bar{u}^1 = c \cosh \frac{u^1}{a}, \qquad x^3 = \int \sqrt{1 - \frac{c^2}{a^2} \sinh^2 \frac{u^1}{a}}\, du^1;$$

$$(49.14) \qquad \text{(ii)} \qquad \bar{u}^1 = c \sinh \frac{u^1}{a}, \qquad x^3 = \int \sqrt{1 - \frac{c^2}{a^2} \cosh^2 \frac{u^1}{a}}\, du^1;$$

$$\text{(iii)} \qquad \bar{u}^1 = c\, e^{u^1/a}, \qquad x^3 = \int \sqrt{1 - \frac{c^2}{a^2} e^{2u^1/a}}\, du^1.$$

In considering these three cases in detail we remark that the integrals in (i) and (ii) are elliptic and consequently in each of these cases x^3 is expressed by means of appropriate elliptic functions.

CASE (i). The maximum and minimum values of $\sinh^2 \frac{u^1}{a}$ are a^2/c^2 and 0, respectively and consequently the maximum and minimum values of the radius $\bar{u}^1$ are $\sqrt{a^2 + c^2}$ and c respectively. At points of a maximum parallel the tangents to the meridian curve are perpendicular to the axis of rotation and at points of a minimum parallel they are parallel to this axis as follows from the value of $dx^3/d\bar{u}^1$. Since x^3 is periodic the surface consists of a succession of spool-like zones, see Fig. 19, the maximum parallels being cuspidal edges. These pseudospherical surfaces are said to be of the *hyperbolic* type.

Case (ii).   In order that the surface be real, $c^2 \leqq a^2$, a restriction not necessary in either of the other cases.   If we put $c = a \sin \theta$, the maximum and minimum values of $\cosh^2 \dfrac{u^1}{a}$ are $\csc^2 \theta$ and 1 respectively, and the corresponding values of the radius $\bar{u}^1$ are $a \cos \theta$ and 0.   The tangents to the meridians at points of the maximum circle are perpendicular to the axis of rotation and at points for which $\bar{u}^1 = 0$, the tangents make the angle θ with the axis.   The surface is made of a succession of zones

Fig. 19   Fig. 20

Fig. 19. A pseudospherical surface of revolution of the hyperbolic type
Fig. 20. A pseudospherical surface of revolution of the elliptic type

similar in shape to hour glasses. Fig. 20 represents such a zone, the maximum parallel being a cuspidal edge. These pseudospherical surfaces are said to be of the *elliptic* type.

Case (iii).   If we make the substitution $\sin\theta = \dfrac{c}{a}\, e^{u^1/a}$ in (49.14) (iii), we obtain

$$\bar{u}^1 = a \sin \theta, \qquad x^3 = a[\cos \theta - \log (\csc \theta + \cot \theta)].$$

From this result we have that $\dfrac{dx^3}{d\bar{u}^1} = \cot\theta$, and consequently θ is the angle which the meridian makes with the axis of rotation. One finds that the length of the segment of a tangent to a meridian from the point of contact to the axis of rotation is a, and consequently the meridian curve is a tractrix (see §6, Ex. 11). These pseudospherical surfaces are said to be of the *parabolic* type. They are called *pseudospheres*. See Fig. 21.

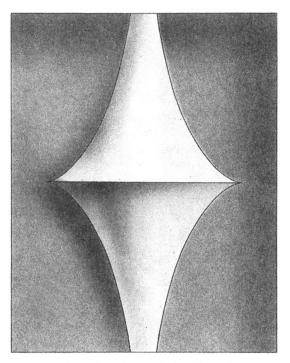

Fig. 21. A pseudosphere (pseudospherical surface of revolution of the parabolic type)

From (49.4) we find that the geodesic curvature of the parallels of the surfaces with the linear elements (49.13) are given by the respective expressions

$$\text{(i)} \quad -\frac{1}{a}\tanh\frac{u^1}{a}, \qquad \text{(ii)} \quad -\frac{1}{a}\coth\frac{u^1}{a}, \qquad \text{(iii)} \quad -\frac{1}{a}.$$

Since no one of these expressions can be transformed into any other, if u^1 is replaced by $u^1 + c$, where c is a real constant, it follows that two pseudospherical surfaces of revolution of different types are not applicable to one another with merdians in correspondence, whereas we have just found this to be the case between different types of spherical surfaces of revolution.

The results (49.11) and (49.13) constitute another proof of the first part of theorem [31.1], and more particularly we have

[49.1] *The linear element of any surface of constant curvature $1/a^2$ or $-1/a^2$ is reducible to the respective forms* (i), (ii), (iii) *of* (49.11) *or* (49.13) *according as the coordinate geodesics are* (i) *orthogonal to a geodesic,* (ii) *pass through a point, or* (iii) *are orthogonal to a curve of constant geodesic curvature.*

We proceed now to the consideration of surfaces of constant curvature when the lines of curvature are coordinate. To this end we assume that the Gaussian curvature of the surface is equal to e/a^2, where e is $+1$ as -1 according as the surface is spherical or pseudospherical. If such a surface is referred to an isometric-conjugate net (see §42), we have

$$(49.15) \qquad \frac{d_{11}}{\sqrt{g}} = e\,\frac{d_{22}}{\sqrt{g}} = \frac{1}{a}, \qquad d_{12} = 0.$$

When the expressions for d_{11}, d_{22} in (49.15) are substituted in the Codazzi equations (39.6), the resulting equations are reducible, in consequence of (28.1) and (28.2), to

$$g_{11}\left(\frac{\partial g_{11}}{\partial u^2} - e\,\frac{\partial g_{22}}{\partial u^2} - 2\,\frac{\partial g_{12}}{\partial u^1}\right) + g_{12}\left(\frac{\partial g_{11}}{\partial u^1} - e\,\frac{\partial g_{22}}{\partial u^1} + 2e\,\frac{\partial g_{12}}{\partial u^2}\right) = 0,$$

$$g_{12}\left(\frac{\partial g_{11}}{\partial u^2} - e\,\frac{\partial g_{22}}{\partial u^2} - 2\,\frac{\partial g_{12}}{\partial u^1}\right) + g_{22}\left(\frac{\partial g_{11}}{\partial u^1} - e\,\frac{\partial g_{22}}{\partial u^1} + 2e\,\frac{\partial g_{12}}{\partial u^2}\right) = 0.$$

Since $g \neq 0$, these equations are equivalent to

$$(49.16) \qquad \begin{aligned} &\frac{\partial}{\partial u^1}(g_{11} - eg_{22}) + 2e\,\frac{\partial g_{12}}{\partial u^2} = 0, \\[2mm] &\frac{\partial}{\partial u^2}(g_{11} - eg_{22}) - 2\,\frac{\partial g_{12}}{\partial u^1} = 0. \end{aligned}$$

These equations are satisfied by

$$(49.17) \qquad g_{11} - eg_{22} = \text{const.}, \qquad g_{12} = 0;$$

hence we have

[49.2] *The lines of curvature on a surface of constant Gaussian curvature form an isometric-conjugate net.*

For the case of spherical surfaces the equations (49.17) are satisfied by

(49.18) $g_{11} = a^2 \cosh^2 \theta,$ $g_{12} = 0,$ $g_{22} = a^2 \sinh^2 \theta,$

and from (49.15) we have

(49.19) $d_{11} = d_{22} = a \sinh \theta \cosh \theta,$ $d_{12} = 0.$

When these values are substituted in the Gauss equation (40.10), where K is given by (28.5) and §28, Ex. 1, one obtains

(49.20) $$\frac{\partial^2 \theta}{\partial u^{1^2}} + \frac{\partial^2 \theta}{\partial u^{2^2}} + \sinh \theta \cosh \theta = 0.$$

Conversely, in consequence of theorem [39.1] we have

[49.3] *For each solution of equation (49.20) the quantities (49.18) and (49.19) determine a spherical surface, the lines of curvature being coordinate.*

For the case of pseudospherical surfaces the equations (49.17) are satisfied by

(49.21) $g_{11} = a^2 \cos^2 \theta,$ $g_{12} = 0,$ $g_{22} = a^2 \sin^2 \theta,$

and from (49.15) we have

(49.22) $d_{11} = -d_{22} = a \sin \theta \cos \theta,$ $d_{12} = 0.$

When these values are substituted in the Gauss equation, one obtains

(49.23) $$\frac{\partial^2 \theta}{\partial u^{1^2}} - \frac{\partial^2 \theta}{\partial u^{2^2}} - \sin \theta \cos \theta = 0.$$

Hence we have

[49.4] *For each solution of equation (49.23) the quantities (49.21) and (49.22) determine a pseudospherical surface, the lines of curvature being coordinate.*

EXERCISES

1. The lines of curvature and the asymptotic lines on a surface of constant Gaussian curvature can be found by quadratures (see §48, Ex. 15).

2. When the linear element of a pseudospherical surface is in the form (49.13) (iii), the surface defined by

$$\bar{x}^i = x^i - a \frac{\partial x^i}{\partial u^1}$$

is pseudospherical, and the tangent planes to the two surfaces at corresponding points are perpendicular.

3. The helicoids

$$x^1 = u^1 \cos u^2, \qquad x^2 = u^1 \sin u^2, \qquad x^3 = \int \sqrt{\frac{1}{a^2 - k^2 u^{1^2}} - \frac{h^2}{u^{1^2}} - 1}\, du^1 + hu^2,$$

where a, h, and k are constants, are spherical surfaces.

4. A helicoid whose generating curve is a tractrix is a pseudospherical surface; it is called a *surface of Dini*.

5. For a pseudospherical surface defined by (49.21) and (49.22) the linear element of the spherical representation is

$$d\bar{s}^2 = \sin^2 \theta\, du^{1^2} + \cos^2 \theta\, du^{2^2}.$$

6. The surfaces of center of a pseudospherical surface are applicable to a catenoid.

7. On a surface of constant curvature the area of a geodesic triangle is in a constant ratio to the difference between the sum of the angles of the triangle and two right angles (see theorem [33.2]).

8. When upon a surface there is more than one family of geodesics which together with their orthogonal trajectories form an isothermal system, the surface is of constant curvature.

9. When the equations of the sphere with center at the origin and radius a are written in the form

$$x^1 = a \sin u^1 \cos u^2, \qquad x^2 = a \sin u^1 \sin u^2, \qquad x^3 = a \cos u^1,$$

the linear element is

(i) $$ds^2 = a^2(du^{1^2} + \sin^2 u^1\, du^{2^2}).$$

The equation of any great circle of the sphere is of the form

(ii) $$A \sin u^1 \cos u^2 + B \sin u^1 \sin u^2 + C \cos u^1 = 0,$$

where A, B, C are constants not all zero; this is an equation of the geodesics on any surface of Gaussian curvature $1/a^2$ in terms of the coordinate system for which the linear element is (i).

10. When in equation (ii) of Ex. 9 u^1 and u^2 are expressed as functions of a general parameter t and this equation is differentiated twice with respect to t, and A, B, and C are eliminated from the three equations, one obtains

$$\frac{du^1}{dt}\left[\frac{d^2 u^2}{dt^2} + 2 \cot u^1 \frac{du^1}{dt}\frac{du^2}{dt}\right] - \frac{du^2}{dt}\left[\frac{d^2 u^1}{dt^2} - \sin u^1 \cos u^1 \left(\frac{du^2}{dt}\right)^2\right] = 0;$$

since this is the equation (37.1) for the case when the linear element is (i) of Ex. 9, this is an analytic proof of the fact that equation (ii) is an equation of the geodesics of the surface with this linear element.

11. When the coordinates on a pseudospherical surface are such that the linear element is

$$ds^2 = a^2(du^{1^2} + \sinh^2 u^1\, du^{2^2}),$$

an equation of the geodesics is

(i) $$A \tanh u^1 \cos u^2 + B \tanh u^1 \sin u^2 + C = 0,$$

as may be verified by the process used in Ex. 10.

12. When equation (ii) of Ex. 9 is divided by cos u^1 and the terms multiplying A and B in this equation are equated to x and y and similarly the terms multiplying A and B of equation (i) of Ex. 11, these equations define the correspondence between the surface and the plane in accordance with which geodesics on the surface are represented by straight lines in the plane.

13. When the coordinates on a pseudospherical surface are such that the linear element is of the form (iii) of (49.13), an equation of the geodesics is

$$A\left(a^2 e^{-2u^1/a} + c^2 u^{2^2}\right) + Bcu^2 + C = 0,$$

as may be verified by the process used in Ex. 10; in this case the equations

$$x = cu^2, \qquad y = ae^{-u^1/a}$$

determine a conformal representation of the surface upon the plane such that any geodesic in the surface is represented in the plane by a circle with its center on the x-axis or by a line perpendicular to this axis.

50. MINIMAL SURFACES

In §43 a minimal surface was defined as one whose mean curvature is equal to zero. In this section we establish a property of minimal surfaces which accounts for their name, and then derive other properties of such surfaces.

Consider a surface S in space defined by the equations

$$(50.1) \qquad\qquad x^i = f^i(u^1, u^2).$$

Consider upon this surface a simply connected region R with curve C as contour and for this region consider the integral

$$(50.2) \qquad\qquad I = \iint_R L\, du^1\, du^2,$$

where L is a function of the x's and their first derivatives $\dfrac{\partial x^i}{\partial u^\alpha}$. Let $\omega^i(u^1, u^2)$ be three arbitrary functions such that

$$(50.3) \qquad\qquad \omega^i(u^1, u^2) = 0 \quad \text{along} \quad C.$$

Then

$$(50.4) \qquad\qquad \bar{x}^i = f^i + \epsilon\,\omega^i,$$

where ϵ is an infinitesimal, define another surface $\bar{S}$ nearby S and containing the curve C. When these expressions are substituted in the function L in (50.2), we have for $\bar{S}$ the corresponding integral

$$(50.5) \qquad \bar{I} = \int_C L\left(x^i + \epsilon\omega^i; \frac{\partial x^i}{\partial u^\alpha} + \epsilon\,\frac{\partial\omega^i}{\partial u^\alpha}\right) du^1\, du^2.$$

In order that I shall be a minimum for all the surfaces passing through C it is necessary that the derivative of $\bar{I}$ with respect to ϵ be equal to zero for ϵ equal to zero. From (50.5) we have that this condition is

$$\iint_R \left(\frac{\partial L}{\partial x^i} \omega^i + \frac{\partial L}{\partial x^i_{,\alpha}} \frac{\partial \omega^i}{\partial u^\alpha} \right) du^1 \, du^2 = 0,$$

where $x^i_{,\alpha} \equiv \dfrac{\partial x^i}{\partial u^\alpha}$. This equation may be written in the form

$$\iint_R \omega^i \left(\frac{\partial L}{\partial x^i} - \frac{\partial}{\partial u^\alpha} \frac{\partial L}{\partial x^i_{,\alpha}} \right) du^1 \, du^2 + \iint_R \frac{\partial}{\partial u^\alpha} \left(\omega^i \frac{\partial L}{\partial x^i_{,\alpha}} \right) du^1 \, du^2 = 0.$$

In consequence of (50.3) the second of these integrals is equal to zero and, since the first integral must be equal to zero for arbitrary functions ω^i subject to the condition (50.3), we have

$$(50.6) \qquad\qquad \frac{\partial}{\partial u^\alpha} \frac{\partial L}{\partial x^i_{,\alpha}} - \frac{\partial L}{\partial x^i} = 0.$$

These are the *generalized equations of Euler*. When these conditions are satisfied the surface S is said to be an *extremal* for the integral (50.2).

We consider now the particular case when the integral (50.2) is the integral of area, namely,

$$(50.7) \qquad\qquad I = \iint \sqrt{g} \, du^1 \, du^2.$$

In this case the equations (50.6) are

$$(50.8) \qquad\qquad \frac{\partial}{\partial u^\alpha} \frac{\partial \sqrt{g}}{\partial x^i_{,\alpha}} - \frac{\partial \sqrt{g}}{\partial x^i} = 0,$$

where g is the determinant $| g_{\alpha\beta} |$ and

$$(50.9) \qquad\qquad g_{\alpha\beta} = \sum_i x^i_{,\alpha} x^i_{,\beta}.$$

Now we have

$$\begin{aligned}
\frac{\partial \sqrt{g}}{\partial x^i_{,\alpha}} &= \frac{1}{2\sqrt{g}} \frac{\partial g}{\partial g_{\beta\gamma}} \frac{\partial g_{\beta\gamma}}{\partial x^i_{,\alpha}} = \frac{1}{\sqrt{g}} \left(\frac{\partial g}{\partial g_{\alpha\alpha}} x^i_{,\alpha} + \frac{\partial g}{\partial g_{\alpha\beta}} x^i_{,\beta} \right) \\
&= \sqrt{g} \, (g^{\alpha\alpha} x^i_{,\alpha} + g^{\alpha\beta} x^i_{,\beta}) \qquad\qquad (\beta \neq \alpha) \\
&= \sqrt{g} \, g^{\alpha\gamma} x^i_{,\gamma}.
\end{aligned}$$

From these equations we have in consequence of (28.2)

$$\frac{\partial}{\partial u^\alpha}\left(\frac{\partial \sqrt{g}}{\partial x^i_{,\alpha}}\right) = \sqrt{g}\left[g^{\alpha\gamma}x^i_{,\gamma}\left\{\begin{matrix}\beta\\\beta\alpha\end{matrix}\right\} + x^i_{,\gamma}\left(-g^{\delta\gamma}\left\{\begin{matrix}\alpha\\\delta\alpha\end{matrix}\right\} - g^{\alpha\delta}\left\{\begin{matrix}\gamma\\\delta\alpha\end{matrix}\right\}\right)\right.$$
$$\left. + g^{\alpha\gamma}\frac{\partial^2 x^i}{\partial u^\alpha \partial u^\gamma}\right]$$
$$= \sqrt{g}\,g^{\alpha\gamma}x^i_{,\alpha\gamma} = \sqrt{g}\,g^{\alpha\gamma}d_{\alpha\gamma}X^i,$$

the last expression being a consequence of (38.18).

From (50.9) it follows that $\sqrt{g}$ does not involve x^i; consequently the second term of (50.8) is identically zero. Hence for the equations (50.8) to hold we must have $g^{\alpha\gamma}d_{\alpha\gamma} = 0$, which by (40.11) is the condition that the mean curvature be equal to zero. Hence we have

[50.1] *A minimal surface is an extremal for the integral of area.*

The determination of whether this is the minimum area for a given contour involves derivatives of higher order and is an important problem in the *calculus of variations.* However, any surface satisfying theorem [50.1] is called a minimal surface.

Lagrange using the equation of a surface in the Monge form (see §10) derived as the differential equation of minimal surfaces (see §39, Ex. 1)

$$(1 + p_2^2)r_{11} - 2p_1p_2r_{12} + (1 + p_1^2)r_{22} = 0;$$

in this notation the left-hand member is the expression for mean curvature. Lagrange raised the question of finding the minimal surface for a given contour. Plateau gave a physical realization of this problem by means of a glycerine film, it being a consequence of surface tension of such a film that the surface would assume the form having a minimum area. Accordingly, this problem proposed by Lagrange is now known as the *Plateau problem.* The mathematical solution of this problem for a given contour has been the subject of study by mathematicians up to the present time.

From (40.11) it follows that when minimal curves on a minimal surface are coordinate, these curves form a conjugate net. In this case $g_{11} = g_{22} = d_{12} = 0$. From the first two of these equations it follows from (28.1) that

(50.10) $$\left\{\begin{matrix}1\\12\end{matrix}\right\} = \left\{\begin{matrix}2\\12\end{matrix}\right\} = 0.$$

From these results we have from (38.18) that $\dfrac{\partial^2 x^i}{\partial u^1 \partial u^2} = 0$ and consequently

$$(50.11) \qquad\qquad x^i = U_1^i + U_2^i,$$

where U_1^i and U_2^i are functions of u_1^i and u_2^i respectively. These functions must be such that

$$(50.12) \qquad \sum_i \left(\frac{\partial U_1^i}{\partial u^1}\right)^2 = 0, \qquad \sum_i \left(\frac{\partial U_2^i}{\partial u^2}\right)^2 = 0,$$

which are the conditions that $g_{11} = g_{22} = 0$. Since the first of equations (50.12) expresses that the sum of the squares of three quantities is zero, and since the three quantities $\frac{1}{2}(1 - u^{1^2})$, $\frac{i}{2}(1 + u^{1^2})$, and u^1 possess this property, it follows that the most general solution of the first of equations (50.12) consists of the above expressions multiplied by an arbitrary function of u^1, say, $F(u^1)$. Since the same argument applies to the second of equations (50.12), we have for equations (50.11) satisfying the conditions (50.12)

$$x^1 = \frac{1}{2} \int (1 - u^{1^2})F(u^1)\,du^1 + \frac{1}{2} \int (1 - u^{2^2})H(u^2)\,du^2,$$

$$(50.13) \quad x^2 = \frac{i}{2} \int (1 + u^{1^2})F(u^1)\,du^1 - \frac{i}{2} \int (1 + u^{2^2})H(u^2)\,du^2,$$

$$x^3 = \int u^1 F(u^1)\,du^1 + \int u^2 H(u^2)\,du^2.$$

From these equations we have that the linear element of the surface is

$$(50.14) \qquad ds^2 = (1 + u^1 u^2)^2 F(u^1)H(u^2)\,du^1\,du^2.$$

From this it is seen that for the surface to be real u^1 and u^2 must be conjugate imaginary and H must be the function conjugate to F. It was in order to effect this result that the negative sign was used before the second integral in the second of equations (50.13).

From (50.13) we have for the direction cosines of the normal to the surface

$$(50.15) \qquad X^1,\ X^2,\ X^3 = \frac{(u^1 + u^2),\ i(u^2 - u^1),\ (u^1 u^2 - 1)}{1 + u^1 u^2},$$

from which one finds that the linear element of the spherical representation is

$$(50.16) \qquad\qquad d\bar{s}^2 = \frac{4\,du^1\,du^2}{(1 + u^1 u^2)^2}.$$

Also by means of (38.13) one finds

(50.17) $d_{11} = -F(u^1), \qquad d_{12} = 0, \qquad d_{22} = -H(u^2).$

From (50.14) and (50.17) one finds that the differential equations of the lines of curvature and of the asymptotic lines are respectively

(50.18) $F(u^1)du^{1^2} - H(u^2)du^{2^2} = 0,$

(50.19) $F(u^1)du^{1^2} + H(u^2)du^{2^2} = 0.$

From the form of these equations we have

[50.2] *When the minimal lines of a minimal surface are coordinate, the equations of its lines of curvature and of its asymptotic lines can be found by quadratures.*

Equations (50.13) are known as the equations of Enneper.* From these equations it is seen that the problem of minimal surfaces is that of functions of a single complex variable (see Ex. 5).

Weierstrass,† who also derived equations (50.13), remarked that these equations can be put in a form free of all quadratures. This is done by replacing $F(u^1)$ and $H(u^2)$ by $f'''(u^1)$ and $h'''(u^2)$ respectively, where the primes indicate differentiation, and then integrating by parts. This gives the equations

$$x^1 = \frac{1 - u^{1^2}}{2} f''(u^1) + u^1 f'(u^1) - f(u^1) + \frac{1 - u^{2^2}}{2} h''(u^2)$$
$$+ u^2 h'(u^2) - h(u^2),$$

(50.20) $$x^2 = i\frac{1 + u^{1^2}}{2} f''(u^1) - iu^1 f'(u^1) + if(u^1) - i\frac{1 + u^{2^2}}{2}h''(u^2)$$
$$+ iu^2 h'(u^2) - ih(u^2),$$

$$x^3 = u^1 f''(u^1) - f'(u^1) + u^2 h''(u^2) - h'(u^2).$$

It is clear that the surface so defined is real when f and h are conjugate imaginary functions of the conjugate imaginary coordinates u^1 and u^2. In this case (50.20) may be written

$$x^1 = \Re[(1 - u^{1^2})f''(u^1) + 2u^1 f'(u^1) - 2f(u^1)],$$

(50.21) $$x^2 = \Re i[(1 + u^{1^2})f''(u^1) - 2u^1 f'(u^1) + 2f(u^1)],$$

$$x^3 = \Re[2u^1 f''(u^1) - 2f'(u^1)],$$

* 1864, 2, p. 107.
† 1866, 1, p. 619.

where in these equations $\Re$ denotes the real part of the expression following $\Re$. Equations (50.20) are of particular value in the study of algebraic surfaces, for it is evident that when f and h are algebraic on the elimination of u^1 and u^2 from these equations the resulting equation will be algebraic in x^1, x^2, and x^3. The converse of this is also true.*

EXERCISES

1. A minimal surface is a surface of translation (see §42, Ex. 1); when its equations are of the form (50.13) a necessary and sufficient condition that the generating curves be congruent is that

$$F(u^1) = -\frac{1}{u^{1^4}} H\left(-\frac{1}{u^1}\right).$$

2. When equations (50.11) are equations of a minimal surface, so also are the equations

$$x_a^j = e^{ia} U_1^j + e^{-ia} U_2^j \qquad (j = 1, 2, 3),$$

where $i = \sqrt{-1}$ and a is a constant. These minimal surfaces as a takes different values are called *associate* minimal surfaces; the normals to these surfaces at points with the same coordinates u^α are parallel.

3. When a in Ex. 2 is equal to $\pi/2$, the surface is called the *adjoint* of the surface with the equation (50.11); the lines of curvature on either surface correspond to the asymptotic lines on the other, and the tangents to the curves with the same equation $\varphi(u^1, u^2) = 0$ on the two surfaces are orthogonal.

4. If x_1^i denote the coordinates of the adjoint surface of the surface with the equation (50.11), the equations of Ex. 2 may be written

$$x_a^j = \cos a \, x^j + \sin a \, x_1^j ;$$

the plane determined by the origin of coordinates, a point P on a minimal surface, and the corresponding point on its adjoint contains the corresponding point on every associate surface and the locus of these points is an ellipse with center at the origin.

5. From §44 Ex. 11 it follows that for a minimal surface $\Delta_2 x^i = 0$; when the coordinate curves form an isometric orthogonal net, we have that x^i are solutions of the harmonic equation

$$\frac{\partial^2 \theta}{\partial u^{1^2}} + \frac{\partial^2 \theta}{\partial u^{2^2}} = 0,$$

from which it follows that for any real minimal surface

$$x^i = f^i(u^1 + iu^2) + f_0^i(u^1 - iu^2),$$

where f^i and f_0^i are conjugate imaginary functions, which is in accord with equations (50.13).

---

* 1909, 1, p. 261.

6. When in equations (50.13)

$$F(u^1) = \frac{b}{2} \frac{e^{ia}}{u^{1^2}}, \qquad H(u^2) = \frac{b}{2} \frac{e^{-ia}}{u^{2^2}},$$

where a and b are constants, and one puts $u^1 = e^{\bar{u}^1 + i\bar{u}^2}$ and $u^2 = e^{\bar{u}^1 + i\bar{u}^2}$, one obtains equations of the following form

$$x^1 = b(\cos a \cosh \bar{u}^1 \cos \bar{u}^2 - \sin a \sinh \bar{u}^1 \sin \bar{u}^2)$$

$$x^2 = b(\cos a \cosh \bar{u}^1 \sin \bar{u}^2 + \sin a \sinh \bar{u}^1 \cos \bar{u}^2),$$

$$x^3 = b(-\bar{u}^1 \cos a + \bar{u}^2 \sin a),$$

which are equations of minimal helicoids (see §24, Ex. 6).

7. When $a = 0$ in Ex. 6, the surface is the catenoid (see §24, Ex. 4); when $a = \pi/2$, the surface is the skew helicoid (see §24, Ex. 7).

8. If two minimal surfaces correspond in such manner that at corresponding points the tangent planes are parallel, the minimal curves on the two surfaces correspond; for two such surfaces the locus of the point which divides in constant ratio a line-segment joining corresponding points is a minimal surface.

9. The spherical representation of the lines of curvature of a minimal surface is an isometric orthogonal net (see §43, Ex. 11 and (46.5)).

10. If one family of the lines of curvature on a minimal surface are plane curves, those of the other system are plane curves also (see §46, Ex.4; §47, Ex. 15).

11. Show that the surface

$$x^1 = au^1 + \sin u^1 \cosh u^2,$$

$$x^2 = u^2 + a \cos u^1 \sinh u^2,$$

$$x^3 = \sqrt{1 - a^2} \cos u^1 \cosh u^2$$

is minimal and that its lines of curvature are plane curves.

12. The surfaces of center of a minimal surface are applicable to one another and to the surface of revolution of the evolute of the catenary.

13. The surface for which $F = H = $ const., say 3, is called the *minimal surface of Enneper*; it possesses the following properties:

(a) it is an algebraic surface of the ninth degree whose equation is unaltered when x^1, x^2, x^3 are replaced by x^2, x^1, $-x^3$ respectively;

(b) it meets the plane $x^3 = 0$ in two orthogonal straight lines;

(c) if we put $u^1 = \bar{u}^1 - i\bar{u}^2$, the equations of the surface are

$$x^1 = 3\bar{u}^1 + 3\bar{u}^1 \bar{u}^{2^2} - \bar{u}^{1^3}, \qquad x^2 = 3\bar{u}^2 + 3\bar{u}^{1^2}\bar{u}^2 - \bar{u}^{2^3}, \qquad x^3 = 3\bar{u}^{1^2} - 3\bar{u}^{2^2},$$

and the curves $\bar{u}^1 = $ const., $\bar{u}^2 = $ const. are the lines of curvature;

(d) the lines of curvature are rectifiable unicursal curves of the third order and they are plane curves, the equations of the planes being

$$x^1 + \bar{u}^1 x^3 - 3\bar{u}^1 - 2\bar{u}^{1^3} = 0, \qquad x^2 - \bar{u}^2 x^3 - 3\bar{u}^2 - 2\bar{u}^{2^3} = 0;$$

(e) the lines of curvature are represented on the unit sphere by a double family of circles whose planes form two pencils with perpendicular axes which are tangent to the sphere at the same point;

(f) the asymptotic lines are twisted cubics;

(g) the sections of the surface by the planes $x^1 = 0$ and $x^2 = 0$ are cubics, which are double curves on the surface and the locus of the double points of the lines of curvature;

(h) the associate minimal surfaces are positions of the original surface rotated through the angle $-a/2$ about the x^3-axis, where a has the meaning of Ex. 2.

(i) the surface is the envelope of the plane normal, at the midpoint, to the join of any two points, one on each of the focal parabolas

$$x^1 = 4\bar{u}^1, \qquad x^2 = 0, \qquad x^3 = 2\bar{u}^{1^2} - 1; \qquad x^1 = 0, \qquad x^2 = 4\bar{u}^2 \qquad x^3 = 1 - 2\bar{u}^{2^2};$$

the planes normal to the two parabolas at the extremities of the join are the planes of the lines of curvature through the point of contact of the given plane.

BIBLIOGRAPHY

This bibliography contains only the books and memoirs which are referred to in the text.

1776 (1) MEUSNIER, J. B. *Mémoire sur la courbure des surfaces.* Published in 1785, in Acad. des sci., Paris, Mém. de math. et de phys. présentés par divers sçavans, vol. 10, pp. 477–510.

1827 (1) GAUSS, K. F. *Disquisitiones generales circa superficies curvas. Werke,* Göttingen, 1880, vol. 4, pp. 217–258.

1848 (1) BONNET, O. *Mémoire sur la théorie générale des surfaces.* Paris, École polytechnique, Jour., vol. 32, pp. 1–146.

1853 (1) BONNET, O. *Sur une propriété de maximum relative à la sphère.* Nouvelles annales, vol. 12, pp. 433–437.

1856 (1) MAINARDI, G. *Sulla teoria generale delle superficie.* Istituto Lombardo, Giornale, n. s., vol. 9, pp. 385–398.

1859 (1) LAMÉ, C. *Leçons sur les coordonnées curvilignes et leurs diverses applications.* Paris, Mallet-Bachelier. 368 pp.

1860 (1) *Bonnet, O. Mémoire sur l'emploi d'un nouveau système de variables dans l'étude des propriétés des surfaces courbes.* Jour. de math. pures, ser. 2, vol. 5, pp. 153–266.

1864 (1) BELTRAMI, E. *Ricerche di analisi applicata alla geometria.* Giornale di matem., vol. 2, pp. 267–282, 297–306, 331–339, 355–375. (This and 1865 (1) are also in his *Opere,* vol. 1.)

    (2) ENNEPER, A. *Analytisch-geometrische Untersuchungen.* Zeits. für Math. und Physik, vol. 9, pp. 96–125.

1865 (1) BELTRAMI, E. *Ricerche di analisi applicata alla geometria.* Giornale di matem., vol. 3, pp. 15–22, 33–41, 82–91, 228–240, 311–314.

1866 (1) WEIERSTRASS, K. *Untersuchungen über Flächen deren mittlere Krümmung überall gleich Null ist.* Preussische Akad., Monatsberichte, pp. 612–625. (Also in his *Mathematische Werke,* vol. 3.)

1869 (1) CHRISTOFFEL, E. B. *Über die Transformation der homogenen Differentialausdrücke zweiten Grades.* Jour. für reine und angew. Math., vol. 70, pp. 46–70.

    (2) BELTRAMI, E. *Teoria fondamentale degli spazii di curvatura costante.* Annali di matem., ser. 2, vol. 2, pp. 232–255.

    (3) CODAZZI, D. *Sulle coordinate curvilinee d'una superficie e dello spazio.* Annali di matem., ser. 2, vol. 2, pp. 269–287.

1870 (1) DINI, U. *Sopra un problema che si presenta nella teoria generale delle rappresentazione geografiche di una superficie su di un' altra.* Annali di matem., ser. 2, vol. 3, pp. 269–293.

1887 (1) DARBOUX, G. *Leçons sur la théorie générale des surfaces.* Paris, Gauthier-Villars, vol. 1.

1888 (1) LELIEUVRE, M. M. C. *Sur les lignes asymptotiques et leur représentation sphérique.* Bulletin des sciences math., vol. 12, pp. 126–128.

1889 (1) DARBOUX, G. *Leçons sur la théorie générale des surfaces.* Paris, Gauthier-Villars, vol. 2.

1901 (1) RICCI-CURBASTRO, G. *and* LEVI-CIVITA T. *Méthodes de calcul différ-
ential absolu et leurs applications.* Math. annalen, vol. 54, pp. 125–
201, 608.

1902 (1) BIANCHI, L. *Lezioni di geometria differenziale,* 2 ed. Pisa, Spoerri.
vol. 1.

1904 (1) FINE, H. B. *A college algebra.* Boston, Ginn. 595 pp.

    (2) RICCI-CURBASTRO, G. *Direzioni e invarianti principali in una varietà
qualunque.* R. Istituto Veneto. Atti, vol. 63, pp. 1233–1239.

1907 (1) BÔCHER, M. *Introduction to higher algebra.* New York. Macmillan.
321 pp.

1909 (1) EISENHART, L. P. *Treatise on the differential geometry of curves and
surfaces.* Boston, Ginn. 474 pp.

1910 (1) VEBLEN, O., *and* YOUNG, J. W. *Projective geometry.* Boston Ginn,
vol. 1.

    (2) DARBOUX, G. *Leçons sur les systèmes orthogonaux.* . . . Paris, Gauthier-
Villars. 567 pp.

1917 (1) LEVI-CIVITA, T. *Nozione di parallelismo in una varietà qualunque e
conseguente specificazione geometrica della curvatura riemanniana.*
Circolo matem. di Palermo. Rendic., vol. 42, pp. 173–205.

1922 (1) BIANCHI, L. *Sul parallelismo vincolato di Levi-Civita nella metrica degli
spazi curvi.* Accad. delle sci. 'fis. e. matem., Naples, Rendic., ser. 3,
vol. 28, pp. 150–171.

1925 (1) GOURSAT, E. *Cours d'analyse mathématique,* 5. éd. Paris, Gauthier-
Villars, vol. 1.

1926 (1) EISENHART, L. P. *Riemannian geometry.* Princeton university press.
262 pp.

    (2) SYNGE, J. L. *On the geometry of dynamics.* Royal Soc. London, Phil.
Trans. A, vol. 226, pp. 31–106.

1927 (1) FINE, H. B. *Calculus.* New York, Macmillan. 421 pp.

    (2) LEVI-CIVITA, T. *The absolute differential calculus.* London, Blackie.
450 pp.

1931 (1) McCONNELL, A. J. *Applications of the absolute differential calculus.*
London, Blackie. 318 pp.

Index

299

Milton Keynes UK
Ingram Content Group UK Ltd.
UKHW010024170724
445591UK00005B/157

9 780691 627465